Theory and Practice of Urban Sustainability Transitions

Series editors
Derk Loorbach, Rotterdam, The Netherlands
Hideaki Shiroyama, Tokyo, Japan
Julia M. Wittmayer, Rotterdam, The Netherlands
Junichi Fujino, Tsukuba, Japan
Satoru Mizuguchi, Tokyo, Japan

THE UNIVERSITY OF TOKYO

This book series Theory and Practice of Urban Sustainability Transitions is intended to explore the different dynamics, challenges, and breakthroughs in accelerating sustainability transitions in urban areas across the globe. We expect to find as much different and diverse stories, visions, experiments, and creative actors as there are cities: from metropolises to country towns, from inner city districts to suburbs, from developed to developing, from monocultural to diverse, and from hierarchical to egalitarian. But we also expect to find patterns in processes and dynamics of transitions across this diversity. Transition dynamics include locked-in regimes that are challenged by changing contexts, ecological stress and societal pressure for change as well as experiments and innovations in niches driven by entrepreneurial networks, and creative communities and proactive administrators. But also included are resistance by vested interests and sunken costs, uncertainties about the future amongst urban populations, political instabilities, and the erosion of social services and systems of provision. And finally there are the forming of transformative arenas, the development of coalitions for change across different actor groups, the diffusion and adoption of new practices, and exponential growth of sustainable technologies. For this series we seek this middle ground: between urban and transition perspectives, between conceptual and empirical, and between structural and practical. We aim to develop this series to offer scholars state-of-the-art theoretical developments applied to the context of cities. Equally important is that we offer urban planners, professionals, and practitioners interested or engaged in strategic interventions to accelerate and guide urban sustainability transition frameworks for understanding and dealing with on-going developments, methods, and instruments.

This book series will lead to new insights into how cities address the sustainability challenges they face by not returning to old patterns but by searching for new and innovative methods and instruments that are based on shared principles of a transitions approach. Based on concrete experiences, state-of-the-art research, and ongoing practices, the series provides rich insights, concrete and inspiring cases as well as practical methods, tools, theories, and recommendations. The book series, informed by transition thinking as it was developed in the last decade in Europe, aims to describe, analyse, and support the quest of cities around the globe to accelerate and stimulate such a transition to sustainability.

To sum up, the book series aims to:

- Provide theory, case studies, and contextualized tools for the governance of urban transitions worldwide
- Provide a necessary and timely reflection on current practices of how transition management is and can be applied in urban contexts worldwide
- Further the theorizing and conceptual tools relating to an understanding of urban sustainability transitions
- Provide best practices of cities across countries and different kinds of cities as well as across policy domains in shaping their city's path towards sustainability

More information about this series at http://www.springer.com/series/13408

Nadja Kabisch • Horst Korn
Jutta Stadler • Aletta Bonn

Editors

Nature-based Solutions to Climate Change Adaptation in Urban Areas

Linkages between Science, Policy and Practice

Editors

Nadja Kabisch
Helmholtz-Centre for Environmental
  Research – UFZ
Leipzig, Germany

Humboldt-Universität zu Berlin
Berlin, Germany

German Centre for Integrative Biodiversity
  Research (iDiv) Halle-Jena-Leipzig
Leipzig, Germany

Jutta Stadler
Federal Agency of Nature Conservation
  (BfN)
Isle of Vilm, Germany

Horst Korn
Federal Agency of Nature Conservation
  (BfN)
Isle of Vilm, Germany

Aletta Bonn
Helmholtz-Centre for Environmental
  Research – UFZ
Leipzig, Germany

Friedrich Schiller University Jena
Jena, Germany

German Centre for Integrative Biodiversity
  Research (iDiv) Halle-Jena-Leipzig
Leipzig, Germany

ISSN 2199-5508   ISSN 2199-5516   (electronic)
Theory and Practice of Urban Sustainability Transitions
ISBN 978-3-030-10417-7   ISBN 978-3-319-56091-5   (eBook)
DOI 10.1007/978-3-319-56091-5

Printed on acid-free paper

This Springer imprint is published by Springer Nature
The registered company is Springer International Publishing AG
The registered company address is: Gewerbestrasse 11, 6330 Cham, Switzerland

Foreword

Today, we are facing increasing challenges from climate change and urbanization. Already, half of the earth's population lives in urban areas, and projections suggest that this share will increase up to 66% by mid of this century. This urban expansion will heavily draw on natural resources, including open space, and will have severe effects on ecosystems and the services they provide. Cities are the first to experience impacts from climate change. Rising temperatures, heat waves, extreme precipitation events, flooding and droughts are causing economic losses, social insecurity and affecting health and human well-being.

Traditionally, urban planners and practitioners in land and resource management have relied on conventional engineering solutions to adapt to climate change, but this may not always be cost-effective, sufficient or sustainable. Nature-based solutions can address societal challenges from climate change and urbanization in a sustainable way. By using ecosystem services, nature-based solutions are innovative solutions that use natural elements to achieve environmental and societal goals. They offer significant potential to provide energy and resource-efficient responses to climate change, and to enhance our natural capital. Nature-based solutions provide additional multiple benefits to city residents such as improvements in health and wellbeing, and improvements of the local green economy.

This volume brings together a wealth of knowledge on the effectiveness of nature-based solutions in addressing climate change adaptation from diverse but inter-related fields of study. Importantly, research from the natural and social sciences is combined and results are interlinked with urban governance and local participation. This volume clearly demonstrates the importance of taking a systemic approach to combine knowledge from different fields, such as urban planning, nature conservation, urban engineering, governance or social justice and public health to address complex issues in a sustainable way. This integrated view to sustainable urban development is also emphasized in the 2030 Agenda for Sustainable Development, the New Urban Agenda adopted at the United Nations' HABITAT III conference, and supported by the European Commission's research and innovation policy on nature-based solutions. It is now time to seize opportunities to act. By understanding the value of nature-based solutions to climate change adaptation for society and by developing the policies, research and practice to implement them, we can contribute to enhancing the preparedness of cities and their communities to meet environmental and societal challenges now and in the future.

European Commission, DG Research and Innovation,   Marco Fritz
Sustainable Management of Natural Resources

Acknowledgements

Synthesising knowledge in this volume from different disciplines and sectors about nature-based solutions has been a very productive and fruitful collaboration of all contributing authors. The trans- and interdisciplinary approach to this book brought together 54 experts from the natural and social sciences as well as from urban policy and planning from across eleven countries. Working on this volume has been a very creative, inspiring and rewarding process.

We are grateful to all contributors for joining the stimulating discussion process and hope this dialogue will continue. Thanks also to all reviewers, who provided helpful and constructive comments for all chapters of this book. We would also like to extend our gratitude to all practitioners and policy advisers, who have contributed to the case study research in this volume. Their efforts and active collaboration made this synthesis possible.

We are especially grateful to Margaret Deignan from the Springer publishing team and to our Springer project coordinator Sofia Priya Dharshini.V for their helpful guidance.

The project developed out of the successful European conference 'Nature-Based Solutions to Climate Change in Urban Areas and Their Rural Surroundings: Linkages between Science, Policy and Practice' on 17–19 November 2015 in Bonn, Germany (for a detailed conference documentation, see http://www.bfn.de/23056+M52087573ab0.html). The conference was organised by the German Federal Agency of Nature Conservation (BfN) and the climate change interest group of the Network of European Nature Conservation Agencies (ENCA) in collaboration with the Helmholtz-Centre for Environmental Research-UFZ and the German Centre for Integrative Biodiversity Research (iDiv) Halle-Jena-Leipzig. More than 230 European experts from 27 countries convened to discuss the importance of nature--based solutions to climate change in urban areas and their rural surroundings at this conference. The large number of presented papers and posters illustrated the highly topical and relevant nature of this field in science, policy and practice and fuelled stimulating debate.

The editors have used their best endeavours to ensure URLs provided for external websites are correct and active at the time of going to press. However, the publisher has no responsibility for websites and cannot guarantee that contents will remain live or appropriate.

Leipzig, Germany	Nadja Kabisch
Isle of Vilm, Germany	Horst Korn
Isle of Vilm, Germany	Jutta Stadler
Leipzig, Germany	Aletta Bonn

Contents

Chapter 1
Nature-Based Solutions to Climate Change Adaptation in Urban Areas—Linkages Between Science, Policy and Practice

Nadja Kabisch, Horst Korn, Jutta Stadler, and Aletta Bonn

Abstract Climate change presents one of the greatest challenges to society today. Effects on nature and people are first experienced in cities as cities form microcosms with extreme temperature gradients, and by now, about half of the human population globally lives in urban areas. Climate change has significant impact on ecosystem functioning and well-being of people. Climatic stress leads to a decrease in the distribution of typical native species and influences society through health-related effects and socio-economic impacts by increased numbers of heat waves, droughts and flooding events. In addition to climate change, urbanisation and the accompanying increases in the number and size of cities are impacting ecosystems with a number of interlinked pressures. These pressures include loss and degradation of natural areas, soil sealing and the densification of built-up areas, which pose additional significant challenges to ecosystem functionality, the provision

N. Kabisch (✉)
Department of Ecosystem Services, Helmholtz Centre for Environmental Research – UFZ, Permoserstraße 15, 04318 Leipzig, Germany

Department of Geography, Humboldt-Universität zu Berlin,
Unter den Linden 6, 10099 Berlin, Germany

German Centre for Integrative Biodiversity Research (iDiv) Halle-Jena-Leipzig,
Deutscher Platz 5e, 04103 Leipzig, Germany
e-mail: nadja.kabisch@ufz.de; nadja.kabisch@geo.hu-berlin.de

H. Korn • J. Stadler
Federal Agency of Nature Conservation (BfN), Isle of Vilm, Germany
e-mail: horst.korn@bfn.de; jutta.stadler@bfn.de

A. Bonn
Department of Ecosystem Services, Helmholtz Centre for Environmental Research – UFZ, Permoserstraße 15, 04318 Leipzig, Germany

Institute of Ecology, Friedrich Schiller University Jena,
Dornburger Str. 159, 07743 Jena, Germany

German Centre for Integrative Biodiversity Research (iDiv) Halle-Jena-Leipzig,
Deutscher Platz 5e, 04103 Leipzig, Germany
e-mail: aletta.bonn@ufz.de

© The Author(s) 2017
N. Kabisch et al. (eds.), *Nature-based Solutions to Climate Change Adaptation in Urban Areas*, Theory and Practice of Urban Sustainability Transitions,
DOI 10.1007/978-3-319-56091-5_1

of ecosystem services and human well-being in cities around the world. However, nature-based solutions have the potential to counteract these pressures. Nature-based solutions (NBS) can foster and simplify implementation actions in urban landscapes by taking into account the services provided by nature. They include provision of urban green such as parks and street trees that may ameliorate high temperature in cities or regulate air and water flows or the allocation of natural habitat space in floodplains that may buffer impacts of flood events. Architectural solutions for buildings, such as green roofs and wall installations, may reduce temperature and save energy. This book brings together experts from science, policy and practice to provide an overview of our current state of knowledge on the effectiveness and implementation of nature-based solutions and their potential to the provision of ecosystem services, for climate change adaptation and co-benefits in urban areas. Scientific evidence to climate change adaptation is presented, and a further focus is on the potential of nature-based approaches to accelerate urban sustainability transitions and create additional, multiple health and social benefits. The book discusses socio-economic implications in relation to socio-economic equity, fairness and justice considerations when implementing NBS.

Keywords Nature-based solutions • Climate change • Urbanisation • Climate change adaptation • Cities

1.1 Background

Climate change presents one of the greatest challenges to society today. Effects on nature and people are first experienced in cities (White et al. 2005) as cities form microcosms with extreme temperature gradients, and by now, about half of the human population globally lives in urban areas (United Nations, Department of Economic and Social Affairs 2014). Already, climate change has significant impact on biodiversity and ecosystem functioning through threatening current habitat conditions due to heat and water stress (European Environment Agency 2012). Climatic stress already leads inter alia to a decrease in the distribution of typical native species and facilitates the establishment of alien invasive species (Knapp et al. 2010). Influences of climate change on society include health-related effects and socio-economic impacts induced by increased numbers of heat waves, droughts and flooding events (European Environment Agency 2016). In addition to climate change, urbanisation and the accompanying increases in the number and size of cities are impacting ecosystems, as urbanisation is driving a significant conversion of rural to urban landscapes (Seto et al. 2011). A number of interlinked pressures, such as loss and degradation of natural areas, soil sealing and the densification of built-up areas pose additional significant challenges to ecosystem functionality and human well-being in cities around the world. These processes may lead to biodiversity loss (for an overview, see Goddard et al. 2010) and a reduction of functions and services that urban ecosystems provide (Haase et al. 2014). However, urban green and blue spaces have the potential to counteract these pressures by providing habitats for a range of species (Niemela 1999; Goddard et al. 2010) and a number of environmental and cultural benefits while contributing to climate change adaptation and mitigation (Kabisch et al. 2015; Kabisch et al. 2016a; see Box 1.1 for definitions).

With regard to urban green and blue spaces, nature-based solutions (NBS) can foster and simplify implementation actions in urban landscapes by taking into account the services provided by nature (Secretariat of the Convention on Biological Diversity 2009). The concept of NBS evolved over the last years and was shaped by several actors (e.g. IUCN and the EU Commission; see Box 1.2 for definition of NBS). The concept of NBS is particularly embedded in the wider discussions on climate change adaptation, ecosystem services and green infrastructure (Kabisch et al. 2016a). Examples of NBS include provision of urban green such as parks and street trees that may ameliorate high temperature in cities (Gill et al. 2007; Bowler et al. 2010) or regulate air and water flows. Allocation of natural habitat space in floodplains may buffer impacts of flood events. Furthermore, architectural solutions for buildings, such as green roofs and wall installations for temperature reduction and related energy savings through reduced cooling loads (Castleton et al. 2010), can contribute to NBS. Importantly, by integrating NBS in urban landscapes, multiple benefits related to climate change adaptation and mitigation are increasingly recognised as influential determinants of human health and well-being (Barton and Grant 2006; Hartig et al. 2014). They relate to the provision and improved availability of urban green spaces and may result in better mental and physical health (Keniger et al. 2013). In addition, NBS may, in many cases, present more efficient and cost-effective solutions than more traditional technical approaches (European Commission 2015). In policy and practice, NBS complement concepts like green infrastructure or ecosystem-based mitigation and adaptation. To date, an increasing number of NBS projects have been implemented. Nevertheless, we are just at the beginning of systematically analysing their (long-term) effects, effectiveness for climate change adaptation and mitigation and provision of co-benefits. Still,

---

Box 1.1 Definition of Climate Change as well as Mitigation and Adaptation to Climate Change (European Environment Agency 2012)

Climate change is defined as any change in climate over time, resulting from natural variability or human activity.

Mitigation to climate change refers to anthropogenic interventions to reduce anthropogenic forces of the climate system. Climate change mitigation strategies include those to reduce greenhouse gas emissions and sources and enhancing greenhouse gas sinks.

Adaptation to climate change is defined as the adjustment in natural or human systems such as urban areas in response to actual or expected climatic stimuli or their effects. Climate change adaptation strategies should moderate harm or exploit beneficial opportunities of climate change.

Box 1.2. Definition of Nature-Based Solutions by IUCN and the European Commission

IUCN defines **nature-based solutions (NBS)** as: '… actions to protect, sustainably manage and restore natural or modified ecosystems, which address societal challenges (e.g., climate change, food and water security or natural disasters) effectively and adaptively, while simultaneously providing human well-being and biodiversity benefits' (p. xii) (Cohen-Shacham et al. 2016).

The European Commission understands: '… **nature-based solutions** to societal challenges as solutions that are inspired and supported by nature, which are cost-effective, simultaneously provide environmental, social and economic benefits and help build resilience. Such solutions bring more, and more diverse, nature and natural features and processes into cities, landscapes and seascapes, through locally adapted, resource-efficient and systemic interventions' (European Commission 2016).

knowledge is needed on measuring effectiveness and how the available evidence can be translated into management strategies and policy instruments.

1.2   Scope of the Book

This book brings together experts from science, policy and practice to provide an overview of our current state of knowledge on the effectiveness and implementation of NBS and their potential to the provision of ecosystem services, for climate change mitigation and adaptation and co-benefits in urban areas. Scientific evidence to climate change adaptation and mitigation is presented, and a further focus is on the potential of nature-based approaches to accelerate sustainability transitions and to create additional, multiple health and social benefits. The book also discusses socio-economic implications in relation to socio-economic equity, fairness and justice considerations when implementing NBS.

Furthermore, the chapters address tools to embed NBS in practice and policy, e.g. through partnership and community approaches between practice (e.g. urban gardening initiatives including allotment gardens), business and policy. As NBS are multifaceted, the book naturally has a strong interdisciplinary and transdisciplinary scope. The evidence reviewed and presented also feeds into recommendations for creating synergies between ongoing policy processes, scientific programmes and practical implementation of climate change mitigation and adaptation actions in European urban areas.

The book provides the current state of knowledge drawing from interdisciplinary research in urban ecology, urban planning, urban sociology and public health. The book also captures in-depth expertise and experience from policy and practice concerned with urban land development, as well as conservation and enhancement of biodiversity and ecosystem services provision. While the focus is on NBS to foster climate change adaptation, the chapters also highlight important multiple co-benefits for human health, quality of life and well-being analysed through interdisciplinary approaches. The book includes papers on new concepts and methods to dealing with the challenges emerging from pressures of climate change and urbanisation—that is, the need for sustainable green space development through NBS at different scales, from single patches to a city wide scale. Many chapters highlight the importance of urban planning on green infrastructure development and biodiversity conservation management within cities and provide pointers to move forward.

Focussing on relevant and up-to-date topics, the contributions of this book relate to the following essential main fields of interdisciplinary socio-environmental science:

1. Theory and management approaches related to nature-based solutions for climate change adaptation
2. Analysis of urban ecosystem services provided through multifunctional urban green spaces
3. Assessment of co-benefits of nature-based solutions to human health and well-being
4. Considerations of environmental justice and social equity related to nature-based solutions implementation
5. Nature-based solutions from a transition theory perspective
6. Municipal governance and socio-economic aspects of implementing nature-based solutions

These topics were intensively discussed at the European conference 'Nature-Based Solutions to Climate Change in Urban Areas and Their Rural Surroundings - Linkages Between Science, Policy and Practice' that took place in Bonn, Germany, from 17 to 19 November 2015 (Kabisch et al. 2016b). The conference was organised by the German Federal Agency for Nature Conservation (BfN), the Helmholtz Centre for Environmental Research-UFZ, the German Centre for Integrative Biodiversity Research (iDiv) Halle-Jena-Leipzig and the Network of European Nature Conservation Agencies (ENCA).

This book contributes to an increased understanding of how NBS can help to adapt to climate change through the provision of urban ecosystem services, of possibilities and limitations to their performance, and of how urban governance can use this understanding for a successful urban planning in growing cities under global change.

1.3 Structure and Contents of the Book

This book is divided into four main parts developing the case for adopting NBS for climate change adaptation. In addition, co-benefits and the implementation challenges of NBS as planning and management tool in urban development are presented.

1. Part I: Setting the Scene—Climate Change and the Concept of Nature-Based Solutions
2. Part II: Evidence for Nature-Based Solutions to Adapt to Climate Change in Urban Areas
3. Part III: Health and Social Benefits of Nature-Based Solutions in Cities
4. Part IV: Policy, Governance and Planning Implications for Nature-Based Solutions

The various chapters provide up-to-date scientific background information, address policy-related issues and lay out pressing urban land-use planning and management questions. Chapters provide specific examples and applications of NBS in cities with case studies, mainly from Europe but also North American and Chinese settings. Chapters further identify knowledge gaps. Their content is presented below.

1.3.1 Part I: Setting the Scene—Climate Change and the Concept of Nature-Based Solutions

The first part presents an overview of the concept of NBS and places it in the context of other relevant concepts such as green infrastructure, ecosystem-based adaptation (EbA) and ecosystem services in urban areas. The part discusses how different interpretations of the NBS concept result in multiple ways of describing and promoting it by a wide range of interested stakeholders.

To set the scene, *Tobias Emilsson and Åsa Ode Sang* provide an extended overview on climate change impacts on urban areas in Europe with specific focus on urban heat, energy and flooding. The overview also introduces climate change mitigation and adaptation options through urban green and blue spaces as an NBS in urban areas. Important potential planning aspects are discussed. Stephan Pauleit and co-authors discuss main features of the NBS concept in relation to overlaps and differences with other concepts, such as ecosystem-based adaptation (EbA), urban green infrastructure (UGI) and ecosystem services (ESS), which have all recently gained prominence in academic debates and are increasingly referred to in policy making. With this regard, *Erik Andersson and co-authors* present the idea of a double insurance value of urban ecosystems, which can be seen as one step towards governance processes that better take into account the complexity of the systems we live in and the multifaceted nature of 'hazards'. Using real-world examples from climate change-induced weather extremes, the authors illustrate that insurance consists of two components: first, functioning ecosystems can insure human societies against external disturbances, and second, these habitats need to be resilient themselves in the face of future disturbances that might affect their functioning. *Niki Frantzeskaki and co-authors* provide case study evidence that NBS are practices that transition initiatives in cities can put in place in order to intervene in their place and change the urban fabric. Focussing on three case study examples, the authors can show that NBS have transformative social impact contributing to social innovation in cities. In particular, the chapter highlights different ways how NBS as practices of transition initiatives in cities can get scaled up and hence contribute to accelerating sustainability initiatives.

1.3.2 Part II: Evidence for Nature-Based Solutions to Adapt to Climate Change in Urban Areas

Chapters in the second part of the book discuss the evidence for effectiveness of NBS also in comparison to technology-based solutions. In particular, the significance of biodiversity and its elements in cities and their rural surroundings for the adaptation to climate change and in providing ecosystem services is assessed.

In a first paper, *Yaella Depietri and Timon McPhearson* refer to the role of urban ecosystems in disaster risk reduction. They underline that evidence of the role of healthy ecosystems in disaster risk reduction is still scarce. By referring to cases in Northern America and in Europe, the authors discuss the role of green, blue and grey infrastructures as well as mixed approaches for climate change adaptation in cities in order to illustrate the different opportunities available for urban areas. In their chapter *Vera Enzi and colleagues* develop the case for architectural solutions and refer to green roof and wall technologies as part of the urban green infrastructure network and as an integrative NBS strategy to adapt to climate change. In particular, the chapter provides an overview about small-scale regulating ecosystem services as microclimatic benefits and impacts of green roofs and walls to city residents, which can be implemented even in densely settled areas. Using best practice examples, authors further show how ecologically improved green roof and wall systems can contribute to urban biodiversity. How urban green space further provides important regulating ecosystem services is shown by *Francesc Baró and Erik Gómez-Baggethun* by synthesising existing knowledge and using data of green space assessments carried out in Europe. They highlight in particular the role of NBS regarding global and local climate regulation as well as air quality improvement using the case study city of Barcelona. *McKenna Davis and Sandra Naumann* introduce sustainable urban drainage systems as an NBS to manage flood risk and to minimise the potential impact of floods on the environment and people. In particular, authors assess if sustainable urban drainage systems (SUDS) are cost-effective and offer long-term drainage alternatives to traditional drainage systems. *Dagmar Haase* also refers to flooding in urban areas, here highlighting the maintenance of natural urban habitats such as wetlands and riparian forest as an NBS to buffer climate change-induced flooding effects. *Dagmar Haase* shows the different additional ecosystem services through NBS measures, such as provision of recreation opportunities for urban residents and important habitat for wildlife.

1.3.3   Part III: Health and Social Benefits of Nature-Based Solutions in Cities

The third part of this book deals with the potential provision of multiple benefits when applying NBS to climate change adaptation. In particular, multiple benefits of urban green spaces related to health and social justice for urban residents are critically discussed. In this context, urban gardens are presented as one green infrastructure element that can be maintained by private individuals and provides multiple co-benefits.

Matthias Braubach and co-authors provide a comprehensive overview on the scientific literature of how urban green spaces can affect the health of urban residents and present epidemiological evidence of public health benefits of green spaces. In their review, the authors address three urban health dimensions, namely environmental conditions and related health outcomes, urban equity and vulnerability as well as resilience to extreme climate conditions related to climate change. Complementing the previous chapter, *Nadja Kabisch and Matilda Annerstedt van den Bosch* show how residents' health is linked to urban green space availability and discuss this in light of environmental justice concerns using the case study of Berlin, Germany. The link between the social effects, environmental justice and green implementation projects as NBS in cities is further critically discussed by *Annegret Haase*. In particular, potential weaknesses of NBS related to social inclusion and cohesion are explained, and the need to fully consider potential drawbacks and repercussions of the implementation and development of urban NBS particularly for lower-income communities is discussed. Using one particular component of the urban green infrastructure network, *Ines Cabral and co-authors* explore the contribution of allotments and community gardens as multifunctional NBS to achieve societal as well as environmental goals, using the case studies of Lisbon (Portugal), Leipzig (Germany), Manchester (UK) and Poznan (Poland). Furthermore, ecosystem services provided by allotment gardens are identified and analysed.

1.3.4   Part IV: Policy, Governance and Planning Implications for Nature-Based Solutions

The last part focusses on policy, governance and planning implications of NBS. In particular, good practice examples of efficient and successful governance approaches are shown, and new actor-networks created by NBS are discussed. This part also shows how NBS might be assessed economically and how economic valuation and related aspects may provide justification to the introduction of NBS in cities. In addition, the chapters in this part discuss new tools and instruments to invest in working with nature for people, to empower people and to encourage multi sectoral partnerships.

Reflecting on institutional aspects and challenges of the implementation of NBS projects, *Chantal van Ham and Helen Klimmek* highlight the need for increased and improved collaboration between sectors and stakeholders as well as for a sound evidence base of the economic, social and environmental benefits of NBS in order to foster increasing uptake of NBS in urban areas. The authors analyse current policies and practices for implementing NBS and highlight those which used unconventional but creative partnerships between policymakers, the private sector and civil society, which resulted in innovative and financeable NBS. The chapter draws from international examples of NBS implementation projects and reports and summarises successes as well as tensions presented in these pioneering solution partnerships. *Christine Wamsler and co-authors* introduce the concept of mainstreaming climate change adaptation to foster sustainable urban development and resilience, in particular mainstreaming ecosystem- or nature-based solutions into urban governance and planning. They also address challenges of NBS implementation through urban governance. In this approach adaptation mainstreaming is considered as the inclusion of climate risk considerations in sector policies and practices. Authors introduce an integrated framework that illustrates potential mainstreaming measures and strategies at different levels of governance by using case studies from Germany and Portugal. The spread of implementation of NBS as driver for innovation in a country with a transition background is discussed *by Jakub Kronenberg and co-authors* at the case of Poland. They analysed different groups of stakeholders using different example implementation projects across the country. The last chapter in this part refers to socio-economic and financial aspects of NBS implementation projects. *Nils Droste and co-authors* highlight the difficulties of resource allocation to NBS implementation as municipal revenues are mostly dedicated to single policy goals and predefined sectorial purposes, thereby leaving little room for cross-sectoral investments such as NBS. The authors identify policy instruments with the potential to foster NBS as well as difficulties in leveraging finance to implement NBS.

The book is complemented with a conclusion chapter by *the editors*. The editors summarise the main challenges for research, urban governance and management described in the chapters and highlight opportunities for future developments, thus leading to overall recommendations for NBS implementation.

We hope this book provides important pointers to the flourishing debate on NBS in urban environments and illustrates good practice with demonstration case studies, so it can fuel further advances in science, policy and practice. Many of the themes have applications beyond urban system with a focus on solutions for sustainable management and conservation in a changing world.

Acknowledgements We thank the external reviewers for providing critical and helpful comments on earlier versions of this chapter and on the book proposal. This work was supported by the German Federal Agency for Nature Conservation (BfN) with funds of the German Federal Ministry for the Environment, Nature Conservation, Building and Nuclear Safety (BMUB) through the research project 'Conferences on Climate Change and Biodiversity' (BIOCLIM, project duration from 2014 to 2017, funding code: 3514 80 020A).

References

Barton H, Grant M (2006) A health map for the local human habitat. J R Soc Promot Heal 126:252–253. doi:10.1177/1466424006070466

Bowler DE, Buyung-Ali L, Knight TM, Pullin AS (2010) Urban greening to cool towns and cities: a systematic review of the empirical evidence. Landsc Urban Plan 97:147–155. doi:10.1016/j.landurbplan.2010.05.006

Castleton HF, Stovin V, Beck SBM, Davison JB (2010) Green roofs; building energy savings and the potential for retrofit. Energ Buildings 42:1582–1591. doi:10.1016/j.enbuild.2010.05.004

Cohen-Shacham E, Walters G, Janzen C, Maginnis S (2016) Nature-based solutions to address societal challenges. Gland, Switzerland

European Commission (2015) Towards an EU research and innovation policy agenda for nature-based solutions & re-naturing cities. Final Report of the Horizon 2020 Expert Group on Nature-Based Solutions and Re-Naturing Cities

European Commission (2016) Policy topics: nature-based solutions. https://ec.europa.eu/research/environment/index.cfm?pg=nbs. Accessed 11 Sept 2016

European Environment Agency (2012) Climate change, impacts and vulnerability in Europe 2012. An indicator-based report

European Environment Agency (2016) Building resilient cities key to tackling effects of climate change. Copenhagen, Denmark

Gill SE, Handley JF, Ennos AR, Pauleit S (2007) Adapting cities for climate change: the role of the green infrastructure. Built Environ 33:115–133

Goddard M, Dougill AJ, Benton TG (2010) Scaling up from gardens: biodiversity conservation in urban environments. Trends Ecol Evol 25:90–98. doi:10.1016/j.tree.2009.07.016

Haase D, Larondelle N, Andersson E, Artmann M, Borgström S, Breuste J, Gomez-Baggethun E, Gren Å, Hamstead Z, Hansen R, Kabisch N, Kremer P, Langemeyer J, Rall EL, McPhearson T, Pauleit S, Qureshi S, Schwarz N, Voigt A, Wurster D, Elmqvist T (2014) A quantitative review of urban ecosystem service assessments: concepts, models, and implementation. Ambio 43:413–433. doi:10.1007/s13280-014-0504-0

Hartig T, Mitchell R, de Vries S, Frumkin H (2014) Nature and health. Annu Rev Public Health 35:207–228. doi:10.1146/annurev-publhealth-032013-182443

Kabisch N, Qureshi S, Haase D (2015) Human–environment interactions in urban green spaces — a systematic review of contemporary issues and prospects for future research. Environ Impact Assess Rev 50:25–34. doi:10.1016/j.eiar.2014.08.007

Kabisch N, Frantzeskaki N, Pauleit S, Naumann S, Davis M, Artmann M, Haase D, Knapp S, Korn H, Stadler J, Zaunberger K, Bonn A (2016a) Nature-based solutions to climate change mitigation and adaptation in urban areas: perspectives on indicators, knowledge gaps, barriers, and opportunities for action. Ecol Soc 21:art39. doi:10.5751/ES-08373-210239

Kabisch N, Stadler J, Korn H, Duffield S, Bonn A (2016b) Proceedings of the European conference on nature-based solutions to climate change in urban areas and their rural surroundings. BfN-Skripten. German Federal Agency of Conservation, Bonn

Keniger L, Gaston K, Irvine K, Fuller R (2013) What are the benefits of interacting with nature? Int J Environ Res Public Health 10:913–935. doi:10.3390/ijerph10030913

Knapp S, Kuehn I, Stolle J, Klotz S (2010) Changes in the functional composition of a Central European urban flora over three centuries. Perspect Plant Ecol Evol Syst 12:235–244

Niemela J (1999) Ecology and urban planning. Biodivers Conserv 8:119–131

Secretariat of the Convention on Biological Diversity (2009) Connecting biodiversity and climate change mitigation and adaptation: report of the second Ad Hoc technical expert group on biodiversity and climate change, Montreal

Seto KC, Fragkias M, Güneralp B, Reilly MK (2011) A meta-analysis of global urban land expansion. PLoS One 6:1–9. doi:10.1371/Citation

United Nations, Department of Economic and Social Affairs PD (2014) World urbanization prospects: the 2014 revision, Highlights (ST/ESA/SER.A/352)

White P, Pelling M, Sen K, Seddon D, Russel S, Few R (2005) Disaster risk reduction: a development concern. A scoping study on links between disaster risk reduction, poverty and development. London, Glasgow

Part I
Setting the Scene: Climate Change and the Concept of Nature-Based Solutions

Chapter 2
Impacts of Climate Change on Urban Areas and Nature-Based Solutions for Adaptation

Tobias Emilsson and Åsa Ode Sang

Abstract This chapter outlines the general impacts and direct consequences climate change is likely to have on urban areas in Europe and how nature-based solutions (NBS) could increase our adaptive capacity and reduce the negative effects of a changing climate. The focus is on urban temperatures while we will also include effects on hydrological, ecological and social factors. We also discuss challenges for planning and design of successful implementation of NBS for climate change adaptation within urban areas.

Keywords Urban design • Ecosystem services • Urban temperatures • Strategic planning • Vegetation maintenance • NBS implementation • Modelling techniques • Collaborative processes

2.1 Introduction

With the current process of climate change, Europe is expected to face major challenges in order to adapt to and mitigate the consequences of severe weather conditions (Kreibich et al. 2014). Year 2016 has seen new temperature records for each month, with July 2016 being the hottest month since temperature started to be recorded according to NASA measurements (NOAA 2016). An increase in temperature can cause discomfort, economical loss, migration and increased mortality rates on a global level (Haines et al. 2006). In addition, there are predicted increases in extreme weather events (e.g. heat and cold waves, floods, droughts, wildfires and windstorms) with several parts of Europe predicted to be exposed to multiple climate hazards (Forzieri et al. 2016).

Next to a changing climate both in Europe and globally, there is an ongoing urbanisation process. In year 2007, half of the world's population lived in urban

T. Emilsson (✉) • Å. Ode Sang
Department of Landscape Architecture, Planning and Management, Swedish University of Agricultural Sciences, Alnarp, Sweden
e-mail: tobias.emilsson@slu.se; asa.sang@slu.se

N. Kabisch et al. (eds.), *Nature-based Solutions to Climate Change Adaptation in Urban Areas*, Theory and Practice of Urban Sustainability Transitions, DOI 10.1007/978-3-319-56091-5_2

areas, and it is predicted that by 2050, 66% of the world's population will live in urban areas (UN 2014). The urban climate often differs from the surrounding rural countryside as it is generally more polluted, warmer, rainier and less windy (Givoni 1991). This suggests that the effect of climate change with the predicted increase in temperature and more extreme weather events will be experienced to a greater extent in urban areas compared to the surrounding landscape. The changing climate might also exaggerate the negative effects of urbanisation already experienced, such as increased urban temperatures and flooding (Semadeni-Davies et al. 2008).

Still, increasing urban densities are seen as a way forward towards sustainable urban development. Across Europe, there is presently a trend for densification as a planning approach for sustainable development to foster efficient use of resources, efficient transport systems and a vibrant urban life (e.g. Haaland and van den Bosch 2015). Development often takes place on areas that are often viewed as underutilised land (such as green space) or through redevelopment on previous industrial estates (van der Waals 2000). However, this approach has also been challenged for its threat to urban green spaces (Haaland and van den Bosch 2015) since together with urban brown fields they potentially have an important role for offering climate change adaptation solutions. The creation, re-establishment, improvement and upkeep of existing vegetation systems and the development of an integrated urban green infrastructure network could provide a valuable asset, in which to incorporate establishment of new nature-based solutions (NBS) to deal with local effects on climate change. The dual inner urban development could here be seen as a constructive way forward (BfN 2008). The approach combines a densification of existing built-up areas with a mixture of conservation actions, thereby boosting the presence, quality and usability of green spaces and enhancing other green infrastructure such as street trees, green walls and roofs (BfN 2008).

Within this chapter, we review (1) the general impacts and consequences of climate change for urban areas in Europe, (2) climate change adaptation possibilities using nature-based solutions (NBS) and (3) some challenges for planning and design for successful implementation of NBS within urban areas. The review focusses on urban temperatures and includes hydrological, ecological and social factors. The review is aimed at setting a baseline for future possible research on planning alternatives for climate change adaptation and providing general guidelines and support for the professional planning community working with climate change adaptation.

2.2 General Impact and Consequences of Climate Change for Urban Areas in Europe

Climate change will have far-reaching impacts and consequences for urban Europe. The impact will range from direct impact of increasing temperatures and changed precipitation dynamics to indirect effects resulting from perturbations and climate change-linked events elsewhere.

2.2.1 Effect on Urban Temperatures

Changing urban temperatures are driven both by large-scale climatic changes and ongoing urbanisation (Fujibe 2009). There is agreement that the current changing climate has to be kept well below an average global increase of 2 °C (EC 2007; UNFCCC 2015) to avoid major future climate-driven catastrophes (Lenton et al. 2008). The urban temperature is dependent on global development but is in general highly influenced by, e.g. the urban heat island (UHI) effect which is seen as a major problem of urbanisation (e.g. Gago et al. 2013; Taha 1997). There are three parameters of urbanisation that have direct bearing on UHI according to Taha (1997), namely, (1) increasing amount of dark surfaces such as asphalt and roofing material with low albedo and high admittance, (2) decreasing vegetation surfaces and open permeable surfaces such as gravel or soil that contribute to shading and evapotranspiration and (3) release of heat generated through human activity (such as cars, air-condition, etc.). These factors are not equally distributed across the city, and hence, certain areas will experience the UHI to a higher degree. The effect will, for example, be higher for areas with a high degree of built-up land and little green space than for leafy suburbs and hence will affect the population differently within an urban area.

The urban climate itself is suggested to increase the heat stress experienced by people during periods of high temperature, particularly during the night, when the UHI is largest (Pascal et al. 2005). Studies suggest that there is an adaptation factor in relation to heat and that early season heat waves or heat waves in regions where hot weather is infrequent have more negative consequences (Anderson and Bell 2011). This suggests that for parts of Europe that previously have not experienced periods with dangerously high temperature people are less adapted to deal with the increase in temperature.

2.2.2 Effect on Urban Hydrology

With a changing climate, the frequency of flood peaks is predicted to increase. Estimations point towards an average doubling of severe flood peaks with a return period of 100 year within Europe by 2045 (Alfieri et al. 2015). In addition, this is matched by a rise in sea level that, together with a predicted increase in windstorm frequency, will lead to an increase in coastal flooding (Nicholls 2004). As most of the urban areas within Europe are situated either on floodplains or along the coast, these two types of flooding will have a major impact across European cities. Climate driven increasing sea levels in certain areas of Europe will also translate into more frequent basement flooding (Arnbjerg-Nielsen et al. 2013).

The impact of a changing climate will differ across the continent whereby Northern Europe is expected to experience more annual mean precipitation as compared to Southern and Central European countries that are projected to experience a reduction in rainfall (Stagl et al. 2014; Olsson et al. 2009). Several models have pointed in a direction of decreasing total summer precipitation and increasing intensity of storms interspersed with drought. Increasing high-precipitation events will mean that the current urban drainage system will exceed its capacity more frequently, causing economic loss, increased discomfort and even loss of lives (Semadeni-Davies et al. 2008). Increasing urban temperatures will also have a strong influence on evapotranspiration that is largely limited by precipitation. Thus, there might be increased evapotranspiration in areas with more precipitation but also increased durations of drought in areas with reduced precipitation. In northern regions there is also an expected seasonal change in precipitation with more winter precipitation falling as rain and higher spring temperatures, leading to increased winter runoff and a reduction in late season snowmelt (Madsen et al. 2014).

2.2.3 Indirect Effects on Urban Habitats and Biodiversity

Climate change will influence several factors of importance to habitat quality and development of urban biodiversity. The projected change in temperatures, rainfall, extreme events and enhanced CO_2 concentrations will influence a range of factors related to single species (e.g. physiology), population dynamics, species distribution patterns, species interactions and ecosystem services, as a result of spatial or temporal reorganisation (Bellard et al. 2012). Increasing urban temperatures and changed precipitation dynamics will influence species community development through limiting water availability during the growing season as well as changing the nutrient dynamics. Especially northern or alpine regions will be severely impacted due to enhanced temperature changes, e.g. as more common species will be able to colonise niches that were otherwise restricted to specialised species (Dirnböck et al. 2011).

Urban areas already have in many cases a higher plant richness compared to their natural counterparts (Faeth et al. 2011) due to influx of alien plant material, more nutrient-rich systems, a larger habitat heterogeneity and more continuous land use or directed management (Kowarik 2011). With a change in the urban climate, there is likely to be a change in invasiveness of alien species (Crossman et al. 2011) as well as an increase in the spread of disease and pests (Wilby and Perry 2006).

2.3 Climate Change Adaptation Possibilities Using Green Infrastructure and Nature-Based Solutions

Adaptation to actual or expected climate change effects involves a range of measures or actions that can be taken to reduce the vulnerability of society and to improve the resilience capacity against expected changing climate. Possible adaptation measures to handle climate change can take many forms and be effective at a range of spatial and temporal scales, proactively planned or as a results of socio political drivers such as new planning regulations, market demand or even social pressure (Metz et al. 2007).

2.3.1 Urban Green Infrastructure (UGI) and Nature-Based Solutions (NBS)

Vegetation can indeed play an important role in moving the urban climate closer to a pre-development state. Urban green infrastructure (UGI) and nature-based solutions (NBS) are fundamental concepts in this work with emphasis on the role that nature can play in providing multiple services to the urban population (Pauleit et al. this volume). UGI is a concept that stems from planning, and hence the focus is on the strategic role for integrating green spaces and their associated ecosystem services within urban planning at multiple scales (Benedict and McMahon 2006). NBS is according to Pauleit et al. (this volume) broad in its definition and scope, with a broad view on 'nature', and an emphasis on participatory processes in creation and management. The European Commission and Directorate-General for Research and Innovation (EC DG 2015) defines NBS as 'living solutions inspired by, continuously supported by and using nature, which are designed to address various societal challenges in a resource-efficient and adaptable manner and to provide simultaneously economic, social, and environmental benefits'. NBS is by Pauleit et al. (this volume) proposed to be seen as an umbrella term that incorporates UGI as well as ecosystem-based adaption and ecosystem services.

2.3.2 Reducing Urban Temperature Through Green or Blue Infrastructure and NBS

Urban temperatures can be strategically handled through a network of planned urban green space. This includes the selection of appropriate surfaces, their spatial organisation and management.

Studies have shown that urban parks have a cooling effect in the range of 1 °C during the daytime, with indications that larger parks have a larger effect as well as systems including trees (Bowler et al. 2010). The surface type will also influence the cooling effect of the blue or green infrastructure. For instance, surface temperatures of water is lower compared to vegetated areas which in turn are markedly cooler than streets and roofs (Leuzinger et al. 2010). This means that there is a larger cooling effect per unit surface water as compared to a vegetated park system (Žuvela-Aloise et al. 2016). This effect varies with time of the day, with largest differences between park and water bodies during daytime. Several studies therefore suggest that in order to maximise the use of space for urban cooling more focus should be placed on inclusion of water bodies as well as concentrating these surfaces in the city centres as compared to an alternative approach with smaller parks distributed over the city in general (Žuvela-Aloise et al. 2016; Skoulika et al. 2014). There is also a substantial seasonality in the effect of urban vegetation, with stronger effects in summer than early spring. While these broad differences in cooling occur, there is also variation found linked to the level of soil sealing and amount of vegetation, which could explain microclimatic effects (Lehmann et al. 2014).

The effect and importance of vegetation systems are also dependent on the organisation of the urban fabric such as structure and type of building (Lehmann et al. 2014). The potential for temperature reduction through the use of vegetation has been shown to be larger in densely built-up area as compared to more sparse developments, with variation due to prevailing wind direction and time of day. The model follows a saturation model where the first installations are of greatest importance with each additional surface area contributing to a lesser extent (Žuvela-Aloise et al. 2016).

Individual urban trees can have an effect on urban temperatures by contributing to reducing UHI. The climatic performance is dependent on the tree characteristics such as leaf organisation and canopy shape, where sparse crowns with large leaves have higher cooling capacity (Leuzinger et al. 2010). Novel types of vegetation systems such as green roofs and green walls can also alter the energy balance of urban areas something that is discussed in more depth by Enzi et al. (this volume). The direct advantage of these systems is that they can be added as a complement to existing blue and green infrastructure and that they make it possible to utilise spaces that normally are not green (see Enzi et al., this volume). Green walls have indeed been shown to reduce wall temperatures (Cameron et al. 2014) and street canyon temperatures with close to 10 °C during the day in hot and dry climates (Alexandri and Jones 2008). The performance of the vegetation depends on species composition with different species having varied cooling capacity and different modes of cooling, i.e. evaporative or shade cooling (Cameron et al. 2014), as well as management variables such as irrigation and water levels in the substrate (Song and Wang 2015; Hunter et al. 2014).

2.3.3 Selection and Management of Urban Vegetation Under Changing Climatic Conditions

It is important to remember that a changing climate will have positive and negative effects on the existing plant material, but in many cases, it will experience increasing stress and consequently lower survival and performance rates. The selection of the right tree is important to achieve high temperature efficiency at the same time as having limited maintenance needs and fulfilling other ecosystem services such as habitat creation and delivering aesthetical values (Rahman et al. 2015). The current selection of plant material as well as planting design has to be adjusted to accommodate a changing climate. A moderate planting design, for example, with tree distances of 7.5 m in combination with permeable pavement or bare soil extending to the canopy extension can achieve good cooling and low water stress (Vico et al. 2014). Changed rainfall patterns might exaggerate the need for irrigation during extended drought periods, something that will be stressed when using higher planting densities or surfaces with low permeability. Xeric trees will have higher performance in relation to cooling and survival under water-limited situations and can also contribute to urban cooling through shading but does not have the same effect as other vegetation types such as perennial plantations and in particular lawns when it comes to increasing humidity (Song and Wang 2015).

Stressed, unhealthy or declining vegetation cover will also cause reduced ecosystem function. Speak et al. (2013) showed that green roofs can lower the air temperature above the system with approximately 1 °C. The effect was increased at night by 50% coinciding with the time when UHI is the strongest. Sections where vegetation cover had declined were warmer during the daytime, highlighting the importance of maintenance and upkeep and the design and installation of quality green systems (Speak et al. 2013; Klein and Coffman 2015). Yaghoobian and Srebric (2015) came to similar conclusions showing that the green roof performance, i.e. surface temperature decreases, is connected with increasing plant coverage. A high plant cover will lead to reduced solar radiation uptake due to high albedo, shading and vegetation system evapotranspiration. In a declining vegetation system, the albedo will be worse, especially if a dark-coloured substrate is used and the efficiency of the green roof is only dependent on evaporation. Thus, it is fundamental that these nature-based solutions are designed in a way that maintain a good plant cover over time, installed and maintained to actually deliver the ecosystem services that they are supposed to deliver. There is also some evidence that the vegetation composition and species or functional diversity can impact on the level of evapotranspiration and reduction of urban stormwater (Lundholm et al. 2010). Some of the most common succulent species can have high survival rates on green roofs and commonly make up for a substantial part of the total cover, but due to their water-preserving physiological adaptations, they have rather low evapotranspiration rates and consequently a lower cooling capacity. Using plant traits to select plants from natural dryland habitats that have optimised water-use strategies for evaporation during wet periods at the same time as being drought tolerant could be a way to optimise green roof cooling capacity (Farrell et al. 2013).

Vegetation can also be used to change the energy balance of buildings directly (see also Enzi et al., this volume). Modelling results show high reduction in energy use as well as reduced maximum temperatures in buildings close to the vegetation as compared to a traditional sunblocking material such as blinds and panels (Stec et al. 2005). The maximum temperature reduction deduced from green roof vegetation has been shown to be close to 20 °C lower as compared to using blinds or physical shading panels. In modern buildings, the insulation is generally much thicker making the surface characteristics of the outer layer less important (Castleton et al. 2010). However, roofs retrofitted with green roofs can have a substantial positive effect on winter energy cost if installed on poorly insulated buildings and if thicker substrate depths are used (Berardi 2016).

2.3.4   Green Infrastructure, NBS and Urban Hydrology

Green infrastructure and nature-based solutions such as green roofs, rain gardens and bioswales have been shown to reduce local flooding, economical loss and discomfort at storm events with medium or frequent return periods. Still, it is important to remember that these small-scale installations have little impact on the large-scale catastrophic rain events such as river flooding, seaside flooding or very intense cloud bursts that pose the greatest danger to urban infrastructure and communities. Thus, there is a need to work on multiple spatial scales to adapt to changing precipitation dynamics focussing both on the installation of local solutions and developing zoning regulations for housing developments as well as planning for safer proactively planned flooding areas forming an integrated and multifunctional urban drainage system (Fletcher et al. 2015; Burns et al. 2012).

There has been a rapidly increasing body of research on the efficiency and function of individual installations (see also Davies and Naumann, this volume; Enzi et al., this volume). Green roofs have been shown to have large effects on annual stormwater runoff but also on peak flows (Bengtsson 2005; Stovin et al. 2013; Stovin 2010). Thin green roofs have a limited storage capacity meaning that these systems have reduced efficiency on very long or intense rain events (Bengtsson 2005). Green roofs and other vegetated systems might influence the water quality of runoff water negatively if conventional fertilisers are used or if they contain nutrient-rich compost without addition of substances such as biochar (Beecham and Razzaghmanesh 2015; Gong et al. 2014; Beck et al. 2011). Bioswales, biofilters or rainbeds or other types of planted retention beds are alternative solutions to handle stormwater on ground if space is available. Ground-based systems can be built with thicker substrates as compared to roofs, which simplifies the use of large perennials, shrubs and small trees. Functionally, these systems also have a potential for infiltration and evapotranspiration (Daly et al. 2012; Muthanna et al. 2008).

2.4   Planning and Design Aspects of Green Infrastructure and Nature-Based Solutions for Adapting to Climate Change

The introduction and enhancement of UGI often provide a local effect for the microclimate both by providing a 'cool island' effect (Oliveira et al. 2011) and contributing to an overall global climate effect through the binding of CO_2 (Nowak and Crane 2002).

From a planning perspective, it is interesting to pose the question on where and which NBS to implement when prioritising resources. In the previous section, we have shown that qualities such as vegetation type as well as amount and level of soil sealing have important bearing on the effect of climate regulations and adaptation measurement. When planning and implementing NBS, these are important considerations to take into account together with existing local conditions. Several studies have further shown that urban morphology plays an important role for explaining climatic effects (Oliviera et al. 2011; Jamei et al. 2016).

When it comes to the allocation of where to invest in NBS for climate change adaptation, it is important to look at the urban area on a strategic level, taking into account the character of the urban morphology as well as information on population details. The following questions are important in order to ensure the most cost-effective, highest gain and to take into account environmental justice (see also A. Haase, this volume) with regard to mitigating the negative effects of climate change: (1) Where does the UHI have the largest impact? (2) Where do vulnerable population groups live (e.g. old people as well as high density of population)? (3) Where is a current lack of green and blue infrastructure? Here, strategic documents such as green infrastructure plans could provide a valuable tool for working with NBS on a strategic level. Norton et al. (2015) present a novel approach through using a hierarchial process for how to prioritise and strategically select NBS (in this case green open spaces, shade trees, green roofs and vertical greening systems) to mitigate high temperature, taking into account the relationship between urban morphology, UGI and temperature mitigation.

There is an abundance of different modelling techniques available, differing in complexity and accuracy that could aid a strategic planning and design of NBS for climate change adaptation (Deak-Sjöman and Sang 2015). However, to ensure environmental justice, there are also strong calls for involving the local population in different processes of co-planning, co-design and co-management. Pauleit et al. (this volume) identify this as a key component of the NBS concept as it also has the potential to ensure the viability of the different solutions and to provide processes to site adaptation. Through the inclusion of scenario and impact modelling techniques in a collaborative process, it is possible to implement NBS that are both climate effective and ensuring environmental justice (see Fig. 2.1).

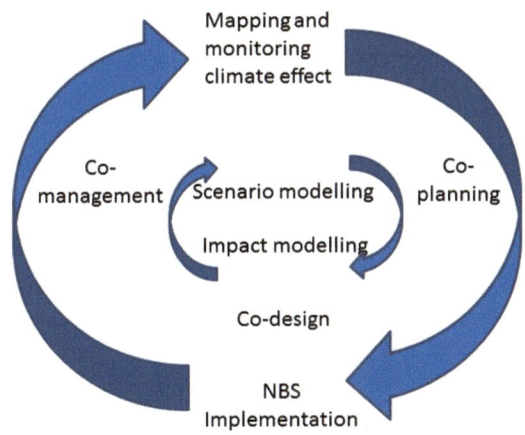

Fig. 2.1 Process for implementing NBS in a collaborative process with integration of modelling techniques

However, while the modelling techniques are available, the skills needed might not be present within local authorities, as shown in a recent survey of Swedish municipalities (Sang and Ode Sang 2015). This hinders the use of modelling techniques for analysing potential climate effects in more iterative and strategic processes through exploring alternative solutions as well as accumulative climate effects by introducing different green space interventions across the urban area.

2.5 Conclusion

Nature-based solutions have a key role to play in achieving a future compact city that is liveable and sustainable. Vegetation in different forms can contribute to various degrees to climate adaptation, depending on NBS type and quality as well as climatic and socio-ecological contexts. Through integrating modelling techniques with collaborative processes, we could ensure a strategic planning of green space interventions that are climate effective and ensure environmental justice.

References

Alexandri E, Jones P (2008) Temperature decreases in an urban canyon due to green walls and green roofs in diverse climates. Build Environ 43:480–493

Alfieri L, Burek P, Feyen L, Forzieri G (2015) Global warming increases the frequency of river floods in Europe. Hydrol Earth Syst Sci 19(5):2247–2260

Anderson GB, Bell ML (2011) Heat waves in the United States: mortality risk during heat waves and effect modification by heat wave characteristics in 43 U.S. communities. Environ Health Perspect 119(2):210–218

Arnbjerg-Nielsen K, Willems P, Olsson J, Beecham S, Pathirana A, Bülow Gregersen I, Madsen H, Nguyen VTV (2013) Impacts of climate change on rainfall extremes and urban drainage systems: a review. Water Sci Technol 68(1):16–28

Beck DA, Johnson GR, Spolek GA (2011) Amending greenroof soil with biochar to affect runoff water quantity and quality. Environ Pollut 159(8–9):2111–2118

Beecham S, Razzaghmanesh M (2015) Water quality and quantity investigation of green roofs in a dry climate. Water Res 70:370–384

Bellard C, Bertelsmeier C, Leadley P, Thuiller W, Courchamp F (2012) Impacts of climate change on the future of biodiversity. Ecol Lett 15(4):365–377

Benedict MA, McMahon ET (2006) Green infrastructure: linking landscapes and communities. Island Press, Washington, DC

Bengtsson L (2005) Peak flows from thin sedum-moss roof. Nord Hydrol 36(3):269–280

Berardi U (2016) The outdoor microclimate benefits and energy saving resulting from green roofs retrofits. Energ Buildings 121:217–229

BfN (2008) Stärkung des Instrumentariums zur Reduzierung der Flächeninanspruchnahme. Position paper of the Federal Agency for Nature Conservation, Germany. Available at: http://www.bfn.de/fileadmin/MDB/documents/themen/siedlung/positionspapier_flaeche.pdf

Bowler DE, Buyung-Ali L, Knight TM, Pullin AS (2010) Urban greening to cool towns and cities: a systematic review of the empirical evidence. Landsc Urban Plan 97(3):147–155

Burns MJ, Fletcher TD, Walsh CJ, Ladson AR, Hatt BE (2012) Hydrologic shortcomings of conventional urban stormwater management and opportunities for reform. Landsc Urban Plan 105(3):230–240

Cameron RWF, Taylor JE, Emmett MR (2014) What's 'cool' in the world of green façades? How plant choice influences the cooling properties of green walls. Build Environ 73:198–207

Castleton HF, Stovin V, Beck SBM, Davison JB (2010) Green roofs; building energy savings and the potential for retrofit. Energ Buildings 42(10):1582–1591

Crossman ND, Bryan BA, Cooke DA (2011) An invasive plant and climate change threat index for weed risk management: integrating habitat distribution pattern and dispersal process. Ecological Indicators, Spatial information and indicators for sustainable management of natural resources 11(1): 183–198

Daly E, Deletic A, Hatt BE, Fletcher TD (2012) Modelling of stormwater biofilters under random hydrologic variability: a case study of a car park at Monash University, Victoria (Australia). Hydrol Process 26(22):3416–3424

Deak-Sjöman J, Sang N (2015) Flood and climate modelling for urban ecosystem services. In: Sang N, Ode Sang Å (eds) A review on the state of the art in scenario modelling for environmental management, Report 6695. Swedish Environmental Agency, Stockholm, pp 131–162

Dirnböck T, Essl F, Rabitsch W (2011) Disproportional risk for habitat loss of high-altitude endemic species under climate change. Glob Chang Biol 17:990–996

EC (2007) Communication from the Commission to the Council, the European Parliament, the European Economic and Social Committee and the Committee of the Regions – Limiting Global Climate Change to 2 Degrees Celsius – The Way Ahead for 2020 and beyond. http://eur-lex.europa.eu/legal-content/EN/TXT/HTML/?uri=CELEX:52007DC0002&from=EN

EC DG (2015) Towards an EU research and innovation policy agenda for nature-based solutions & re-naturing cities final report of the horizon 2020 expert group on "Nature-based solutions and re-naturing cities": (Full Version). Publications Office of the European Union, Luxembourg. http://dx.publications.europa.eu/10.2777/765301

Faeth SH, Bang C, Saari S (2011) Urban biodiversity: patterns and mechanisms. Ann N Y Acad Sci 1223(1):69–81

Farrell C, Szota C, Williams NSG, Arndt SK (2013) High water users can be drought tolerant: using physiological traits for green roof plant selection. Plant Soil 372(1–2):177–193

Fletcher TD, Shuster W, Hunt WF, Ashley R, Butler D, Arthur S, Trowsdale S et al (2015) SUDS, LID, BMPs, WSUD and more – the evolution and application of terminology surrounding urban drainage. Urban Water J 12(7):525–542

Forzieri G, Feyen L, Russo S, Vousdoukas M, Alfieri L, Outten S, Migliavacca M, Bianchi A, Rojas R, Cid A (2016) Multi-hazard assessment in Europe under climate change. Clim Chang 137(1–2):105–119

Fujibe F (2009) Detection of urban warming in recent temperature trends in Japan. Int J Climatol 29(12):1811–1822

Gago EJ, Roldan J, Pacheco-Torres R, Ordóñez J (2013) The city and urban heat islands: a review of strategies to mitigate adverse effects. Renew Sust Energ Rev 25:749–758

Givoni B (1991) Impact of planted areas on urban environmental quality: a review. Atmos Environ Part B, Urban Atmos 25(3):289–299

Gong K, Wu Q, Peng S, Zhao X, Wang X (2014) Research on the characteristics of the water quality of rainwater runoff from green roofs. Water Sci Technol 70(7):1205–1210

Haaland C, Konijnendijk van den Bosch C (2015) Challenges and strategies for urban green-space planning in cities undergoing densification: a review. Urban For Urban Green 14(4):760–771

Haines A, Kovats RS, Campbell-Lendrum D, Corvalan C (2006) Climate change and human health: impacts, vulnerability and public health. Public Health 120(7):585–596

Hunter AM, Williams NSG, Rayner JP, Aye L, Hes D, Livesley SJ (2014) Quantifying the thermal performance of Green Façades: a critical review. Ecol Eng 63:102–113

Jamei E, Rajagopalan P, Seyedmahmoudian M, Jamei Y (2016) Review on the impact of urban geometry and pedestrian level greening on outdoor thermal comfort. Renew Sust Energ Rev 54:1002–1017

Klein PM, Coffman R (2015) Establishment and performance of an experimental green roof under extreme climatic conditions. Sci Total Environ 512–513:82–93

Kowarik I (2011) Novel urban ecosystems, biodiversity, and conservation. Environmental pollution, selected papers from the conference Urban Environmental Pollution: overcoming obstacles to sustainability and quality of Life (UEP2010), 20–23 June 2010, Boston, USA, 159 (8–9): 1974–1983

Kreibich H, Bubeck P, Kunz M, Mahlke H, Parolai S, Khazai B, Daniell J, Lakes T, Schröter K (2014) A review of multiple natural hazards and risks in Germany. Nat Hazards 74(3):2279–2304

Lehmann I, Mathey J, Rößler S, Bräuer A, Goldberg V (2014) Urban vegetation structure types as a methodological approach for identifying ecosystem services – application to the analysis of micro-climatic effects. Ecol Indic 42:58–72

Lenton TM, Held H, Kriegler E, Hall JW, Lucht W, Rahmstorf S, Schellnhuber HJ (2008) Tipping elements in the earth's climate system. Proc Natl Acad Sci 105(6):1786–1793

Leuzinger S, Vogt R, Körner C (2010) Tree surface temperature in an urban environment. Agric For Meteorol 150(1):56–62

Lundholm J, MacIvor JS, MacDougall Z, Ranalli M (2010) Plant species and functional group combinations affect green roof ecosystem functions. PLoS One 5(3):e9677

Madsen H, Lawrence D, Lang M, Martinkova M, Kjeldsen TR (2014) Review of trend analysis and climate change projections of extreme precipitation and floods in Europe. J Hydrol 519(PD):3634–3650

Metz B, Davidson OR, Bosch PR, Dave R, Meyer LA (2007) Contribution of working group III to the fourth assessment report of the intergovernmental panel on climate change. https://www.ipcc.ch/publications_and_data/ar4/wg3/en/contents.html

Muthanna TM, Viklander M, Thorolfsson ST (2008) Seasonal climatic effects on the hydrology of a rain garden. Hydrol Process 22(11):1640–1649

Nicholls RJ (2004) Coastal flooding and wetland loss in the 21st century: changes under the SRES climate and socio-economic scenarios. Glob Environ Chang 14(1):69–86

NOAA (2016) State of the climate: global analysis for July 2016. August. http://www.ncdc.noaa.gov/sotc/global/201607

Norton BA, Coutts AM, Livesley SJ, Harris RJ, Hunter AM, Williams NSG (2015) Planning for cooler cities: a framework to prioritise green infrastructure to mitigate high temperatures in urban landscapes. Landsc Urban Plan 134:127–138

Nowak DJ, Crane DE (2002) Carbon storage and sequestration by urban trees in the USA. Environ Pollut 116(3):381–389

Oliveira S, Andrade H, Vaz T (2011) The cooling effect of green spaces as a contribution to the mitigation of urban heat: a case study in Lisbon. Build Environ 46(11):2186–2194

Olsson J, Berggren K, Olofsson M, Viklander M (2009) Applying climate model precipitation scenarios for urban hydrological assessment: a case study in Kalmar City, Sweden. Atmospheric research, 7th international workshop on precipitation in urban areas, 92 (3), 364–375

Pascal M, Laaidi K, Ledrans M, Baffert E, Caserio-Schönemann C, Le Tertre A, Manach J, Medina S, Rudant J, Empereur-Bissonnet P (2005) France's heat health watch warning system. Int J Biometeorol 50(3):144–153

Rahman MA, Armson D, Ennos AR (2015) A comparison of the growth and cooling effectiveness of five commonly planted urban tree species. Urban Ecosyst 18(2):371–389

Sang N, Ode Sang Å (2015) A review on the state of the art in scenario modelling for environmental management, Report 6695. Swedish Environmental Agency, Stockholm

Semadeni-Davies A, Hernebring C, Svensson G, Gustafsson LG (2008) The impacts of climate change and urbanisation on drainage in Helsingborg, Sweden: suburban stormwater. J Hydrol 350:114–125

Skoulika F, Santamouris M, Kolokotsa D, Boemi N (2014) On the thermal characteristics and the mitigation potential of a medium size urban park in Athens, Greece. Landsc Urban Plan 123:73–86

Song J, Wang Z-H (2015) Impacts of mesic and xeric urban vegetation on outdoor thermal comfort and microclimate in phoenix, AZ. Build Environ 94(Part 2):558–568

Speak AF, Rothwell JJ, Lindley SJ, Smith CL (2013) Reduction of the urban cooling effects of an intensive green roof due to vegetation damage. Urban Clim 3:40–55

Stagl J, Mayr E, Koch H, Hattermann FF, Huang S (2014) Effects of climate change on the hydrological cycle in central and eastern Europe. In: Rannow S, Neubert M (eds) Managing protected areas in central and Eastern Europe under climate change, Advances in global change research 58. Springer, Dordrecht, pp 31–43

Stec WJ, van Paassen AHC, Maziarz A (2005) Modelling the double skin Façade with plants. Energ Buildings 37(5):419–427

Stovin V (2010) The potential of green roofs to manage urban stormwater. Water Environ J 24(3):192–199

Stovin V, Poë S, Berretta C (2013) A modelling study of long term green roof retention performance. J Environ Manag 131:206–215

Taha H (1997) Urban climates and heat islands: albedo, evapotranspiration, and anthropogenic heat. Energ Buildings 25(2):99–103

UN (2014) World urbanization prospects: the 2014 revision, highlights (ST/ESA/SER.A/352). Department of Economic and Social Affairs, Population Division

UNFCCC (2015) The Paris Agreements. United Nations framework convention on climate change. https://unfccc.int/resource/docs/2015/cop21/eng/l09r01.pdf

van der Waals J (2000) The compact city and the environment: a review. Tijdschr Econ Soc Geogr 91(2):111–121

Vico G, Revelli R, Porporato A (2014) Ecohydrology of street trees: design and irrigation requirements for sustainable water use. Ecohydrology 7(2):508–523

Wilby RL, Perry GLW (2006) Climate change, biodiversity and the urban environment: a critical review based on London, UK. Prog Phys Geogr 30(1):73–98

Yaghoobian N, Srebric J (2015) Influence of plant coverage on the total green roof energy balance and building energy consumption. Energ Buildings 103:1–13

Žuvela-Aloise M, Koch R, Buchholz S, Früh B (2016) Modelling the potential of green and blue infrastructure to reduce urban heat load in the city of Vienna. Clim Chang 135(3–4):425–438

Chapter 3
Nature-Based Solutions and Climate Change – Four Shades of Green

Stephan Pauleit, Teresa Zölch, Rieke Hansen, Thomas B. Randrup, and Cecil Konijnendijk van den Bosch

Abstract 'Nature-based -solutions' (NbS) aim to use nature in tackling challenges such as climate change, food security, water resources, or disaster risk management. The concept has been adopted by the European Commission in its research programme Horizon 2020 to promote its uptake in urban areas and establish Europe as a world leader of NbS. However, the concept has been defined vaguely. Moreover, its relationships with already existing concepts and approaches to enhance nature and its benefits in urban areas require clarification.

Notably, ecosystem-based adaptation (EbA), urban green infrastructure (UGI) and ecosystem services (ESS) have gained prominence in academic debates and are increasingly referred to in policy-making. In this chapter main features of each of the concepts, as well as overlaps and differences between them are analysed based on a review of key literature.

NbS is the most recent and broadest of the four concepts. Therefore, it may be considered as an umbrella to the other concepts but with a distinct focus on deployment of actions on the ground. EbA is a subset of NbS that is specifically concerned with climate change adaptation via the use of nature. As a planning approach, UGI, on the other hand, can provide strategic guidance for the integration of NbS into

S. Pauleit (✉)
Centre for Urban Ecology and Climate Adaptation (ZSK) and Chair for Strategic Landscape Planning and Management, Technical University of Munich, Munich, Germany
e-mail: pauleit@wzw.tum.de

T. Zölch • R. Hansen
Technical University of Munich, Munich, Germany
e-mail: teresa.zoelch@tum.de; hansen@tum.de

T.B. Randrup
Swedish University of Agricultural Sciences, Uppsala, Sweden

Norwegian University of Life Sciences, Ås, Norway
e-mail: thomas.randrup@slu.se

C. Konijnendijk van den Bosch
Department of Forest Management, University of British Columbia, Vancouver, BC, Canada
e-mail: cecil.konijnendijk@ubc.ca

© The Author(s) 2017
N. Kabisch et al. (eds.), *Nature-based Solutions to Climate Change Adaptation in Urban Areas*, Theory and Practice of Urban Sustainability Transitions, DOI 10.1007/978-3-319-56091-5_3

29

developing multifunctional green space networks at various scales. Finally, ESS value the benefits that humans derive from urban nature. ESS can support policy making for prioritising strategies and actions to maximise the benefits of NbS and can thus be considered as a kind of connecting concept between the other concepts. Overall, it is concluded that NbS is a powerful metaphor which, however, critically depends on UGI and ESS for its further definition and systematic uptake in urban areas.

Keywords Nature-based solutions • ecosystem-based adaptation • green infra-structure • ecosystem services

3.1 Introduction

Improving quality of life in cities, reducing their ecological footprint, and adapting them to climate change are three fundamental challenges that need to be urgently addressed (UN 2010). (Re-)integrating nature and natural processes into built areas is increasingly considered as a solution to these challenges (Handley et al. 2007).

This notion is not entirely new and can be traced back to the writings and works of eminent scholars and practitioners of planning, landscape architecture and urban ecology, such as Patrick Geddes (Welter and Lawson 2000), Ebenezer Howard (1902), Frederick Law Olmsted (Eisenman 2013), Ian McHarg (1969), Michael Hough (2004), Anne W. Spirn (1984) and Herbert Sukopp (Sukopp and Wittig 1998). However, an almost explosive emergence of statements, visions and concepts for "eco-urbanism" can be observed over the past two decades (Beatley 2000, 2011; Register 2006; Newman et al. 2009; Mostafavi and Doherty 2010; Lehmann 2010).

Some concepts for ecologically-oriented urban development have primarily enriched the academic discourse, such as "landscape urbanism" (e.g., Waldheim 2006), while other concepts have been conceived for, or found their way into the realm of policy making. Four of the latter concepts – nature-based solutions (NbS) (Balian et al. 2014), ecosystem-based adaptation (EbA) (Munang et al. 2013), green infrastructure (GI) (Benedict and MacMahon 2006), and ecosystem services (ESS) (MEA 2005) – are at the focus of this chapter. These four concepts have been selected because they have gained prominence in academic debates and are increasingly referred to in policy-making. Moreover, nature-based solutions is the core concept of this book, while ecosystem-based adaptation and green infrastructure are widely discussed and increasingly used in both planning and the climate change communities (Davies et al. 2015; Wamsler 2015; Zölch et al. submitted). Ecosystem services, in turn, are probably the most widely used concept of the four to strengthen the role of nature in decision-making (Haase et al. 2014). Therefore, these concepts appear to hold particular potential for informing and hence advancing the practice of landscape planning and landscape architecture. Of the concepts, only that of green infrastructure has had a clear link to the urban context from the start (Benedict and MacMahon 2002). Meanwhile all four concepts are now applied in urban set-tings (Gómez-Baggethun et al. 2013; Brink et al. 2016). Moreover, as all four concepts are still fairly new, we assume that they reflect current framing of environmental problems and solutions to these.

In this chapter we hypothesize that the four concepts are closely interrelated, partly overlapping and partly complementing each other. Furthermore, all of these concepts have a broad scope and they have been interpreted and taken up differently in academic debate and in practice (e.g., Davies et al. 2015; Hansen et al. 2015; Wamsler and Pauleit 2016). Hence, this chapter aims to characterise the four concepts to identify and discuss their commonalities and differences, as well as the relations between them. In doing so, the chapter will contribute to a well-informed use of the four concepts and a critical debate for their advancement.

3.2 Approach

This article is based on a selective, scoping literature review to identify the most relevant texts about the four concepts in focus, i.e., nature-based solutions, ecosystem-based adaptation, green infrastructure and ecosystem services. For all of the four concepts the scientific database ISI Web of Knowledge as well as Google were searched with different keyword combinations:

- Concept + urban
- Concept + climate change adaptation
- Concept + urban + climate change adaptation

In ISI Web of Knowledge, the search term combinations had to be refined depending on the hit rate and accuracy of the results. From the results of this search, the first 10 displayed as "newest" and the 10 "most cited" were scanned for suitability (title and abstract).

The search was repeated with Google adding "PDF" to each keyword combination to also include policy documents of international relevance. If it was likely that a source included content related to one of the four concepts, it was included in the literature review. Seminal literature, i.e., documents repeatedly referenced in the reviewed sources, was added through snowballing. The scoping review was undertaken in April 2016.

3.3 Nature-Based Solutions in Comparison with Other Concepts

3.3.1 Nature-Based Solutions

3.3.1.1 Definitions of the Concept and Its Origin

The concept of 'Nature-based solutions' (NbS) was introduced towards the end of the 2000s by the World Bank (MacKinnon et al. 2008) and IUCN (2009) to highlight the importance of biodiversity conservation for climate change mitigation and adaptation. NbS were put forward by IUCN in the context of the climate change negotiations in Paris "as a way to mitigate and adapt to climate change, secure water, food and energy supplies, reduce poverty and drive economic growth." (IUCN 2014). IUCN suggested seven principles as comprising the core of this concept, including cost efficiency, harnessing both public and private funding, ease of communication, and replicability of solutions (van Ham 2014). Notably, these principles highlight the role of NbS to address global challenges. More recently, the European Commission defined NbS as "actions which are inspired by, supported by or copied from nature" (EC 2015). Thus, NbS puts an explicit emphasis on linking biodiversity conservation with goals for sustainable and climate resilient development (Balian et al. 2014; Eggermont et al. 2015), and represent innovative, implementable 'solutions'.

Moreover, it is highlighted that NbS can be cost-effective and that benefits range from environmental protection to creating jobs and stimulating innovation for a green economy. Particular weight is placed on combining policy influence with actions on the ground to implement NbS (IUCN nd).

The European Commission adopted the concept of nature-based solutions for its research programme Horizon 2020 (EC 2015) with an explicit focus on urban areas. In preparation of the programme, a working group of scientists and policy makers elaborated on the concept (EC 2015, Annex 1:24). In short, Maes and Jacobs (2015:3) defined NbS "as any transition to a use of ecosystem services with decreased input of non-renewable natural capital and increased investment in renewable natural processes".

3.3.1.2 Main Features and Elements of NbS

Main features of NbS can be broadly summarised in four points:

First, NbS is broad in definition and scope. While the concept is rooted in climate change mitigation and adaptation, it is understood as an umbrella term for simultaneously addressing several policy objectives. Biodiversity conservation and enhancement of ecosystem services are considered as the basis for finding solutions to major challenges, ranging from climate change and disaster risk reduction to addressing poverty and promoting a green economy. The goal to simultaneously further economic growth and sustainability via NbS has been particularly stressed by the European Commission (Maes and Jacobs 2015, EC 2016).

Second, the concept is broad in terms of "nature". The report by the European Commission's expert group (EC 2015:38 ff.) lists 310 actions as examples of NbS, ranging from the protection and expansion of forest areas to capture gaseous pollutants, planting wind breaks for soil conservation to the protection of urban green spaces or planting of green roofs for various benefits such promotion of biodiversity, carbon storage and stormwater retention. Despite this breadth of concept, Eggermont et al. (2015) and Maes and Jacobs (2015) distinguished NbS from conventional engineering approaches for being multifunctional, conserving and adding to the stock of natural capital, and being adaptable and contributing to the overall resilience of landscapes.

Third, integrative and governance-based approaches to the creation and management of NbS are embraced (van Ham 2014). Therefore, the concept is distinguished from more traditional and top-down conservation, e.g., via protected areas towards finding solutions that aim to meet the needs of a diverse range of stakeholders. For this purpose, participatory approaches to co-design, co-creation and co-management ('co-co-co') of nature-based solutions are advocated (EC 2016).

Fourth, the concept of NbS is action-oriented. While IUCN recognises the need for linking policy with action on the ground, the latter is emphasised (MacKinnon et al. 2008, IUCN n.d.). However, the Horizon 2020 work programme for 2016–2017 seeks for systemic solutions to the development and implementation of NbS (EC 2016). This will require that attention is placed on regulatory frameworks, planning systems and economic instruments. Concurrently, Horizon 2020 expects large-scale pilot and demonstration projects that may serve as reference points for the upscaling of NbS across Europe and beyond.

3.3.2   Ecosystem-Based Adaptation

3.3.2.1   Definitions of the Concept and Its Origin

The concept of EbA is defined as "the use of biodiversity and ecosystem services as part of an overall adaptation strategy to help people adapt to the adverse effects of climate change" (CBD 2009:41). Its main focus relates to sustainable management, conservation and restoration of ecosystems with the objective to provide services supporting humans' adaptation to climate change (CBD 2009, Munang et al. 2013). Accordingly it is embedded into the concepts of ecosystem services and climate change adaptation (Chong 2014, Wamsler et al. 2014). Besides adaptation benefits, multiple social, economic and cultural co-benefits for local communities are also taken into account.

EbA first entered the stage in 2008 during the United Nations Framework Convention for Climate Change (UNFCCC), when the concept was included into the Bali Action Plan (IUCN 2008; Girot et al. 2012). It was then applied primarily with a geographical focus on the global South, but later also in the global North (Vignola et al. 2009, Brink et al. 2016, Andrade et al. 2011). Today the concept is considered to be valid for both developing and developed countries, and EbA is widely used internationally (Munang et al. 2013, Naumann et al. 2011a). The European Climate Change Adaptation Strategy, for example, encourages its implementation (EC 2013b).

To date, the EbA concept has mainly been applied in the sectors of agriculture and forestry (Doswald et al. 2014, Vignola et al. 2009), but in urban areas and sectors such as urban planning the interest in EbA as a cost-efficient, comprehensive and multifunctional approach is rising (Brink et al. 2016). In cities, EbA includes the design and improvement of green and blue infrastructures (Doswald and Osti 2011). EbA measures are referred to as use of urban ecosystems providing ecosystem services that benefit climate adaptation (Zandersen et al. 2014, Geneletti and Zardo 2016). However, pathways and supporting legislation for its systematic integration into urban planning is missing and there is limited evidence available on the actual uptake of the concept in municipal policies and plans (Wamsler et al. 2014, Wamsler 2015, Geneletti and Zardo 2016). Thus, academics advocate for mainstreaming EbA into urban planning, i.e., incorporating its principles into relevant policies and planning tools across sectors (e.g., Ojea 2015, Wamsler et al. 2014, Geneletti and Zardo 2016).

3.3.2.2 Main Features and Elements of EbA

Similar to the NbS concept, EbA is applied at different scales and in different sectors, and its implementation integrates various stakeholders, from national and regional governments to local communities, companies and NGOs, as well as involving multiple academic fields (Brink et al. 2016, Vignola et al. 2009). In cities, EbA measures can span from the micro-scale at building or small garden level to the macro-scale at city level (Geneletti and Zardo 2016; EC 2013a).

Moreover, EbA is promoted not only for environmental but also socio-economic benefits (Geneletti and Zardo 2016). Therefore, a people-centred approach is seen as main focus of EbA. Reid (2016) even argues that it is mutually supportive with community-based approaches aiming for societal benefits at the local level as compared to overall advances in economy and that this increases the potential for commitment of governments. Hence, the concept should be used with bottom-up and participatory approaches as well as ensure local interests and cultural prerequisites (Girot et al. 2012).

EbA focuses primarily on climate change adaptation and is hence more limited in scope than NbS. EbA is an integral part of overall adaptation strategies and encompasses adaptation policies and measures (CBD 2009, Andrade et al. 2011). Adaptation efforts then lead to co-benefits that extend beyond adaptation such as biodiversity conservation (Munang et al. 2013, Jones et al. 2012). According to a recent review, many EbA case studies concentrate on bio-geophysical assessment criteria for building an evidence-base for their adaptation potential (Brink et al. 2016), whereas a lack of quantitative estimates has been identified before (Jones et al. 2012, Doswald et al. 2014).

3.3.3 Green Infrastructure

3.3.3.1 Definitions of the Concept and Its Origin

The concept of GI stands for interconnected networks of all kinds of green spaces "that support native species, maintain natural ecological processes, sustain air and water resources and contribute to the health and quality of life" (Benedict and McMahon 2006:281). GI emerged from a growing concern of uncontrolled urban sprawl in the US in the 1990s (Benedict and McMahon 2002, Walmsley 2006). A new approach was called for whereby GI should not deal with space that was left after building or with infrastructure development but rather actively influence spatial planning by identifying ecologically valuable land as well as suitable areas for development (Benedict and McMahon 2002). GI planning should improve open space protection by offering an integrative and proactive approach (McDonald et al. 2005). GI is strongly connected to spatial planning and rooted in both landscape architecture and landscape ecology (Fletcher et al. 2014).

Initially, the GI approach resembled the approach to ecological networks that had been applied in Europe by being mainly described as a network of core habitats, stepping stones and corridors of areas with high nature value (Jongman 2004). Later, in the US, GI gained attention as a concept for sustainable stormwater management, promoted by the US Environmental Protection Agency. Here GI is often used interchangeably with approaches such as Low Impact Development (LID) or Sustainable Urban Drainage Systems (SUDS) (Fletcher et al. 2014).

Concurrently, GI is conveyed as a broader concept that can contribute to human well-being in many ways, including by strengthening stormwater management (Rouse and Bunster-Ossa 2013), not only in the US but also in Europe. European policy aims at mainstreaming GI in spatial planning and territorial development (EC 2013c) to reach the aims of the Biodiversity Strategy 2020 (EC 2013d). Conserving biodiversity is an important aim of the European GI Strategy, but GI is supposed to also contribute to various policy aims, including improving human health and well-being, achieving a more sustainable use of natural capital, and supporting the development of a green economy (EC DG Environment 2012; EC 2013d). However, GI is an elusive concept, being described depending upon the author(s) as either ecological networks or emphasising the benefits for human well-being (Mell 2009).

Overall, GI is often described as contributing to the same policy aims as NbS. Compared to EbA, the connection between GI and climate change adaptation is less in focus. Climate change adaptation is often just one of several policy aims GI is supposed to contribute to (EC 2013d; Lafortezza et al. 2013, Lovell and Taylor 2013). For instance, biodiversity conservation, promotion of human health and well-being, and social cohesion were more frequently mentioned than climate adaptation as goals in relevant policy documents in a comparative European study of strategic greenspace planning (Davies et al. 2015).

3.3.3.2   Main Features and Elements of GI

Compared to NbS and EbA, the GI concept has already found its way into the practice of spatial planning in urban areas. It has been applied by cities across the globe, most notably in the US and UK, but also including cities such as Barcelona and Lyon, and regions such as the Alpine Carpathian Corridor in Slovakia & Austria (Naumann et al. 2011a, Davies et al. 2015).

GI appears to be well-suited for urban planning for several reasons. First, the GI concept includes a spatial layer and criteria for GI components. GI is usually described as comprising a broad range of environmental features (see Table 3.1) and

Table 3.1 Green infrastructure components (EC 2013d)

Hubs: Core areas of high biodiversity value such as protected areas (e.g., Natura 2000 sites) and non-protected core areas with large healthy functioning ecosystems

Corridors and stepping stones: natural features like small watercourses, ponds, hedgerows, woodland strips

Restored habitats to reconnect or enhance existing natural areas (e.g., restored reedbed or wild flower meadow)

Artificial features such as eco-bridges, fish ladders or green roofs,to enhance ecosystem services or assist wildlife movement

Buffer zones that improve the general ecological quality and permeability of the landscape to biodiversity (e.g., wildlife-friendly farming)

Multi-functional zones with compatible land uses that support multiple land uses in the same spatial area (e.g., food production and recreation)

as existing at different spatial scales (e.g., national, regional, local) (EC 2013c). These components are required to be of high quality and be part of an interconnected network (EC 2013d).

Second, GI is based on a number of principles that can be applied in spatial planning. While multifunctionality and connectivity are most prominently described as core principles (Hansen and Pauleit 2014), additional principles have been proposed for the content and process of GI planning (see Table 3.2). Like NbS and EbA, GI is supposed to maintain and promote ecosystem services and deliver multiple benefits for humans (Lafortezza et al. 2013, Fletcher et al. 2014). For instance, in the European policy context, GI is defined as a green space network "designed and managed to deliver a wide range of ecosystem services". GI shall help to enhance and synergize benefits provided by nature in contrast to mono-functionally planned "grey" infrastructure (EC 2013c).

Table 3.2 Green infrastructure planning principles (Hansen and Pauleit 2014, based on Benedict and McMahon 2006, Kambites and Owen 2006, Pauleit et al. 2011).

Despite its different origin, the GI approach shares many features with the concepts of NbS and EbA. Starting with multifunctionality and the provision of multiple ecosystem services: the principle of multifunctionality further requires the involvement of a variety of stakeholders, such as private businesses, planning authorities, conservationists, the public and a range of policymakers (Naumann et al. 2011b, EC DG Environment 2012, Lovell and Taylor 2013). Therefore, like NbS and EbA, GI is supposed to be based on participatory planning processes that include a broad variety of community groups (Lovell and Taylor 2013).

Table 3.2 Green infrastructure planning principles

Approaches addressing the green structure
Integration: Green infrastructure planning considers urban green as a kind of infrastructure and seeks the integration and coordination of urban green with other urban infrastructures in terms of physical and functional relations (e.g., built-up structure, transport infrastructure, water management system).
Multi-functionality: Green infrastructure planning considers and seeks to combine ecological, social and economic/abiotic, biotic and cultural functions of green spaces.
Connectivity: Green infrastructure planning includes physical and functional connections between green spaces at different scales and from different perspectives.
Multi-scale approach: Green infrastructure planning can be used for initiatives at different scales, from individual parcels to community, regional and state. Green infrastructure should function at multiple scales in concert.
Multi-object approach: Green infrastructure planning includes all kinds of (urban) green and blue space; e.g., natural and semi-natural areas, water bodies, public and private green space like parks and gardens.
Approaches addressing governance processes
Strategic approach: Green infrastructure planning aims for long-term benefits but remains flexible for changes over time.
Social inclusion: Green infrastructure planning stands for communicative and socially-inclusive planning and management.
Transdisciplinarity: Green infrastructure planning is based on knowledge from different disciplines such as landscape ecology, urban and regional planning, and landscape architecture and developed in partnership with different local authorities and stakeholders.

Hansen and Pauleit (2014), based on Benedict and McMahon (2006), Kambites and Owen (2006) and Pauleit et al. (2011)

3.3.4 Ecosystem Services

3.3.4.1 Definitions of the Concept and Its Origin

Nature is the basis for the production of food, clean water and fresh air; natural elements such as trees and other vegetation act as filter for air pollution and can reduce the risk of flooding by runoff retention and infiltration. Also, nature has significant influence on humans, for example by providing restorative settings, for educational purposes, offering inspiration and promoting creativity. The provision of these services to humans by nature has been captured in the concept of ESS. Thus, ESS is basically a categorization of the broad range of 'benefits people obtain from ecosystems' (MEA 2005:V).

The concept emerged in the late 1970s when ecosystem functions beneficial to humans where termed as "services" in order to raise public awareness for biodiversity conservation (Gómez-Baggethun et al. 2010). Since then, the literature on ecosystem services has strongly expanded, which has also led to a rich debate on the concept, with particular emphasis on methods for assessment and valuation of ecosystem services, but also addressing ethical issues and how the concept can be mainstreamed into policy making (e.g., Cowling et al. 2008, Thompson 2008, Norgaard 2010, Vierikko and Niemela 2016). The United Nations led Millennium Ecosystem Assessment (MEA) report was a milestone with regard to the latter, as it represented the first ever global assessment of ecosystem services to policy makers.

Since the application of the ESS concept to the urban context at the end of the 1990s (Bolund und Hunhammar 1999), specific urban ESS research has led to a quick rise in publications (Haase et al. 2014). Goméz-Baggethun et al. (2013) concluded in a major review of urban ecosystem services, that the concept can play a critical role in reconnecting cities to the biosphere and in reducing the ecological footprint and ecological debt of cities, while enhancing resilience, health, and quality of life of their inhabitants. Moreover, the economic advantages of applying an ESS approach have been widely described, e.g., by Elmqvist et al. (2015) who stated that the benefits of investing in actively restoring rivers, lakes and woodlands occurring in urban areas may not only be ecologically and socially desirable, but also economically advantageous.

3.3.4.2 Main Features and Elements of ESS

The ESS concept developed out of a growing concern that the benefits humans derive from nature are not adequately reflected, if at all, in conventional economics (Goméz-Baggethun et al. 2010, Lelea et al. 2013). Therefore, the ESS concept can be considered as an attempt to redress this balance by the systematic assessment of demands for and supply of all kinds of services that ecosystems generate. To this end, the most popular current definition of ESS is that of the MEA: "the functions and products of ecosystems that benefit humans, or yield welfare to society" (MEA 2005). The MEA divides ecosystem services into four basic categories which in turn can be comprised of a large number of individual ecosystem services (Table 3.3).

Table 3.3 Categories of ecosystem services

Provisioning services

Goods, such as food or freshwater, that ecosystems provide and humans consume or use. In an urban context this may include, for instance, food production on urban and peri-urban farmland, on rooftops, in backyards and in community gardens.

Regulatory services

Services, such as flood reduction and water purification that healthy natural systems, such as wetlands, can provide. In an urban context this may include reduction of temperature and air pollution by vegetation through shading, by absorption of heat through evapotranspiration, and by removing pollution through leaves.

Cultural services

Intangible benefits, such as aesthetic enjoyment or contributing to identity of place that nature often provides. In urban areas recreation and aesthetics is probably the most significant service the natural environment serves to humans. This includes green infrastructure at large, including parks and other public as well as private green spaces.

Supporting services

Basic processes and functions, such as soil formation and nutrient cycling, that are critical to the provision of the first three types of ecosystem services.

Based on MEA (2005), Barthel et al. (2010), Gómez-Baggethun and Barton (2013)

Table 3.3 Categories of ecosystem services (Based on MEA 2005; Barthel et al. 2010; Gómez Baggethun and Barton 2013)

Although there is ample evidence of overall prevailing benefits of urban nature, it needs to be noted that ecosystems can also generate disservices, such as the clogging of gutters by leaves, production of allergenic pollen, or by enhancing the spread of diseases (e.g., via ticks or mosquitos) (Lyytimäki et al. 2008, Escobedo et al. 2011).

Increasing attention has also been given to exploring the potential synergies and trade-offs between various ESS (Raudsepp-Hearne et al. 2010, Martínez-Harms and Balvanera 2012). Urban river restoration, for instance, may not only reduce the risk of urban flooding but can at the same time restore typical floodplain habitats and provide new opportunities for recreational access to the river (Oppermann 2005). Enhancing the recreational capacity of a park, on the other hand, may lead to pressures on its biodiversity through more users and associated disturbances, and thus generate a trade-off (Chace and Walsh 2006). Therefore, approaches for multifunctional green infrastructure are sought that create synergies while avoiding trade-offs between the provision of different ESS (Hansen and Pauleit 2014).

Assessments of ESS provide policy-makers and practitioners with a comprehensive framework for building on and enhancing, rather than replacing, traditional approaches to solving environmental challenges. However, these assessments of ESS are an attempt to showcase services derived from nature for the benefit of humans rather than a tool for the long-term sustainment of these benefits. Therefore, there seems to be a tension between studying the values of ESS and communication of these on the one hand, and using the concept in practical planning and management on the other (e.g., Albert et al. 2014). In this regard, EbA and GI can become important bridging concepts to integrate ESS in urban development.

3.4   Discussion

3.4.1   Foundations of the Four Concepts

NbS, EbA, GI and ESS are four concepts that have been introduced in the past two decades to strengthen the role of nature in its widest meaning in policy-making – from the global to the site level. Table 3.4 provides a comparative overview of these concepts. They have co-evolved and are widely overlapping in terms of their scope and definition of nature. On the one hand, they are motivated by the concern to better protect nature, and specifically biodiversity, in a human-dominated world. On the other hand, the use of nature is considered as an option to complement, improve or even replace traditional engineering approaches, for example, for stormwater management. Therefore, all four concepts are clearly focusing on human interests, aiming to assert the environmental, social and economic benefits that people gain from nature. Moreover, they are problem-focused and they require inter- and transdisciplinary approaches. For instance, EbA is considered to tackle the challenge of climate change adaptation from multiple academic fields and concepts, e.g.,

Table 3.4 Comparison of the four concepts

Concept	Roots/origin and definition	Current focus	Governance focus	Use in urban context	Application in (planning) practice
NbS	New concept, definition still under debate/development Rooted in climate change mitigation and adaptation	Dealing with multiple societal challenges; biodiversity seen as central to solution	Integrative and governance-based approaches are embraced	Urban focus from the start	Still needs to be developed, but has a strong action focus (problem solving)
EbA	Rather new concept, with definition which is still debated Rooted in climate change adaptation	Climate change adaptation	People-centred approach; bottom-up and participatory approaches are called for	Focus initially mostly on wider agriculture and forestry, but now increasingly also urban	Still needs to be developed
GI	Concept with a history of about two decades; in Europe more recent; definition quite well established but also divergent Rooted in controlling urban sprawl, ecological network creation, but also stormwater management	Broad socio-ecological focus, with major role for landscape architecture and landscape ecology	Participatory planning processes are favoured	Well established	Very well established
ESS	Longest history and definition well established, although still debated Rooted in biodiversity conservation	Biodiversity conservation by (economic) valuation of services provided by nature	Focus on governance aspects, participation	Urban ESS have been in focus only more recently	Partly established, but needs operationalisation through other concepts (such as GI, NbS)

ecology, nature conservation, risk management and development, while NbS should address alternative ways to deal with broader societal challenges, such as unemployment and crime (Brink et al. 2016). Importantly, the four concepts aim to better integrate nature conservation into the economy without fundamentally challenging the economic system. Moreover, they highlight the need for community involvement in the management of natural capital, and to this end, they advocate the inclusion of a broad range of relevant actors in decision making.

3.4.2 Commonalities and Differences

Due to the breadth and the vagueness of their definitions, it is difficult to establish clear differences between the four concepts analysed in this chapter. Figure 3.1 suggests, however, that relationships can be observed between these concepts.

3.4.2.1 NbS vs. EbA

NbS, the most recent of the concepts, can be considered as an umbrella for the other three concepts while EbA may be considered as a subset of NbS for climate change adaptation (Naumann et al. 2014). Moreover, the concept of NbS is characterized by

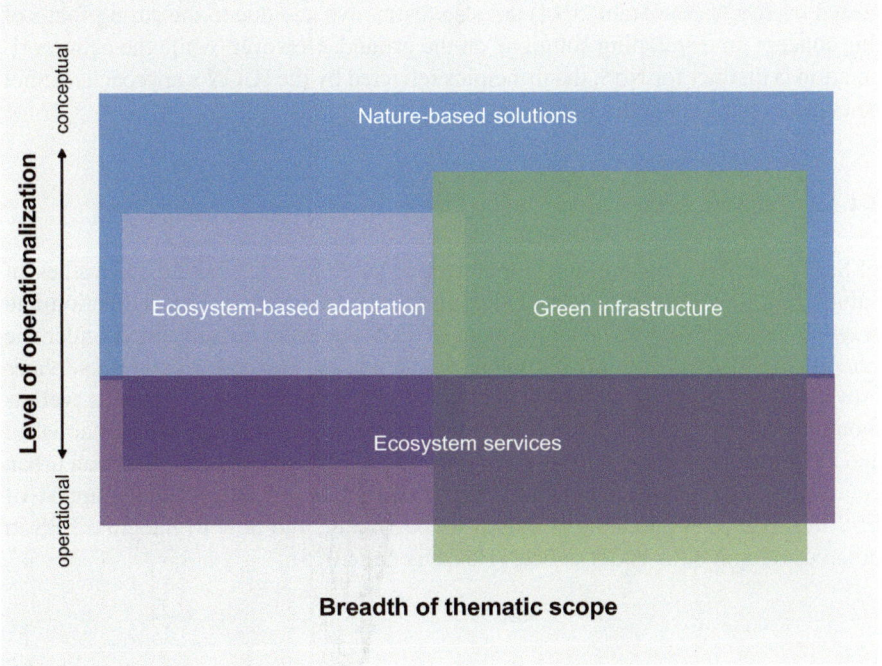

Fig. 3.1 Illustration of thematic scope and current level of operationalization of the four concepts

its orientation towards solutions, including the creation of new ecosystems (Eggermont et al. 2015).

NbS, but also EbA and GI should adhere to the principle of multifunctionality (Eggermont et al. 2015, Doswald et al. 2014, Davies et al. 2015). This would distinguish these approaches from mono-functional engineering solutions but also from e.g., intensive farming landscapes where the main focus is to generate agricultural products and further benefits such as biodiversity and recreation are not adequately considered. Multifunctionality means not only that, for instance, NbS deliver more than one ecological, social or economic function, but also that synergies between these functions should be sought while at the same time minimising trade-offs. How this can be achieved is rarely specified (Hansen et al. 2016). However, it has been suggested that the concept of ESS supports a systematic consideration of different functions respective services in GI planning as it defines a broad range of these services and provides tools for their assessment and valuation (Hansen and Pauleit 2014).

3.4.2.2   NbS vs. GI

While multifunctionality may thus be considered as linking between the four concepts, some differences between NbS and GI can also be observed. The role of biodiversity for developing solutions to global challenges is at the core of NbS but not necessarily in GI planning (Davies et al. 2015). Further, NbS principles suggested by IUCN (van Ham 2014) are also distinctive and due to the strong focus of this concept on developing solutions on the ground. However, while the action orientation is distinct for NbS, the principles reflected by the IUCN's approach, cannot be considered as generally agreed to.

3.4.2.3   NbS vs. ESS

ESS can support devising and implementing NbS by establishing the values of nature, and thus providing further definition of its substance. The distinction and assessment of a potentially large range of ESS that may be subsumed under the categories of supporting, provisioning, regulating and cultural ecosystem services provide a necessary foundation for defining and targeting policy goals as well as monitoring their outcomes. The ESS concept is now theoretically well established and a wide range of tools have been developed for ESS assessment, also in an urban context (e.g., Gómez-Baggethun and Barton 2013). However, systematic uptake of ESS in urban policy making is still at its beginning, and how to integrate ESS in urban development is under debate (Hansen et al. 2015).

3.4.3   Applicability in Urban Planning

It has been suggested that GI may support the uptake of ESS in urban planning (Hansen and Pauleit 2014), while Jones et al. (2012) call for recognizing urban GI as an important category of EbA strategies that are specifically appropriate within cities worldwide. Similar to NbS, GI is also a broad concept as recent work has shown (Rouse and Bunster-Ossa 2013, Davies et al. 2015) but it has its roots in planning and thus adds a spatial perspective to the concepts of NbS and EbA. Its application to urban settings may be considered as an approach for strategic planning of NbS, which is founded in principles of multifunctionality and connectivity (e.g., Hansen et al. 2016). GI can thus help to integrate NbS, EbA and not least ESS into the realm of urban planning with its established repertoire of instruments. In turn, national policies, such as the US Clean Water Act, have proven to be strong drivers for mainstreaming the GI concept into urban development (Rouse and Bunster-Ossa 2013). Consequently, so-called 'stormwater GI' has become more and more widespread in the USA. Swedish policies for ESS, on the other hand, have been shown to drive the adoption of EbA at municipal level (Wamsler and Pauleit 2016).

Furthermore, expanding governance-based approaches for GI may also advance the development and implementation of NbS via activities initiated by civil society at large (Buijs et al. 2016). In turn, GI may benefit from closely connecting it to the NbS and EbA discourses to re-emphasise the importance of biodiversity.

From this, we argue that NbS and EbA can make a significant change to current practice of urban development but core principles of the concepts should be more clearly articulated. A mere re-labelling of business as usual under the new concept of NbS would risk to discredit these concepts, on the other hand (Reid 2016). Therefore, the rapidly developing body of theory and methods of ESS and GI as well as evidence from their application should be recognised in further developing NbS as a concept. Moreover, emphasising the links to GI as planning approach and ESS as an approach to assess nature's benefits can promote systematic integration of NbS in urban development (Vierikko and Niemelä 2016; Hansen and Pauleit 2014. Conversely, it has been suggested that the NbS approach complements the ESS framework because it promotes (and relies more on) biodiversity to increase the resistance and resilience of soci-ecological systems to global changes (Eggermont et al. 2015).

3.5   Conclusion

Human activities have generated major environmental changes, which have accelerated from the 1950s onwards. Consequences, e.g., in terms of a changed climate with increasing stormwater events, as well as larger and increased soil and air pollution rates, are eminent today (e.g., NASA 2015). Both evidence and a general understanding exist for the fact that nature is both impacted by environmental changes and offers opportunities to solve a number of the challenges modern urban populations experience in their daily life.

In this context, this chapter reviewed four concepts: nature-based solutions (NbS), ecosystem-based adaptation (EbA), green infrastructure (GI) and ecosystem services (ESS). These concepts represent the dominant discourse on human-nature relationships in Western societies. They aim for better protecting and integrating nature into human development and, even more, for harnessing the values of nature for increasing human well-being. As hypothesised at the outset of this paper, the literature shows that the four concepts are interrelated. They build by and large on the same principles, such as multifunctionality and participation, but some differences can be observed in terms of breadth of concepts and their implementation in planning and practice. Based on the present analysis, it is suggested that NbS is an umbrella concept for EbA, GI and ESS. EbA is more specifically emphasising nature's role for climate change adaptation and it can be considered as a subcategory of NbS. GI is a concept that emerged in planning; it can help to develop strategic approaches for systematically integrating NbS and EbA into urban development at various scales. Finally, ESS provides means for measuring and valuing nature's benefits. While the other concepts are more practical and solution oriented, the concept of ESS is a more abstract one with a very strong focus on valuation.

The four concepts presented in this chapter should not be considered as competing but rather as complementary and mutually reinforcing. It would be difficult and even counterproductive to attempt providing sharp and narrow definitions for the concepts of NbS, EbA, and GI, as they would lose their flexibility to be applied in different local contexts and bridge between different actors. However, core principles such as multifunctionality, connectivity, being adaptive and adopting socially inclusive approaches to their implementation need to be operationalised if these concepts should make a substantial change to current urban development practices (see Ahern 2007, and Hansen et al. 2016, for a more elaborate discussion of core principles with reference to GI, Maes and Jacobs 2015, with reference to NbS, and Reid 2016, for EbA.) Findings from recent studies suggest that there is still ample scope for further research in this regard (e.g., Davies et al. 2015). The journey for the transformation of urban areas towards sustainability and resilience by means of large-scale implementation of NbS has only just begun.

Acknowledgments We are greatly indebted to the two reviewers whose comments substantially helped to improve the paper. Writing of this chapter was supported by the Bavarian State Ministry of the Environment and Consumer Protection [project number TLK01U- 63929] and by the European Union's Research and Innovation funding programme for 2007-2013 FP 7 (FP7-ENV.2013.6.2-5-603567).

References

Ahern J (2007) Green Infrastructure for cities. The spatial dimension. In: Novotny V (ed) Cities of the future. Towards integrated sustainable water and landscape management. IWA Publications, London, pp 267–283

Albert C, Aronson J, Fürst C, Opdam P (2014) Integrating ecosystem services in landscape planning: requirements, approaches, and impacts. Landsc Ecol 29:1277–1285

Andrade A, Córdoba R, Dave R, Girot P, Herrera FB, Munroe R, Oglethorpe J, Pramova E, Watson J, Vergara W (2011) Draft principles and guidelines for integrating ecosystem-based approaches to adaptation in project and policy design: a discussion document. CEM/IUCN, CATIE, Nairobi

Balian E, Eggermont H, Le Roux X (2014) Outcomes of the strategic foresight workshop "Nature-based solutions in a BiodivERsA context", Brussels, June 11–12, 2014. BiodivERsA Report. Available via http://www.biodiversa.org/671. Accessed 31 Jan 2015

Barthel S, Folke C, Colding J (2010) Social-ecological memory in urban gardens: retaining the capacity for management of ecosystem services. Glob Environ Chang 2082:255–265

Beatley T (2000) Green urbanism. Learning from European cities. Island Press, Washington, DC

Beatley T (2011) Biophilic cities. Island Press, Washington, DC

Benedict MA, McMahon ET (2002) Green infrastructure: smart conservation for the 21st century. Renewable Resour J 20:12–17

Benedict MA, McMahon ET (2006) Green infrastructure. Linking landscapes and communities. Island Press, Washington, DC

Bolund P, Hunhammar S (1999) Ecosystem services in urban areas. Ecol Econ 29(2):293–301

Brink E, Aalders T, Adam D, Feller R, Henselek Y, Hoffmann A, Ibe K, Matthey-Doret A, Meyer M, Negrut NL, Rau A-L, Riewerts B, von Schuckmann L, Tornros S, von Wehrden H, Abson DJ, Wamsler C (2016) Cascades of green: a review of ecosystem-based adaptation in urban areas. Global Environ Chang Hum Policy Dimens 36:111–123

Buijs A, Elands B, Havik G, Ambrose-Oji B, Gerőházi É, van der Jagt A, Mattijssen T, Møller MS, Vierikko K (2016) Innovative governance of urban green spaces – learning from 18 innovative examples across Europe. EU FP7 project GREEN SURGE (ENV.2013.6.2–5-603567), Deliverable 6.2. [Online] Available via http://greensurge.eu/working-packages/wp6/files/Innovative_Governance_of_Urban_Green_Spaces_-_Deliverable_6.2.pdf. Accessed 6 Mar 2016

CBD (Secretariat of the Convention on Biological Diversity) (2009) Connecting biodiversity and climate change mitigation and adaptation: report of the second Ad Hoc technical expert group on biodiversity and climate change, Technical series no. 41. Canadian Electronic Library, Montreal

Chace JF, Walsh JJ (2006) Urban effects on native avifauna, a review. Landsc Urban Plan 74:46–69

Chong J (2014) Ecosystem-based approaches to climate change adaptation: progress and challenges. Int Environ Agreements 14(4):391–405

Cowling RM, Egoh B, Knight AT, O'Farrell PJ, Reyers B, Rouget M, Roux DJ, Welz A, Wilhelm-Rechman A (2008) An operational model for mainstreaming ecosystem services for implementation. PNAS 105(28):9483–9488

Davies C, Hansen R, Rall E, Pauleit S, Lafortezza R, De Bellis Y, Santos A, Tosics I (2015) The status of European green space planning and implementation based on an analysis of selected European city-regions. EU FP7 project GREEN SURGE, Deliverable D5.1, Available via www.greensurge.eu. Accessed 10 Aug 2016

Doswald N, Osti M (2011) Ecosystem-based approaches to adaptation and mitigation – good practice examples and lessons learned in Europe. BfN-Skripten, Vol. 306. Bundesamt für Naturschutz, Bonn.

Doswald N, Munroe R, Roe D, Giuliani A, Castelli I, Stephens J, Moller I, Spencer T, Vira B, Reid H (2014) Effectiveness of ecosystem-based approaches for adaptation: review of the evidence-base. Climate Dev 6(2):185–201

EC (European Commission) (2013a) Thematic issue: ecosystem-based adaptation. Science for environment policy. European Commission, Brussels

EC (2013b) Communication from the Commission to the European Parliament, the Council, the European Economic and Social Committee and the Committee of the Regions. An EU strategy on adaptation to climate change. Brussels

EC (2013c) Communication from the Commission to the European Parliament, the Council, the European Economic and Social Committee and the Committee of the Regions. Green Infrastructure (GI) — Enhancing Europe's Natural Capital. COM(2013) 249 final

EC (2013d) Building a green infrastructure for Europe. Publications office of the European Union, Luxembourg. Available via http://ec.europa.eu/environment/nature/ecosystems/docs/green_infrastructure_broc.pdf. Accessed 20 July 2016.

EC (2015) HORIZON 2020 –Work Programme 2016–2017: 12. Climate action, environment, resource efficiency and raw materials. European Commission Decision C (2016) 1349 of 9 March 2016 EC (2015) Towards an EU Research and Innovation policy agenda for Nature-Based Solutions & Re-Naturing Cities. Final Report of the Horizon 2020 Expert Group on 'Nature-Based Solutions and Re-Naturing Cities' (full version). European Commission, Directorate-General for Research and Innovation. Available via http://bookshop.europa.eu/en/towards-an-eu-research-and-innovation-policy-agenda-for-nature-based-solutions-re-naturing-cities-pbKI0215162/. Accessed 10 Aug 2016

EC (2016) Horizon 2020 work programme 2016–2017. 12. Climate action, environment, resource efficiency and raw materials. European Commission Decision C(2016)4614 of 25 July 2016. Available via http://ec.europa.eu/research/participants/data/ref/h2020/wp/2016_2017/main/h2020-wp1617-climate_en.pdf. Accessed 10 Aug 2016

EC DG Environment (2012) The multifunctionality of green infrastructure. Brussels. Available via http://ec.europa.eu/environment/nature/ecosystems/docs/Green_Infrastructure.pdf, Accessed 10 Aug 2016

Eggermont H, Balian E, Azevedo JMN, Beumer V, Brodin T, Claudet J, Fady B, Grube M, Keune H, Lamarque P, Reuter K, Smith M, van Ham C, Weisser WW, Le Roux X (2015) Nature-based solutions: new influence for environmental management and research in Europe. Gaia 24(4):243–248

Eisenmann TS (2013) Frederick law olmsted, green infrastructure, and the evolving city. J Plan Hist 12(4):287–311

Elmqvist T, Setälä H, Handel SN, van der Ploeg S, Aronson J, Blignaut JN, Gómez-Baggethun E, Nowak DJ, Kronenberg J, de Groot R (2015) Benefits of restoring ecosystem services in urban areas. Curr Opin Environ Sustain 14:1010–1108

Escobedo FJ, Kroeger T, Wagner JE (2011) Urban forests and pollution mitigation: analyzing ecosystem services and disservices. Environ Pollut 159(8–9):2078–2087

Fletcher TD, Shuster W, Hunt WF, Ashley R, Butler D, Arthur S, Trowsdale S, Barraud S, Semadeni-Davies A, Bertrand-Krajewski J-L, Mikkelsen PS, Rivard G, Uhl M, Dagenais D, Viklander M (2014) SUDS, LID, BMPs, WSUD and more – the evolution and application of terminology surrounding urban drainage. Urban Water J 12:1–18

Geneletti D, Zardo L (2016) Ecosystem-based adaptation in cities: an analysis of European urban climate adaptation plans. Land Use Policy 50:38–47

Girot P, Ehrhart C, Oglethorpe J (2012) Integrating community and ecosystem-based approaches in climate change adaptation responses. Ecosystem livelihoods adaptation network. Availabe via http://careclimatechange.org/files/adaptation/ELAN_IntegratedApproach_150412.pdf. Accessed 31 July 2016

Gómez-Baggethun E, Barton DN (2013) Classifying and valuing ecosystem services for urban planning. Ecol Econ 86:235–245

Gómez-Baggethun E, de Groot R, Lomas PL, Montes C (2010) The history of ecosystem services in economic theory and practice: from early notions to markets and payment schemes. Ecol Econ 69:1209–1218

Gómez-Baggethun E, Gren Å, Barton DN, Langemeyer J, McPhearson T, O'Farrell P, Andersson E, Hamstead Z, Kremer P (2013) Urban ecosystem services. In: Elmqvist T, Fragkias M, Goodness J, Güneralp B, Marcotullio P, McDonald R, Parnell S, Haase D, Sendstad M, Seto K, Wilkinson C (eds) Urbanization, biodiversity and ecosystem services: challenges and opportunities. Springer, Dordrecht, pp 175–251

Haase D, Larondelle N, McPhearson T, Schwarz N, Hamstead Z, Kremer P, Artmann M, Wurster D, Breuste J, Borgström S, Jansson A, Elmqvist T, Andersson E, Langemeyer J, Gómez-Baggethun E, Kabisch N, Rall EL, Pauleit S, Hansen R, Voigt A, Qureshi S (2014) A quantitative review of urban ecosystem services assessment, concepts, models and implementation. Ambio 43(4):413–433

Handley J, Pauleit S, Gill S (2007) Landscape, sustainability and the city. In: Benson JF, Roe M (eds) Landscape and sustainability, 2nd edn. Spon, London, pp 167–195

Hansen R, Pauleit S (2014) From multifunctionality to multiple ecosystem services? A conceptual framework for multifunctionality in green infrastructure planning for urban areas. Ambio 43(4):516–529

Hansen R, Frantzeskaki N, McPhearson T, Rall E, Kabisch N, Kaczorowska A, Kain J-H, Artmann M, Pauleit S (2015) The uptake of the ecosystem services concept in planning discourses of European and American cities. Ecosyst Serv 12:228–246

Hansen R, Werner R, Santos A, Luz A, Száraz L, Tosics I, Vierikko K, Davies C, Rall E, Pauleit S (2016) Advanced urban green infrastructure planning and implementation. EU FP7 project GREEN SURGE, Deliverable D5.2. Available via www.greensurge.eu. Accessed 10 Aug 2016

Hough M (2004) Cities and natural process, 2nd edn. Routledge, London

Howard E (1902) Garden cities of tomorrow. Swan Sonnenscheind & Co. Ltd., London

IUCN (2009) No time to lose: make full use of nature-based solutions in the post-2012 climate change regime. Position paper on the fifteenth session of the conference of the parties to the United Nations Framework Convention on Climate Change (COP 15). IUCN, Gland

IUCN (2014) Nature-based solutions. Available via http://www.iucn.org/about/union/secretariat/offices/europe/european_union/key_issues/nature_based_solutions. Accessed 30 Apr 2016

IUCN (n.d.) Pioneering nature's solutions to global challenges. Available via https://cmsdata.iucn.org/downloads/iucn_english_brochure.pdf. Accessed 30 Apr 2016

IUCN (International Union for Conservation of Nature) (2008) Ecosystem-based adaptation: an approach for building resilience and reducing risk for local communities and ecosystems. A submission by IUCN to the Chair of the AWG-LCA with respect to the Shared Vision and Enhanced Action on Adaptation. UNFCCC, Gland

Jones HP, Hole DG, Zavaleta ES (2012) Harnessing nature to help people adapt to climate change. Nat Clim Chang 2:504–509

Jongman RHG (2004) The context and concept of ecological networks. In: Jongman RHG, Pungetti G (eds) Ecological networks and greenways: concept, design, implementation. Cambridge University Press, Cambridge, pp 7–33

Kambites C, Owen S (2006) Renewed prospects for green infrastructure planning in the UK. Plan Pract Res 21:483–496

Lafortezza R, Davies C, Sanesi G, Konijnendijk CC (2013) Green Infrastructure as a tool to support spatial planning in European urban regions. iForest 6:102–108

Lehmann S (2010) The principles of green urbanism. Earthscan, London

Lelea S, Springate-Baginskib O, Lakerveldc R, Debd D, Dashe P (2013) Ecosystem services: origins, contributions, pitfalls, and alternatives. Conserv Soc 11(4):343–358

Lovell ST, Taylor JR (2013) Supplying urban ecosystem services through multifunctional green infrastructure in the United States. Landsc Ecol 28:1447–1463

Lyytimäki J, Petersen LK, Normander B, Bezák P (2008) Nature as a nuisance? Ecosystem services and disservices to urban lifestyle. Environ Sci 5(3):161–172

MacKinnon K, Sobrevila C, Hickey V et al (2008) Biodiversity, climate change and adaptation: nature-based solutions from the Word Bank portfolio. World Bank, Washington, DC

Maes J, Jacobs S (2015) Nature-based solutions for Europe's sustainable development. Conserv Lett. doi:10.1111/conl.12216

Martínez-Harms MJ, Balvanera P (2012) Methods for mapping ecosystem service supply. A review. Int J Biodiver Sci Ecosystem Serv Manag 8:17–25

McDonald LA, Allen WL, Benedict MA, O'Conner K (2005) Green infrastructure plan evaluation frameworks. J Conser Plann 1:6–25

McHarg I (1969) Design with nature. The Natural History Press, Garden City/New York

MEA (2005) Millennium assessment report. Ecosystems and human well-being: synthesis. Island Press, Wahsington, DC

Mell IC (2009) Can green infrastructure promote urban sustainability? Proc ICE – Eng Sustain 162:23–34

Mostafavi M, Doherty G (2010) Ecological urbanism. Lars Müller Publ, Zurich

Munang R, Thiaw I, Alverson K, Mumba M, Liu J, Rivington M (2013) Climate change and ecosystem-based adaptation: a new pragmatic approach to buffering climate change impacts. Curr Opin Environ Sustain 5(1):67–71

NASA (2015) Global temperature. Accessible via http://climate.nasa.gov/vital-signs/global-temperature. Accessed 10 Aug 2016

Naumann S, Anzaldua G, Berry P, Burch S, Davis M, Frelih Larsen A, Gerdes H, Sanders M (2011a) Assessment of the potential of ecosystem-based approaches to climate change adaptation and mitigation in Europe. Oxford University Centre for the Environment, Oxford

Naumann S, Davis M, Kaphengst T, Pieterse M, Rayment M (2011b) Design, implementation and cost elements of Green Infrastructure projects. Final report to the European Commission, DG Environment, Contract no. 070307/2010/577182/ETU/F.1. Available via http://ec.europa.eu/environment/enveco/biodiversity/pdf/GI_DICE_FinalReport.pdf. Accessed 10 Aug 2016

Naumann S, Kaphengst T, McFarland K, Stadler J (2014) Nature-based approaches for climate change mitigation and adaptation. The challenges of climate change – partnering with nature. German Federal Agency for Nature Conservation (BfN), Ecologic Institute, Bonn

Newman P, Beatley T, Boyer H (2009) Resilient cities. Island Press, Washington, DC

Norgaard RN (2010) Ecosystem services: from eye-opening metaphor to complexity blinder. Ecol Econ 69(6):1219–1227

Ojea E (2015) Challenges for mainstreaming ecosystem-based adaptation into the international climate agenda. Curr Opin Environ Sustain 14(0):41–48

Oppermann B (2005) Redesign of the River Isar in Munich, Germany. Getting coherent quality for green structures through competitive process design ? In: Werquin AC, Duhem B, Lindholm G, Oppermann B, Pauleit S, Tjallingii S (eds) Green structure and urban planning. Office for Official Publications of the European Communities, Luxembourg

Pauleit S, Liu L, Ahern J, Kazmierczak A (2011) Multifunctional green infrastructure planning to promote ecological services in the city. In: Niemelä J (ed) Urban ecology. Patterns, processes, and applications. Oxford University Press, Oxford, pp 272–285

Raudsepp-Hearne C, Peterson GD, Bennett EM (2010) Ecosystem service bundles for analyzing tradeoffs in diverse landscapes. Proc Natl Acad Sci U S A 107:5242–5247

Register R (2006) Ecocities. New Society Publishers, Gabriola Island

Reid H (2016) Ecosystem- and community-based adaptation: learning from community-based natural resource management. Climate Dev 8(1):4–9

Rouse DC, Bunster-Ossa IF (2013) Green infrastructure, a landscape approach. American Planning Association, Washington, DC

Spirn AW (1984) The granite garden. Basic Books, New York

Sukopp H, Wittig R (eds) (1998) Stadtökologie, 2nd edn. G Fischer, Stuttgart

Thompson BH Jr (2008) Ecosystem services and natural capital: reconceiving environmental management. New York Univ Environ Law J 17:460–489

UN (2010) World urbanization prospects: the 2009 revision. United Nations Department of Economic and Social Affairs, Population Division, New York

Van Ham C (2014) Pioneering nature-based solutions. Available via www.biodiversa.org/673/download. Accessed 30 Apr 2016

Vierikko K, Niemela J (2016) Bottom-up thinking identifying socio-cultural values of ecosystem services in local blue-green infrastructure planning in Helsinki, Finland. Land Use Policy 50:537–547

Vignola R, Lacatelli B, Martinez C, Imbach P (2009) Ecosystem-based adaptation to climate change: what role for policy-makers, society and scientists? Mitig Adapt Strateg Glob Chang 14:691–696

Waldheim C (ed) (2006) The landscape urbanism reader. Princeton Architectural Press, New York

Walmsley A (2006) Greenways: multiplying and diversifying in the 21st century. Landsc Urban Plan 76:252–290

Wamsler C (2015) Mainstreaming ecosystem-based adaptation: transformation toward sustainability in urban governance and planning. Ecol Soc 20(2). doi:10.5751/ES-07489-200230

Wamsler C, Pauleit S (2016) Making headway in climate policy mainstreaming and ecosystem-based adaptation: two pioneering countries, different pathways, one goal. Clim Chang 135(3–4):71–87

Wamsler C, Luederitz C, Brink E (2014) Local levers for change: mainstreaming ecosystem-based adaptation into municipal planning to foster sustainability transitions. Glob Environ Chang 29(0):189–201

Welter V, Lawson J (eds) (2000) The city *after* Patrick Geddes. Lang, Bern

Zandersen M, Jensen A, Termansen M, Buchholtz G, Munter B, Bruun HG, Andersen AH (2014) Ecosystem based approaches to climate adaptation – urban prospects and barriers, vol 83. Aarhus University, Aarhus

Zölch T, Wamsler C, Pauleit S (submitted) Integrating the ecosystem-based approach into municipal climate change adaptation strategies. Submitted to Landscape and Urban Planning

Chapter 4
Double Insurance in Dealing with Extremes: Ecological and Social Factors for Making Nature-Based Solutions Last

Erik Andersson, Sara Borgström, and Timon McPhearson

Abstract Global urbanisation has led to extreme population densities often in areas prone to problems such as extreme heat, storm surges, coastal and surface flooding, droughts and fires. Although nature based solutions (NBS) often have specific targets, one of the overarching objectives with NBS design and implementation is to protect human livelihoods and well-being, not least by protecting real estate and built infrastructure. However, NBS need to be integrated and spatially and functionally matched with other land uses, which requires that their contribution to society is recognised. This chapter will present an ecologically grounded, resilience theory and social-ecological systems perspective on NBS, with a main focus on how functioning ecosystems contribute to the 'solutions'. We will outline some of the basic principles and frameworks for studying and including insurance value in work towards climate change adaptation and resilience, with a special emphasis on the need to address both internal and external insurance. As we will demonstrate through real world examples as well as theory, NBS should be treated as dynamic components nested within larger systems and influenced by social as well as ecological factors. Governance processes seeking to build urban resilience to climate change in cities and other urban dynamics will need to consider both layers of insurance in order to utilize the powerful role NBS can play in creating sustainable, healthy, and liveable urban systems.

E. Andersson (✉)
Stockholm Resilience Centre, Stockholm University, Stockholm, Sweden
e-mail: erik.andersson@su.se

S. Borgström
School of Architecture and the Built Environment, KTH Royal Institute of Technology, Stockholm, Sweden
e-mail: sara.borgstrom@abe.kth.se

T. McPhearson
Urban Ecology Lab, Environmental Studies, The New School, New York, NY, USA
e-mail: timon.mcphearson@newschool.edu

© The Author(s) 2017   51
N. Kabisch et al. (eds.), *Nature-based Solutions to Climate Change Adaptation in Urban Areas*, Theory and Practice of Urban Sustainability Transitions, DOI 10.1007/978-3-319-56091-5_4

Keywords Extreme weather events • Infrastructure • Resilience • Insurance value •
Social-ecological systems • Vulnerability • Social acceptance • Multifunctionality •
Flexibility • Participation

4.1   Introduction

Societies have suffered from extreme weather events throughout human history, and
despite tremendous technological advances over the last centuries we are today – in
some ways – even more vulnerable than before. Global urbanisation has led to extreme
population densities often in areas prone to problems such as extreme heat, storm
surges, coastal and surface flooding, droughts and fires (Grimm et al. 2000). We are
also increasingly dependent on infrastructure for our daily lives – transportation of
different kinds, communication networks, and power supply chains to mention just a
few. The reduction of risk, not least in the context of a changing climate and the pros-
pect of more frequent weather extremes (Coumou and Rahmstorf 2012), is a primary
target for nature based solutions (NBS). Although NBS often have specific targets,
one of the overarching objectives with NBS design and implementation is to protect
human livelihoods and well-being, not least by protecting real estate and built infrastructure, and unless otherwise specified this is what we mean by 'protection' in this
chapter. To do this, we argue that NBS need to be integrated and spatially and functionally matched with other land uses, which requires that their contribution to society
is recognised.

This chapter will present an ecologically grounded, resilience theory and social-
ecological systems perspective (Box 4.1) on NBS with a main focus on how functioning ecosystems contribute to NBS, rather than green technology or hybrid
systems with often a more minor biological component. We argue that the implications of such a perspective are critical for how we think about and design NBS as a
long term strategy to deal with climate change and, especially, its effects in terms of
extreme events. As we will show, NBS need to meet two criteria to provide long-
term protection: First, they must fit functionally and spatially with the vulnerable
areas (e.g. by providing a barrier between the source of a disturbance and potential
sufferers) and be sufficiently sized to match the magnitude of the disturbance.
Second, the NBS themselves need be resilient to the disturbance and long lag times
between events; they must be perceived as valuable, and thus supported, in times
when this capacity is not actively in use. This is especially true in densely populated
and contested urban landscapes, where other land uses constantly challenge the
preservation and protection of functioning ecosystems (Depietri and McPhearson
2017, Chap. 6, this volume).

We will outline some of the basic principles and frameworks for studying and
including insurance value (Box 4.2) in work towards climate change adaptation and
resilience. As we will demonstrate through real world examples as well as theory,
NBS should be treated as dynamic components nested within larger systems and
influenced by social as well as ecological factors. Since cities and urbanizing regions
are both very vulnerable to weather extremes and other climate change effects, and

Box 4.1 Resilience in Social-Ecological Systems
Resilience, as we here use it, has its roots in a seminal paper published by Holling in 1973 (Holling 1973), where he proposed that ecosystem dynamics are non-linear and that there are certain properties that make it more likely that a system retain its functions and character despite disturbances and changing drivers. A resilient system would thus be one that can absorb shocks, and reorganise, without undergoing fundamental change. As this chapter describes, the first writing on resilience within this school of thought had a strong ecological focus. Later, resilience thinking became a dominant approach within studies of social-ecological systems, adding to the study of the ecological properties of different ecosystems and species assemblages the different connections and interactions between people and the nature they are part of (Berkes and Folke 1998). Following this line, later iterations of resilience have included more social factors and now also explicitly include transformations within a development process as one of the necessary ingredients in a resilient system (Folke 2006). In accordance, we argue that NBS and their resilience are made up of social-ecological components and interactions, making governance a key challenge for resilience building. Further, transformation, in the context of this chapter, means that the embedding system around the NBS may need to shift and change, sometimes profoundly, to make sure that the NBS and the insurance they offer survive fundamental changes like an altered climate and new weather regimes.

Box 4.2 Insurance Value
The societal importance of ecosystems and biodiversity in buffering shocks, climate change induced weather extremes prominently among them, is increasingly examined through the metaphor of insurance value (Baumgärtner 2007; Green et al. 2016). Referring to the insurance value offered by ecosystems suggests that there is a critically important value in the structure and core ecosystem processes responsible for maintaining ecosystem functions and properties (e.g. Perrings 1995; Millennium Ecosystem Assessment 2005). Recognising that multiple definitions exist, we use insurance value to reflect the avoided socioeconomic and wellbeing costs associated with weather related disasters, and insurance itself as the maintenance of ecosystem services provided by social-ecological systems despite variability, disturbance and management uncertainty. With this definition, the insurance value of an ecosystem is closely related to its resilience, self-organizing capacity, and to what extent it may continue to provide flows of ecosystem service benefits over a range of variable environmental conditions (Green et al. 2016). While these aspects are increasingly appreciated in the work on NBS and insurance against climate change, we argue that the field has yet to recognise the need for making the NBS themselves survive over time.

tend not to give full consideration to functioning ecosystems as critical components of climate change adaptation, we focus on examples from such systems. That said, this approach should still be applicable to most or all social-ecological systems. Applying a 'double insurance' thinking, external and internal, can be seen as one step towards governance processes that better take into account the complexity of the systems we live in and the multifaceted nature of urban pressures and disturbances.

4.2 External Insurance

4.2.1 Ecological Foundations

Ecosystems of different kinds have been shown capable of mitigating weather extremes and thereby protecting other parts of urban systems and reducing the impact of disturbances (McPhearson et al. forthcoming). Examples include coastal ecosystems that provide a physical barrier from storm surges (Costanza et al. 2006; Koch et al. 2009), open land and permeable surfaces that protect from flooding through percolation (Farrugia et al. 2013) and urban trees and forests that mitigate heat waves (Jenerette et al. 2011; Depietri et al. 2012). Ecosystems differ from the more inert physical elements of many 'conventional' solutions. While, for example, coastal wetlands may provide coastal protection from storm surge and flooding, similar to sea walls, the wetlands do this as an integral part of an internal dynamics of wetland ecosystems. All ecosystems have their own dynamics, and to understand when, why and where they help us deal with climate change and its consequences requires knowledge of the fundamentals of systems ecology. It is important to get down to the details of understanding the role for species, species interactions, spatial structure, and how these together generate (or not) the ecological contribution to NBS insurance. Ecological character also has bearing on the efficiency of the NBS, and will influence how they compare to alternative solutions (cf. The Royal Society 2014).

Ecosystem function and resilience is an outcome of the organisms present in the system and their interactions – with each other and the physical environment (e.g. Chapin et al. 1997). The *life history traits* of an organism, such as physiology, behaviour, resource use and competitive strategies, determine both the ecological functions it contributes to and how it may respond to different disturbances and pressures (e.g. Mori et al. 2013). These traits comprise characters that affect soil stabilisation, water retention, radiation reflectance and different shielding effects (waves, noise, high winds etc.), among other things. In and of themselves, or in different combinations, these characters are the foundation for the first type of insurance value offered by NBS. For example, the shading and evapotranspiration by urban forest trees, which provide important cooling benefits in increasingly hot urban areas, depend on ecological processes like nutrient retention and cycling traits of soil microbes and invertebrates to maintain health and function of urban trees (Ballinas and Barradas 2016). Similarly, vegetation growing along urban riparian areas directly affect the outflow of water via the reduction of surface flow velocity, the partitioning of precipitation via canopy interception, evaporation and the promotion of infiltration, and via water uptake, storage and transpiration (Brauman et al. 2007; Nepf 2012; Gurnell 2014). Plants can also provide soil stabilization, which allows soil organism communities to develop. Riparian soils then are able provide important ecosystem functions such as nitrogen retention important in areas with significant upstream nitrogen fluxes (Groffman et al. 2003, 2004) and thus reduce the risk of soil erosion and eutrophication.

4.2.2 Vulnerability and Exposure

There is still considerable need for fundamental research on the ecological characteristics that could mitigate the effects of extreme events and climate change. Sound ecological knowledge is the first building block for understanding how to design NBS to meet different needs. These needs can be understood as the 'demand' for protection, and are very much determined by socioeconomic factors and cannot be captured by a biophysical assessment alone. An ecologically suitable NBS will only deliver expected solutions if it is sufficiently sized and adequately located (cf. Andersson et al. 2015). The answers to both these questions need to be sought in the larger social-ecological system – as is clearly demonstrated by the literature on vulnerability and exposure (e.g. Adger et al. 2005; Folke 2006).

According to IPCC "vulnerability describes a set of conditions of people that derive from the historical and prevailing cultural, social, environmental, political, and economic contexts" and "the propensity of exposed elements such as human beings, their livelihoods, and assets to suffer adverse effects when impacted by hazard events" (IPCC 2012, p. 71 and p. 69, respectively), while exposure refers to an "inventory of elements in an area in which hazard events may occur" (ibid., p. 69). While the literature and common usage sometimes conflate the two they do capture different aspects. To be vulnerable you must be exposed to the risk, but exposure alone is not sufficient. Awareness of exposure together with appropriately planned and implemented adaptation measures may at least reduce the vulnerability of people and property.

If NBS are to contribute to making communities less vulnerable they must be sized to match the magnitude of the disturbance and the extent of the exposed area. For example, bioswales and other types of green infrastructure being employed around the world in cities to provide NBS for mitigating surface flooding may fail if overwhelmed by large or extreme precipitation or flood events. Second, they need to be located in the right place. There are several possible spatial relationships between the source of the insurance capacity – the NBS – and where people and property may benefit from it (see e.g. Fisher et al. 2009; Renaud et al. 2016). Most usually, the insurance is achieved by the NBS providing a 'buffer' between the exposed area and the potential risk, e.g. wetlands upstream or along the coast outside a city (see Haase 2017, Chap. 7, this volume). This connection between the NBS and their beneficiaries is mediated by social structures such as built infrastructure and institutions defining access to land and designating land uses as well as socio-economic priorities of management and stewardship.

4.3 Internal Insurance

4.3.1 The Role of Diverse Ecosystems

Of concern in cities facing significant climate challenges is that assumed protection through NBS may fail when the ecosystems themselves are not adequately resistant and resilience to climate and other disturbances (McPhearson et al. 2015). Fundamental for the second level of insurance – the survival of the NBS themselves – is how the NBS respond to different pressures. This can be ecologically captured by the *robustness,* the capacity to cope and continue to deliver the desired function during an event, and by the presence of alternative functional pathways (e.g. different types of vegetation providing shading or soil stabilisation) with differential ability to respond to and cope with different pressures over time. Elmqvist and co-authors (2003) pointed to the importance of *response diversity* for being able to deal with changes, which in the context of this chapter can be understood as NBS being built up by components (organisms or communities) with differences in their responses to disturbances. By making sure that the NBS that are expected to help us cope and adapt to climate change are robust and resilient enough to withstand not only weather extremes and climate change itself, but a spectrum of different disturbances, we increase the probability that they will persist over time and thus be there when the rare, extreme events occur (cf. Mori et al. 2013).

To provide an example, urban forests are being widely used as NBS and generally considered as important to mitigate several weather related hazards – they keep the temperature down during heat waves, they can stabilize soils and help in preventing flooding (Roy et al. 2012; Livesley et al. 2016). However, despite this wide recognition of value, the fact that urban forests themselves may be vulnerable to the same disturbances is less recognised. Most studies that have looked at the benefits of such forests are based upon the species, number and size of trees in the urban landscape and how these shape effect (Nowak et al. 2013; Farrugia et al. 2013; Livesley et al. 2016). Less attention has been dedicated to the benefits of diversity and species specific traits in the composition of the forests, and when discussed it is usually along broad theoretical reasons for why diversity is important (Muller and Bornstein 2010). While tree population diversity is regarded as desirable few studies directly reflect or inform on the capacity of different tree species and species assemblages to deal with environmental stresses (May et al. 2013). For example, while temperature regulation – especially mitigating extreme heat – is one of the most important services trees can contribute, trees are also sensitive to drought, which is often associated with periods of above average temperatures and heat waves (Déry and Wood 2005). Drought induces physiological stress that may either kill the tree or make it vulnerable to pest attacks or disease challenges or fully inhibiting the ability of trees to provide the functions needed for NBS to heat (McDowell et al. 2011).

4.3.2 Spatial Dynamics

Beyond life history traits, *spatial dynamics* have been highlighted as an essential factor for resilience (Webb and Bodin 2008; Allen et al. 2016). In principle, having connections between for example the specimens of a specific type of NBS will usually promote the resilience of the individual NBS as well as the connected network of NBS. Each node in the network can potentially support the other nodes by replenishing locally extirpated populations and facilitating reorganisation (e.g. Hanski and Gilpin 1991; Bengtsson et al. 2003). While perhaps more indirect than the immediate response to disturbance, these processes make it more likely that the NBS survives and are in place to support system resilience when needed. However, connectivity can be problematic as some of the disturbances that challenge NBS may follow the same linkages (Holling et al. 2002; Allen et al. 2014).

For this reason, *modularity* has been suggested as a fluid middle way where smaller groups of nodes within a network are internally well connected but relatively isolated from other groups (e.g. Webb and Bodin 2008). Bioswales are a good example for how modularity can help maintain resilience. Since they are often small and spatially disaggregated, failure of one set of bioswales may not impact bioswales in other locations in the city allowing continued functioning for flood mitigation in places where bioswales did not fail. Similarly, street trees, which tend to be spatially separated, may be able to sustainably provide NBS when faced with disturbances more reliably in some instances than more spatially aggregated trees in urban forests. Tree insect pests, which may become increasingly threatening to NBS (Dale and Frank 2014), can spread more easily among closely growing tree stands. Here again, the principle of modularity where redundant functions are disaggregated can help to ensure continued green infrastructure performance for mitigating effects of climate change in cities.

4.3.3 Public Support: Making Sense of NBS

Ecological factors and features are, however, only one side of making NBS themselves resilient. NBS are embedded in social-ecological systems and social aspects are critical, for example political and economic priorities, human perceptions, norms and values, historical legacies and institutional contexts. Besides promoting governance that supports the necessary ecological underpinnings of NBS, a double insurance thinking approach to NBS also requires flexibility and open-ended designs to handle shifting priorities, and form strategies for long-term public support.

The survival of a NBS over time, especially in systems with a very active presence of people, depends on how people view them, and how they are managed. If the function of a NBS is not appreciated or understood, it risks being replaced by something else with a more apparent value to people. Climate change may make weather extremes more frequent, but there is still quiet periods between the times when NBS are needed and when their function and value are less evident (cf. Andersson et al. 2015). Also, diversity in preferences must be taken into account; an NBS might be attractive to some while disagreeable to others. One example is dense forests, which are of great recreational values to some people while perceived as unsafe and dangerous by others. Wetlands is another; they can be seen as either beautiful, rich bird habitats providing flood protection and water cleaning, or as breeding grounds for parasites caring diseases and environments that create risks for drowning accidents.

One way of engaging with social acceptance is to make sure that the NBS are multi-functional and, most importantly, that they have clear functions also during the periods between extreme events. Multi-functional green structures are today increasingly discussed and also implemented as a solution to the decreasing extent of green space in densifying cities (e.g. Gomez-Baggethun et al. 2013). However, presently this is mostly a matter of designing green structures to meet multiple demands at the same location at the same time, less about safeguarding multiple latent functions that can be activated or utilized when needed. The availability of different functions, and sometimes the magnitude of their effect, is also a question of the institutional set up, where different user rights and policy recommendations can shift and intentionally or unintentionally change functionality by emphasizing, strengthening or suppressing different functions. An example of this is the current trend of urban gardening, which while in many ways adding functionality to many urban green spaces might suppress other functions or the access to these.

In addition to direct utility, ecological quality, and long term resilience, must be better recognised and appreciated if we are to reach a publically supported double insurance. With increasingly participatory approaches to NBS design and management, finding ways to incorporate functions that are not needed at present and that are for a common good (flood prevention) instead of specific interests (urban gardening) is an urgent and challenging necessity.

4.4   Investing in Insurance: Governance Frameworks

How can we design and plan for NBS that provide protection to present large scale disturbances and at the same time also can withstand drastic economic and social fluxes at different scales (e.g. the recent European economic and migration crises) without losing their long-term functionality and protective capacity? The effectiveness of each NBS element is dependent on the interactions with the surroundings as well as remotely connected complex systems. This reinforces the need for governance approaches to shift focus to anticipating, planning for, and navigating change instead of planning for a presumed controlled development driven by a few, often disconnected social and economic parameters (Duit et al. 2010). This calls for increasing connections between levels of governance and across professions, where for example green spaces are not treated as an isolated entity but integrated functional entities in the urban fabric.

4.4.1 Flexibility

An ecologically resilient NBS also needs a resilient governance system which implies a shift towards more adaptive and participatory approaches (Biggs et al. 2015). The need to explore new ways of NBS planning (spatially and practically), knowledge development and transfer, and NBS design and maintenance – all with a long time perspective, is at odds with current often short term, mainstreamed and efficiency driven NBS governance. A more adaptive mode of NBS governance means embracing experimental approaches, where evaluation of goals, measures and outcomes are built into continuous learning (Walters 1986). It also requires a high degree of flexibility and open-endedness where present requirements for functionality and protection are constantly weighed against the long-term capacity of the NBS to respond to yet unknown disturbances and changing demands and needs. This requires political and public acceptance of failures as learning leverages in NBS planning, design and management. One important aspect is to identify and articulate how to sustain options for the future, making sure that an NBS with certain function can also be used to provide other services if needed. For example, an urban green space that is today mostly used for outdoor recreation, but in times of heavy rain can act as a reservoir for water and thereby protect surrounding areas for getting flooded. Another spin to multifunctionality and flexibility is our tendency to build identity and value on continuity (with its fallacy for command and control (Holling and Meffe 1996)). Although challenging, a shift to increased temporal variation in what functions of an NBS are promoted could support resilience in two ways: it would make change rather than permanence part of people's everyday life (and perhaps something that could be of value in itself), and it would encourage more nimble governance where land designations, procedures, policy and institutional framings are less static. Here more research is needed.

4.4.2 Participation

The need for participation in resilient NBS governance goes far beyond public hearings in the planning phase. Reflecting and similar to the ecological principles, diversity in the social components of social-ecological systems is perhaps the most important characteristic. An NBS that is planned and designed in a way that makes it heavily dependent on one actor's flow of resources, organisation, institutional framework and motivation, is very vulnerable to social and economic changes relating to that particular actor. This is evident in several cities in Europe where the public finances have been severely impacted by economic crisis lately and the character of many green spaces has changed, either into vacant lots or into management by volunteering local groups. In heavily human influenced ecosystems new management regimes often lead to an altered functionality (cf. Andersson et al. 2007). This illustrates that functionality and mainstreaming of NBS are closely connected to multi-actor engagement and partnerships that provide a greater diversity of knowledges and practices, ranging from expert to experience driven, from which the experimental approach can draw (Tengö et al. 2014). In order to recognise the system context of the NBS it is necessary to create governance linkages that match these interactions and dependencies (Andersson et al. 2014). The participatory component of resilient governance is also a matter of creating economic insurance, where different financial resources can be activated to sustain functionality over time. Furthermore, it is about relying on multiple actors for continuous knowledge generation, e.g. citizen science (Krasny et al. 2014), knowledge transfer over time (Andersson and Barthel 2016) and practical management. This will in turn be of importance for the social support of the NBS existence and awareness of its changing functional design over time.

4.5 Conclusions

Ecosystems and green and blue infrastructure can provide long-term insurance to climate change, making them an integral part of strategies to meet this multi-faceted challenge (Table 4.1). However, the extent to which they will be able to do this will depend both on their quality and the context they are set in. First, the ability of NBS to provide insurance against impacts of extreme events requires understanding of

Table 4.1 The two levels of insurance. Definitions, key factors underlying the insurance capacity and key aspects of governance that could promote and support them

	Definition	Key factors	Governance
1st level of insurance, external	Capacity to protect the larger system based on regulating ecosystem functions and location relative to vulnerable areas	Ecosystem configuration, spatial location, the nature of exposure	Spatial planning, value recognition, cross boundary considerations and linkages
2nd level of insurance, internal	Robustness during an event and resilience over time, the survival of the NBS itself	Response diversity, multifunctionality, participation and involvement, broad recognition of value	Participative processes, recognition of multiple values, legal frameworks and recommendations that can facilitate flexible use over time

spatial context of ecosystems. If an NBS is not properly positioned, it will do little to mitigate extreme events. Coastal cities need coastal wetlands physically located between them and a storm surge if wetlands are to provide resilience to coastal storms. This first insurance value stems from NBS being entities with physical attributes, but, of course, they are more than that. They are alive, and often constitute complex systems in themselves. Functioning ecosystems interact with the larger social-ecological systems they are embedded in and have their own vulnerabilities and resilience. The second layer of insurance, the survival of the NBS over time, has both ecological and social roots. Ecologically, this can be described and assessed through the functional traits present in an NBS and through its functional linkages to its surroundings, e.g. as part of a blue or green network. Following from these, key principles for building resilience are to promote diversity among response traits and to find an appropriate level of modularity.

On the social side, we have highlighted flexibility in the governance of NBS and recognition and support from the public. NBS governance includes promoting diversity and redundancy, and having more open-ended and adaptive decision-making processes for governing ecosystems for multi-functionality in the face of multiple changes and pressures. Additionally, public support for specific and often mostly dormant NBS (the need for insurance is not constant but linked to occasional events (Andersson et al. 2015)) can be strengthened by managing and planing them to be multifunctional over time, and by making sure that some functions achieve social purposes even in the periods between extreme events. Governance processes seeking to build urban resilience to climate change and other urban dynamics will need to consider both layers of insurance in order to utilize the powerful role NBS can play in creating sustainable, healthy, and liveable urban systems. Though NBS are complex, which this chapter's discussion on a number of less considered aspects of social-ecological complexity of NBS has demonstrated, working with NBS remains an unrealized and high potential opportunity for resilience building that fits well within the power and capacity of planning offices or stewardship bodies to achieve.

References

Adger WN, Arnell NW, Tompkins EL (2005) Successful adaptation to climate change across scales. Glob Environ Chang Part A 15:77–86

Allen CR, Angeler DG, Garmestani AS et al (2014) Panarchy: theory and application. Ecosystems. doi:10.1007/s10021-013-9744-2

Allen CR, Angeler DG, Cumming GS et al (2016) REVIEW: Quantifying spatial resilience. J Appl Ecol 53:625–635. doi:10.1111/1365-2664.12634

Andersson E, Barthel S (2016) Memory carriers and stewardship of metropolitan landscapes. Ecol Indic 70:606–614. doi:10.1016/j.ecolind.2016.02.030

Andersson E, Barthel S, Ahrné K (2007) Measuring social-ecological dynamics behind the generation of ecosystem services. Ecol Appl 17:1267–1278. doi:10.1890/06-1116.1

Andersson E, Barthel S, Borgström S et al (2014) Reconnecting cities to the biosphere: stewardship of green infrastructure and urban ecosystem services. Ambio 43:445–453. doi:10.1007/s13280-014-0506-y

Andersson E, McPhearson T, Kremer P et al (2015) Scale and context dependence of ecosystem service providing units. Ecosyst Serv 12:157–164. doi:10.1016/j.ecoser.2014.08.001

Ballinas M, Barradas VL (2016) The urban tree as a tool to mitigate the urban heat island in Mexico City: a simple phenomenological model. J Environ Qual 45:157–166. doi:10.2134/jeq2015.01.0056

Baumgärtner S (2007) The insurance value of biodiversity in the provision of ecosystem services. Nat Resour Model 20:87–127. doi:10.1111/j.1939-7445.2007.tb00202.x

Berkes F, Folke C (eds) (1998) Linking social and ecological systems: management practices and social mechanisms for building resilience. Cambridge University Press, Cambridge.

Bengtsson J, Angelstam P, Elmqvist T et al (2003) Reserves, resilience and dynamic landscapes. Ambio 32:389–396

Biggs R, Schlûter M, Schoon ML (eds) (2015) Principles for building resilience: sustaining ecosystem services in social-ecological systems. Cambridge University Press, Cambridge

Brauman KA, Daily GC, Duarte TK, Mooney HA (2007) The nature and value of ecosystem services: an overview highlighting hydrologic services. Annu Rev Environ Resour 32:67–98. doi:10.1146/annurev.energy.32.031306.102758

Chapin FS, Walker BH, Hobbs RJ, et al (1997) Biotic control over the functioning of ecosystems. Science (80-) 277:500–504

Costanza R, Mitsch WJ, Day JW Jr (2006) A new vision for New Orleans and the Mississippi delta: applying ecological economics and ecological engineering. Front Ecol Environ 4:465–472

Coumou D, Rahmstorf S (2012) A decade of weather extremes. Nat Clim Chang 2:491–496. doi:10.1038/nclimate1452

Dale AG, Frank SD (2014) The effects of urban warming on herbivore abundance and street tree condition. PLoS One 9:e102996. doi:10.1371/journal.pone.0102996

Depietri Y, Renaud FG, Kallis G (2012) Heat waves and floods in urban areas: a policy-oriented review of ecosystem services. Sustain Sci 7:95–107. doi:10.1007/s11625-011-0142-4

Déry SJ, Wood EF (2005) Observed twentieth century land surface air temperature and precipitation covariability. Geophys Res Lett 32:L21414. doi:10.1029/2005GL024234

Duit A, Galaz V, Eckerberg K, Ebbesson J (2010) Governance, complexity, and resilience. Glob Environ Chang 20:363–368. doi:10.1016/j.gloenvcha.2010.04.006

Elmqvist T, Folke C, Nyström M et al (2003) Response diversity, ecosystem change, and resilience. Front Ecol Environ 1:488–494

Farrugia S, Hudson MD, McCulloch L (2013) An evaluation of flood control and urban cooling ecosystem services delivered by urban green infrastructure. Int J Biodivers Sci Ecosyst Serv Manag 9:136–145. doi:10.1080/21513732.2013.782342

Fisher B, Turner RK, Morling P (2009) Defining and classifying ecosystem services for decision making. Ecol Econ 68:643–653. doi:10.1016/j.ecolecon.2008.09.014

Folke C (2006) Resilience: the emergence of a perspective for social–ecological systems analyses. Glob Environ Chang 16:253–267. doi:10.1016/j.gloenvcha.2006.04.002

Green TL, Kronenberg J, Andersson E, Elmqvist T, Gómez-Baggethun E (2016) Insurance value of green infrastructure in and around cities. Ecosystems 19:1051–1063. doi:10.1007/s10021-016-9986-x

Gomez-Baggethun EP, Gren Å, Barton D et al (2013) Urban ecosystem services. In: Elmqvist T, Fragkias M, Goodness J et al (eds) Global urbanization, biodiversity, and ecosystems – challenges and opportunities cities and biodiversity outlook – scientific analyses and assessments. Springer, Dordrecht, pp 175–251

Grimm NB, Grove JM, Pickett STA, Redman CL (2000) Integrated approaches to long-term studies of urban ecological systems. Bioscience 50:571–583

Groffman PM, Bain DJ, Band LE et al (2003) Down by the riverside: urban riparian ecology. Front Ecol Environ 1:315–321. doi:10.1890/1540-9295(2003)001[0315:DBTRUR]2.0.CO;2

Groffman PM, Law NL, Belt KT et al (2004) Nitrogen fluxes and retention in urban watershed ecosystems. Ecosystems. doi:10.1007/s10021-003-0039-x

Gurnell A (2014) Plants as river system engineers. Earth Surf Process Landforms 39:4–25. doi:10.1002/esp.3397

Hanski I, Gilpin ME (1991) Metapopulation dynamics: empirical and theoretical investigations. Academic, London

Holling CS (1973) Resilience and stability of ecological systems. Annu Rev Ecol Syst 4:1–23.

Holling CS, Meffe GK (1996) Command and control and the pathology of natural resource management. Conserv Biol 10:328–337

Holling CS, Gunderson LH, Peterson GD (2002) Sustainability and panarchies. In: Gunderson LH, Holling CS (eds) Panarchy. Understanding transformations in human and natural systems. Island Press, Washington, DC, pp 63–102

IPCC (2012) Managing the risks of extreme events and disasters to advance climate change adaptation: special report of the intergovernmental panel on climate change. Cambridge University Press, Cambridge/New York

Jenerette GD, Harlan SL, Stefanov WL, Martin CA (2011) Ecosystem services and urban heat riskscape moderation: water, green spaces, and social inequality in Phoenix, USA. Ecol Appl 21:2637–2651. doi:10.1890/10-1493.1

Koch EW, Barbier EB, Silliman BR et al (2009) Non-linearity in ecosystem services: temporal and spatial variability in coastal protection. Front Ecol Environ 7:29–37. doi:10.1890/080126

Krasny ME, Russ A, Tidball KG, Elmqvist T (2014) Civic ecology practices: participatory approaches to generating and measuring ecosystem services in cities. Ecosyst Serv 7:177–186. doi:10.1016/j.ecoser.2013.11.002

Livesley SJ, McPherson GM, Calfapietra C (2016) The urban forest and ecosystem services: impacts on urban water, heat, and pollution cycles at the tree, street, and city scale. J Environ Qual 45:119–124. doi:10.2134/jeq2015.11.0567

May PB, Livesley SJ, Shears I (2013) Managing and monitoring tree health and soil water status during extreme drought in Melbourne, Victoria. Arboricult Urban For 39:136–145

McDowell NG, Beerling DJ, Breshears DD et al (2011) The interdependence of mechanisms underlying climate-driven vegetation mortality. Trends Ecol Evol 26:523–532. doi:10.1016/j.tree.2011.06.003

McPhearson T, Andersson E, Elmqvist T, Frantzeskaki N (2015) Resilience of and through urban ecosystem services. Ecosyst Serv 12:152–156. doi:10.1016/j.ecoser.2014.07.012

McPhearson T, Karki M, Herzog C et al forthcoming Urban ecosystems and biodiversity. In: Rozensweig C, Solecki B, Al E (eds) Urban Climate Change Research Network Second Assessment Report on Climate Change in Cities (ARC3-2). Cambridge University Press, Cambridge.

Millennium Ecosystem Assessment (2005) Living beyond our means: natural assets and human well-being. Island Press, Washington DC.

Mori AS, Furukawa T, Sasaki T (2013) Response diversity determines the resilience of ecosystems to environmental change. Biol Rev 88:349–364. doi:10.1111/brv.12004

Muller R, Bornstein C (2010) Maintaining the diversity of California's municipal forests. J Arboric 36:18–27.

Nepf HM (2012) Hydrodynamics of vegetated channels. J Hydraul Res 50:262–279. doi:10.1080/00221686.2012.696559

Nowak DJ, Hoehn RE III, Bodine AR et al (2013) Assessing urban forest effects and values: Toronto's urban forest. U.S. Department of Agriculture, Forest Service, Northern Research Station, Newtown Square

Perrings C (1995) Biodiversity conservation as insurance. In: Swanson TM (ed) The economics and ecology of biodiversity decline: the forces driving global change. Cambridge University Press, Cambridge, pp 69–78

Renaud FG, Sudmeier-Rieux K, Estrella M, Nehren U (eds) (2016) Ecosystem-based disaster risk reduction and adaptation in practice. Springer, Cham

Roy S, Byrne J, Pickering C (2012) A systematic quantitative review of urban tree benefits, costs, and assessment methods across cities in different climatic zones. Urban For Urban Green 11:351–363. doi:10.1016/j.ufug.2012.06.006

Tengö M, Brondizio ES, Elmqvist T et al (2014) Connecting diverse knowledge systems for enhanced ecosystem governance: the multiple evidence base approach. Ambio. doi:10.1007/s13280-014-0501-3

The Royal Society (2014) Resilience to extreme weather. The Royal Society Science Policy Centre Report 02/14. The Royal Society, London

Walters C (1986) Adaptive management of renewable resources. McGraw Hill, New York

Webb C, Bodin Ö (2008) A network perspective on modularity and control of flow in robust systems. In: Norberg J, Cumming GS (eds) Complex theory a sustain future. Columbia University Press, New York, pp 85–118

Chapter 5
Nature-Based Solutions Accelerating Urban Sustainability Transitions in Cities: Lessons from Dresden, Genk and Stockholm Cities

Niki Frantzeskaki, Sara Borgström, Leen Gorissen, Markus Egermann, and Franziska Ehnert

Abstract Nature based solutions are amongst other practices that transition initiatives work with when intervening in their place and change its fabric. Focusing on the actors establishing, driving and scaling these solutions in and across cities, we come to evince that nature-based solutions have transformative social impact since they mediate new social relations and new social configurations contributing to social innovation in cities, and change nature perception and human-nature relations in urban contexts. We built from evidence in three city-regions that over the past years they saw the proliferation of community-based and policy-based initiatives with the aim to improve sustainability, livability and the aspiration to foster inclusivity and social justice in their cities: the city of Dresden in Germany, the city of Genk in Belgium and the city-region of Stockholm in Sweden. We will elaborate on the different ways nature based solutions as practices of transition initiatives in cities get scaled and contribute to accelerating sustainability transitions in these city-regions. In line with this, we will draw cross-case lessons for urban planning on the tensions transition initiatives that experiment with and institutionalize nature-based solutions in their cities face when actively pursue acceleration strategies and pathways to scale.

N. Frantzeskaki (✉)
Dutch Research Institute For Transitions (DRIFT), Erasmus University Rotterdam, Rotterdam, The Netherlands
e-mail: n.frantzeskaki@drift.eur.nl

S. Borgström
School of Architecture and the Built Environment, KTH Royal Institute of Technology, Stockholm, Sweden
e-mail: sara.borgstrom@abe.kth.se

L. Gorissen
Flemish Institute of Technological Research (VITO), Mol, Belgium
e-mail: leen.gorissen@vito.be

M. Egermann • F. Ehnert
Leibniz Institute of Ecological Urban and Regional Development (IOER), Dresden, Germany
e-mail: m.egermann@ioer.de; f.ehnert@ioer.de

© The Author(s) 2017
N. Kabisch et al. (eds.), *Nature-based Solutions to Climate Change Adaptation in Urban Areas*, Theory and Practice of Urban Sustainability Transitions, DOI 10.1007/978-3-319-56091-5_5

Keywords Cities • Sustainability transitions • Transformations • Scaling • Acceleration • Institutions • Governance • Transition initiatives • Civil society • Learning

5.1 Introduction

Cities are places and spaces of sustainability transitions: where solutions can be created, tested, and scaled (Coenen and Truffer 2012; Wolfram and Frantzeskaki 2016). In urban contexts, we found that actor configurations position themselves through practice and belief and often create new realities worth investigating in view of amounting unsustainability pressures. In our research we cast an eye on transition initiatives, actor collectives led by public, civic, business or partnerships of those, who put in place new ways of doing, thinking and organizing and transform current systems of provision with the aim to actively contribute to environmental sustainability. Understanding how these actor configurations play out with local governments, businesses and how they change the 'rules of the game' in cities is a governance question upon which we reflect. Especially in view of how they can influence the pace of change to more sustainable practices, lifestyles and living in cities of the future (Romero-Lankao 2012).

In recent years, sustainability transition studies have identified transition initiatives as the seeds of local transformation (Seyfang and Smith 2007; Seyfang et al. 2014) looking closely on the ways they create new practices, new narratives and understandings of sustainabilities as well as viewing how they transform infrastructure systems (Reeves et al. 2014). Civil society-led transition initiatives have been examined closely by scholars to assess their transformative potential, and it is the local understanding and local knowledge that civil society has that catalyzes the "tailoring to local context" and consequently leads to a fast-paced realization of new ideas and new approaches for more socially responsible governance (Aylett 2010, 2013, p. 869). Civil society can advocate for more radical and progressive ideas rather than "returning to old ideals" (cf. Calhoun 2012, p. 27). These characteristics of rapid experimentation adapted to the local context make civil society function as a driver of sustainability transitions (Boyer 2015; Burggraeve 2015; Bussu and Bartels 2014; Calhoun 2012; Carmin et al. 2003; Cerar 2014; Christmann 2014; Creamer 2015; Foo et al. 2015; Forrest and Wiek 2015; Fuchs and Hinderer 2014; Garcia et al. 2015; Kothari 2014; Magnani and Osti 2016; Seyfang and Smith 2007; Seyfang and Longhurst 2013; Seyfang et al. 2014; Touchton and Wampler 2014; Verdini 2015; Zajontz and Laysens 2015; Walker et al. 2014; Warshawsky 2014; Wagenaar and Healey 2015).

When investigating how transition initiatives change urban space and urban systems of provision, we found that a great number of those initiatives put in place and experiment with solutions that restore nature, imitate and build upon nature processes as ways to address environmental issues in place-explicit ways, known as nature-based solutions. Nature-based solutions have been defined as living solutions underpinned by natural processes and structures that are designed to address various environmental challenges while simultaneously providing economic, social, and environmental benefits (European Commission 2015). Nature-based solutions as social-spatial interventions have a transformative impact in the relations between people and nature. First, nature-based solutions contribute in the mental and physical health and wellbeing of people in cities (Andersson et al. 2015; Ambrey and Fleming 2014, p. 1298; Bratman et al. 2015; Buchel and Frantzeskaki 2015; Carrus et al. 2015). Reconnecting with nature in cities can contribute to social ties, establishment of sense of community and social cohesion (Kazmierczak 2013). Second, nature-based solutions are systemic ways on locally responding to climate change pressures. So far research has focused on the (potential) insurance value of nature-based solutions that revolves around the restorative capacity of these solutions deeming them effective for climate change adaptation and mitigation (Green et al. 2016; Haase et al. 2012; Kabisch et al. 2016; Mullaney et al. 2015; Andersson et al. Chap. 4 this volume). We add to this understanding is that nature-based solutions can have regenerative impact (Carrus et al. 2015, p. 226).

We argue that for understanding the impact of nature-based solutions in cities, we need to attend to their social production (Ernston 2013). In this way, we will understand how nature-based solutions as social-spatial settings, they mediate the need and ability of actors and communities to establish a positive dependence of place motivating them to restore it (Tidball and Stedman 2012, p. 297). Third, transition initiatives are instrumental in creating and localizing nature-based solutions, moving from a passive experience 'of nature' to an active experience 'with (making) nature'. In this way, transition initiatives experiment with nature-based solutions, learn-by-doing on how to adapt them to city-specific and place-specific situations and geophysical characteristics and create new narratives and understandings of their benefits. As thus, nature-based solutions are seeds of transformation of local practice and local space towards more sustainable ones.

In line with the above, nature-based solutions as social-ecological settings require social actors and social processes to be implemented in cities. In this chapter we examine the way nature-based solutions scale as a social process that includes transition initiatives driving their social production. Specifically, we examine, how transition initiatives as actor configurations that establish, experiment and localise nature-based solutions shift them from 'solutions' to social configurations, making nature-based solutions the new 'urban commons of sustainability' and in this way contributing to accelerating sustainability transitions in cities.

5.2 Understanding the Acceleration Dynamics of Urban Sustainability Transitions

We describe transition initiatives which drive transformative change towards environmental sustainability in an accelerated pace, or simply transition initiatives that actively contribute to accelerating urban sustainability transitions. The latter is understood as contributing to conditions of balance, resilience and interconnected-ness that allows society to satisfy its needs without exceeding the capacity of its supporting ecosystems to continue to regenerate the services to meet those needs and without diminishing biodiversity. The initiatives under study all undertake activities that in one way or another support, promote, maintain, develop or upscale nature-based solutions.

We investigate the way transition initiatives operate in accelerating urban transition initiatives with nature-based solutions with a conceptual framework of acceleration mechanisms developed as the core of a 3 year EU funded research project ARTS (www.acceleratingtransitions.eu). The conceptual framework of the mechanisms driving accelerating of urban sustainability transitions is not to "be taken as a recipe for success" of how transition initiatives should relate and connect but "as handles for reflective practice" that in the context of multi-level governance can instigate accumulation of changes and increase the pace of changes required for urban sustainability transitions to take-off (Bussu and Bartels 2014). Building from the literature of sustainability transitions (Rotmans et al. 2001, p. 17; Van der Brugge and Van Raak 2007; Avelino and Rotmans 2011) and urban governance and with a focus on transformative agency working (Westley et al. 2013 p. 2; Olsson et al. 2014; Cote and Nightingale 2012), we propose five mechanisms that may contribute to acceleration of sustainability transitions and we selectively present the most recent writings that ground our conceptualization and further show the applicability of our conceptualization in understanding accelerating dynamics in cities.

Upscaling is the growth of members, supporters or users of a single transition initiative in order to spread these new ways of thinking, organizing and practicing. Boyer (2015, p. 322) identifies scaling up as a diffusion pathway of grassroots innovations focusing on a scaling of a practice by referring to "application of practice beyond an activist core, to a broader audience". Desa and Koch (2014) identifies the diffusion and spread of one product or practice to greater number of users as a scaling and Staggenborg, and Ogrodnik (2015) point at different ways that transition initiatives meaningfully scale up in broadening participation and engagement of new members and supporters without compromising values and operation models.

Replicating is the take up of new ways of doing, organizing and thinking of one transition initiative by another transition initiative or different actors in order to spread out these new ways. Garcia et al. (2015, p. 96) recognizes replication of innovative practices as a process that contributes to systemic change, further backing our conceptualization as a process for changing the pace of change in transitions. Boyer (2015, p. 322) identifies replication as a diffusion pathway of grassroots innovations addressing "spread through a network of dedicated activists" and later on, as "diffusion of practice with a committed activist network".

Partnering is the pooling and/or complementing of resources, competences, and capacities in order to exploit synergies to support and ensure the continuity of the new ways of doing, organizing and thinking. Partnering is the mechanism that describes the ways transition initiatives act for pooling synergies and leverage resources (Horsford and Sampson 2014, p. 961; Frantzeskaki et al. 2014; Garcia et al. 2015) and establishing partnerships as catalytic for change (Eckerberg et al. 2015). Partnering in the form of public-private partnerships occur to provide specific services "in the realms of public health, development and environment" (Borzel and Risse 2010, p. 120; Healey 2015). Partnering allows collective learning and tackling inequality in place-based projects via new actor collaborations (Devolder and Block 2015) and for seeking resources to sustain the initiative itself (Healey 2015, p. 111; Sagaris 2014).

The majority of research writings on impact of transition initiatives to sustainability transitions focused on site-specific impact and majorly under-examine the impact in context dynamics including governance. In our conceptual framework we include two mechanisms to describe and capture the way transition initiatives play out with the city context dynamics: Embedding and Instrumentalising.

Instrumentalising is tapping into and capitalizing on opportunities provided by the multi-level governance context of the city-region in order to forge resources for the continuity of the operation of the initiative. Chmutina et al. (2014) note that the ability of transition initiatives to capitalize on opportunities also relies on the existence of 'openness' situations within multi-level governance for change and for empowering transition initiatives. Acquiring resources is vital for sustaining the activities of the initiative that can also include regeneration of deprived areas (Fraser and Kick 2014, p. 1447; Healey 2015, p. 115) and for growing beyond a "talking shop" (Forrest and Wiek 2015).

Embedding is the alignment of old and new ways of doing, organizing and thinking in order to integrate them into city-regional governance patterns. Bussu and Bartels (2014) point at embedding engagement and conflict resolution practices from community to the city governance and as a way of "formalizing" community projects and participation methods (also addressed by Barr et al. 2011). Embedding involves the alignment of efforts, strategies and agendas/goals between transition initiatives and local government across scales (Horsford and Sampson 2014, p. 961; Garcia et al. 2015). Embedding is enabled by the recognition from the public sector to alter practices and be more responsive and receptive/dialectic to innovative practices coming from other actors (Boyer 2015, p. 322; Healey 2015; Healey and Vigar 2015; Frantzeskaki and Kabisch 2016). Overall, embedding captures the extent to which transition initiatives strategically shape the context in which they operate (Moss et al. 2014).

5.3 Case Studies

In the ARTS project we conduct transdisciplinary research with transition initiatives in a number of transition city-regions to collectively explore, and assess accelerating dynamics. We selected five transition city-regions in which we see signs of acceleration in the form of accumulated changes in policy, planning and civil society, and a booming number of transition initiatives in place that pressure the status quo with provoking sustainability practices, ideas and ways of organization. These city-regions include: Brighton in the UK, Budapest in Hungary, Dresden in Germany, Genk in Belgium, and Stockholm in Sweden. In this chapter we only present case study findings and cross-case study analysis of three of the city-regions as indicative where a critical mass of transition initiatives practicing and localizing nature-based solutions' exist and actively contribute to accelerating transitions.

5.3.1   The City-Region of Dresden, Germany

Dresden is a growing city of approximately 549.000 inhabitants in 2015 in the East of Germany. It is the capital of the Free State of Saxony. Leaving behind the communist past with the democratic transition of 1989/90, the environmental movement could organize and act freely. Even though the City of Dresden displays many unsustainable patterns, this gave birth to a plethora of local initiatives that promote alternative, sustainable ways of fulfilling human basic needs. The City of Dresden joined the European Climate Alliance in 1994 and committed to reducing its carbon footprint to a sustainable level of 2.5 tonnes of GHG emissions per capita by 2080.

5.3.1.1   Nature-Based Solutions Initiatives in the City-Region Dresden

In 2016, there are almost 100 local initiatives that strive for a sustainable future. They display a considerable diversity of approaches, covering domains such as food production and consumption, building and urban development, mobility and transport, energy production and consumption, biodiversity protection or environmental education. Many of them pursue nature-based solutions and combine them with social and economic innovations to foster a socio-ecological transition.

Urban Gardening Network Dresden   This umbrella organization was established in 2012 to link all the different garden projects within the city-region of Dresden. About two dozen of these kinds of projects with various backgrounds and concepts exist. Among them are community gardens, intercultural and multigenerational gardens, community supported agricultures as well as beekeeping initiatives. The network wants to link the already existing community gardens to support not only green infrastructure within the city, but also to encourage the creation of spaces for learning and experimenting. The objectives of the network are (1) the support of cooperation activities between the (mostly civil society-driven) gardening initiatives and the city-administration, (2) to generate synergies between the urban gardens, like by sharing tools or seeds (3) to exchange know-how and experience, and, (4) to collectively organize workshops, educational programs to raise public awareness for issues related to sustainable food production, green spaces, and biodiversity. The network understands itself as part of a bigger movement, promoting sustainable food production.

Consumer Cooperative for Organic Products This consumer cooperative (VG Verbrauchergemeinschaft für umweltgerecht erzeugte Produkte eG) is one of the biggest local cooperatives for organic food in Germany. With more than 9.000 members, the initiative plays a crucial role in offering Dresden residents a relatively easy option to buy organic food, but also daily goods (detergent, shampoo, etc.). In addition, the VG is well linked with many small sustainable producers in the city-region of Dresden. It helps them to be become much more known (e.g. a shop for natural colours). The VG was established in 1991 with its first rooms located at the Environmental Center Dresden and just a few members. Most of the contacts to the regional suppliers of organic food have existed for years and nowadays they are supplied by more than 80 farms and producers located nearby (production and further processing is provided within a radius of 150 km around the city-region). In 1994, the initiative established an association. In 2005, the economic part of the initiative was transferred into a cooperative structure. Due to the non-profit character of the cooperative, members pay lower prices on regional organic food. This is possible because membership fees cover overheads so that members have to pay only the producer price without an additional add-up of retailing. Nowadays, there are five different shops of the initiative, whereby four of them offer also non-members the opportunity to buy the offered goods, but to a higher retailing price. To re-connect producers and consumers, members can join annual guided tours to visit the suppliers of the VG.

Hufewiesen e.V. The initiative Hufewiesen is named after a unique area of 13 ha forest and meadow landscape located in the historical center of the district of Dresden-Pieschen (populated by about 50.000 inhabitants). Over the course of generations, the residents used the area as farmland. Yet, in 1990, the City Council of Dresden classified the Hufewiesen as prospective building land. Thereafter, the parcels of land were purchased by a real estate company and laid fallow ever since. When the company wanted to build on the entire area, the citizen initiative Hufewiesen formed in order to preserve the land as a common green space that is owned and managed by the citizens themselves ("Bürgergrün") since the Hufewiesen is the last green space in the district Dresden-Pieschen. In so doing, it builds on the idea of commoning. The initiative formed in 2012. It built up a large supporter network and organized a great variety of events and activities. It managed to raise awareness not only among civil society networks, but also in local politics and public administration as well as in the science and business community. In 2014, it started a large participation process, conducting a citizen survey with over 1.200 people, to develop a new concept for the area. The survey clearly showed that the majority of the people wants to preserve the area as a nearby recreation area and a green space accessible to the public. Participants proposed a mosaic of different sustainable uses (e.g. community gardening, children's playground, sports field, etc.). Currently, the initiative debates how to acquire ownership of the land and transform it into a common green space ("Bürgergrün"), for instance by creating a foundation.

5.3.1.2 Accelerating Dresden's Transition with Nature-Based Solutions' Initiatives

The three initiatives described above use nature-based solutions and environmental sustainability in different ways and different fields of actions (gardening, food and nature conservation). Hence, they also contribute to the acceleration of urban sustainability transitions in different ways.

The **Urban Gardening Network** evolved into a hub for acceleration in the food domain. Since the network was established in 2012, many more community gardens were founded in Dresden. This is due to the network's active work on spreading the idea of urban gardening, for instance by organizing the educational program 'Seitentriebe'. It offers manifold courses on cultivation, the conservation of food or herbalist education. All new community gardens participate in this program and spread the idea further. Due to this network character, the community gardeners were able to establish informal contacts and formal cooperation with the city administration (The Office for Green Space) and the Association of Dresden Allotment Gardens.[1] Based on this cooperation, they now jointly plan a 'living lab' in which single former allotment gardens are transformed into community gardens, then existing right next to each other within one site.

The **VG** contributes to acceleration in different ways. First of all, it does so simply by increasing the number of members. It not only managed to upscale its members to over 9.000, but also to attract citizens from different backgrounds, also reaching beyond the "green bubble of already committed people". More importantly, it could induce a co-evolution process of organic farming around the city-region. Most of the VG's suppliers converted from conventional to organic farming as they were encouraged, but also obliged to do so by the VG. These organic farmers benefit from the VG because it creates a constant demand for their organic produce. Without the VG, many of them would not be able to survive at normal market conditions. In so doing, it has a very important stabilizing function for local ecological farmers.

It is worth mentioning that the three presented initiatives, but also many other nature-oriented initiatives in Dresden seek to combine environmentalism with social activism. They put a high emphasis on participation, sharing and commons as values of its own. Against this background, the **Hufewiesen** contributes to promoting NBS by reconnecting citizens with the values of urban nature. They succeeded in gathering a considerable amount of citizens behind the association's objectives of nature protection and nature-based development and in changing the perspective of local politicians. With its initiative, it could shield the urban green from massive building investments so far (Photos 5.1, 5.2, 5.3 and 5.4).

---

[1] This represents about 55.000 allotment gardens in the city-region of Dresden.

Photo 5.1 Community Garden Dresden Johannstadt (Gartennetzwerk – cc license)

Photo 5.2 Guided tour of VG members to their suppliers of organic regional food, Dresden, Germany (VG Busfahrt – Olaf Stiebitz)

Photo 5.3 Hufewiesen I: Air picture of the Hufewiesen area – with this picture the initiative convinced many inhabitants of the value of the site and nature in the city, Dresden, Germany (Hufeweisen – cc license)

Photo 5.4 Hufewiesen II: Once a year, the Hufewiesen initiative invites inhabitants to experience and celebrate the urban green space they would like to protect, Dresden, Germany (Hufeweisen – cc license)

5.3.2   Genk Case, Belgium

Genk is the greenest city of Flanders in Belgium, housing a multicultural population of approximately 65 000 inhabitants. It shares the characteristics of European post-industrial cities that have gone through economic restructuring and are searching for a new identity. The City has an extraordinary history of experimentation on the front of social innovation and is at the moment in the middle of rethinking its economic backbone.

5.3.2.1   Nature-Based Solutions' Initiatives in the City Region

Genk also houses a variety of local volunteers that are active in environmental sustainability. Several transition initiatives aim to change nature perceptions, reinforce human-nature relationships or promote eco-friendly lifestyles and practices. Below we highlight a couple of initiatives that are active in promoting or supporting nature-based solutions:

The Heempark   The Heemparkis a unique example of a Public – Civic Partnership as it combines city supported (Environment and Nature Centre) and volunteer activities (Heempark vzw) in a collaborative and mutually beneficial way. The Heempark was initiated in 1985 by a group of local volunteers who aimed to safeguard the local agricultural heritage through the Heempark model – a combination of demo sites of local agricultural practice with a clear focus on environmental sustainability. When the number of visitors increased and the entire initiative grew over the heads of the volunteers, the City supported with personnel, infrastructure and an educational program for schools and associations. At the moment the Heempark has about 90 members and approximately 35 active volunteers that maintain the gardens and organise a wide variety of activities to promote reconnection of people to nature and increase environmental awareness and engagement. Activities include cooking classes of organic food with autistic children, sessions about herbs, bees and vegetables, making honey, building eco-friendly gardens and so on. Our findings show that the Heempark is a hotspot when it comes to connecting initiatives and reconnecting people to nature. It houses 350 educational groups while attracting approximately 10.000 visitors a year (Photo 5.5).

The Bee Plan   The Bee plan is a multi-stakeholder initiative to turn the city of Genk into a bee friendly city. It originated in 2013 when two concerned citizens organised an open environmental council meeting showcasing the documentary 'More than honey'. Afterwards, the 60 people present brainstormed about what they could do in the city to improve conditions for bees and all ideas were gathered. To take this further, a working group was set up involving beekeepers, city services, environmental organisations and engaged civilians who developed a 'Bee plan' for the city of Genk that was approved by the bench of aldermen in February 2014. The plan aims to strengthen the bee populations in the city region by (1) improving the

Photo 5.5 A volunteer guiding the Sunday herb walk at the Heempark, Genk, Belgium

living conditions for bees on communal land; (2) by engaging citizens to do the same on their land and (3) supporting local beekeepers. The plan thus aims to combine public and civic action, for instance in setting up a voluntary team of 30 bee ambassadors that participate in many public events, in a bee friendly makeover of public spaces, in organizing educational activities such as bees in the library etc. The Bee plan thus originated from a multi-actor brainstorm and developed into a policy plan coordinated by the City through the Heempark but in close cooperation with a wide range of volunteers (e.g. Bee Team) to restore and regenerate nature to place (Photos 5.6, 5.7 and 5.8).

Velt Genk & the Local Organic Allotment Gardens Velt Genk, the Guild for Eco-friendly Lifestyles and Organic Gardening, is a local section of the Flemish Velt vzw organisation. Velt vzw is a non-profit organisation that was founded in 1974 by a group of concerned citizens that wanted to make the agricultural sector healthier and more environmentally friendly by banning chemical pesticides and fertilizers. Gradually, the field of action shifted to private (vegetable) gardening. Velt vzw supported local citizens in Genk in the renewal project to convert the decayed allotment gardens into vibrant organic food production systems. The project, building on strong collaborative partnerships, proved to be so successful that it gave rise to a Flemish fund allowing replication all over Flanders. The local branch, Velt Genk was established in 1978 – thus being one of the oldest divisions of Velt vzw – and has currently

Photo 5.6 A volunteer of the Bee team shows how to make honey at the Heempark, Genk, Belgium

Photo 5.7 Bees in the Library in Genk, Belgium

Photo 5.8 Bee mobile with children in Genk, Belgium

more than 100 members. Volunteers from Velt Genk are involved in several other initiatives associated with environmental sustainability such as the newly established communal garden 'Tuin van Betty', they maintain a demonstration garden at the Heempark, initiate and support vegetable growing in centres for the elderly and local volunteers are active in the Organic allotment garden of Genk Noord. Velt Genk successfully campaigned to ban harmful pesticides from public green management. With their 'Without is healthier' campaign they convinced the mayor to sign an intention declaration spurring eco-friendly public green management.

5.3.2.2 Accelerating Genk's Transition with Nature-Based Solutions' Initiatives

The initiatives described above are all in one way or another bringing nature-based solutions into practice, either by converting vacant land into organic allotment gardens, by turning Genk into a bee friendly city, by setting up and maintaining demo sites showcasing and promoting environmentally friendly practices (e.g. compost schools, bee towers, organic vegetable gardens). They promote reconnection of citizens to nature by organizing a wide range of activities, localize nature to place by involving citizens in the restoration and regeneration of nature in the city. What is more, our findings indicate that these initiatives contribute to accelerating sustainability transitions by spreading nature-based thinking and doing, mobilizing people and money and changing governance institutions. These transformative and community-based initiatives thereby trigger social-ecological reconfigurations that safeguard/promote ecological functions and thus play an instrumental role in fueling systemic change (Gorissen et al. 2017).

5.3.3 Stockholm Case, Sweden

The urbanized parts of the Stockholm city-region consist of an urban mosaic of interwoven green-blue and grey structures. Nature is seen as an important landscape element has a strong foundation in Swedish culture and traditions, which is apparent in the many protected nature and outdoor recreation areas within and in close proximity to Swedish cities. However, besides these protected remains of a more natural landscape, other nature-based solutions have just recently gained interest, especially in relation to climate change challenges, e.g. run off water management. The Stockholm population of approximately 2 million people is increasing rapidly, creating a severe housing shortage and the urbanization is taking place by densification and sprawl with resulting decrease of larger green space, but also an exploration and demand for other kinds of nature based solutions in the densifying city districts. As of now most of the locally based sustainability initiatives are using the blue-green infrastructure as part of their transition agenda.

5.3.3.1 Nature-Based Solutions' Initiatives in the City Region Stockholm

Miljöverkstan Flaten In 2011 it all started around the bathing place Flatenbadet in a nature reserve of high recreational value in southern Stockholm. The place was falling apart but the founders saw that the place could offer much to the local residents. They wanted to build bridges to the site by giving the children living nearby connections to nature. At the same time, they wanted to pay tribute to the location, care for and nurture the nature, and to activate both children and adults. They started a group together with actors in the area for the revitalisation of Flatenbadet as the venue and contacted the City District Administration (CDA, within Stockholm municipality). The CDA had the opportunity to set aside a small amount of money that was used to design a nature trail for school children and also for doing background research and networking to create a meeting place. From these starting seeds Miljöverkstan Flaten was formed and a huge flow of interest was awakened. Today they work extensively with sustainability by place-making, focusing on local kindergartens, school kids, teenagers and recent immigrants in various projects using the area and local resources. Miljöverkstan Flaten view nature as the prime support of strong and tolerant individuals and hence uses nature experiences in a broad sense to handle and proactively work with social and environmental challenges. Nature is present in all their activities, including organizational meetings and partnership building. Their most important growth takes place by networking that has resulted in the creation of new local initiatives, activated youngsters and established a great network linking many parts of the society. In 2016 they became a key partner of a collaborative pilot project initiated by the Stockholm municipality that aims to explore how modern nature protected areas in urbanizing landscapes can be part of a larger transition to urban sustainable development. The Miljöverkstan Flaten use of nature-based solutions is captured by the nature-to-people as well as to localize nature-to-place dimensions.

Stockholm Green Wedges Collaboration From a perceived need to coordinate the 26 local municipalities around the green wedges of Stockholm city-region, NGOs representing environmental protection, tourism, outdoor recreation and cultural heritage formed the network "Protect the green wedges of Metropolitan Stockholm" in 1996. It was triggered by a report about the values of the ten green wedges launched by the Stockholm County Council Planning Division and it was used by the network in local and regional urban development debates. In 2006 the six municipalities and NGOs related to the Rösjö green wedge north of Stockholm were invited to a workshop by the regional branch of Swedish Society for Nature Conservation in order to form a mutual understanding of the green wedge as an ecological unit and hence inspire to cross-border/sector as well as municipal/civic society collaborations. The first steps included joint study visits, workshops, creation of a visitor's map, and building networks with regional authorities and other external key actors and eventually a platform for collaboration was created that also gained political support. At a certain point of concretizing, including work plan and budget, the politicians realized that the municipalities would be the leaders of this initiative. The initiative was then reorganized and the formalization process started, which is generally seen as beneficial. The initiative has gained attention and later became replicated in another green wedge also belonging to one of the initially engaged municipalities. The experiences are now spread to additional green wedges in the city-region and the long term goal of creating a regional green wedge council is coming closer to realization. The first and foremost goal of this initiative is to safeguard a connected regional green structure in the rapidly urbanizing city-region, which can be seen as a traditional nature conservation project. However, the structure that is promoted can on the other hand be seen as the back-bone of many other future NBS in the region since it will support the ecological functionality of smaller structures, e.g. pocket parks and green roofs, whose resilience will be dependent on the vitality of this larger structure. Hence, the green wedge collaborations in Stockholm is primarily about nature-to-people use of nature-based solutions but also has indirect linkages to localize nature-to-place and evidently creates an important ecological support in the urbanized parts of the city in form of restoration and regeneration (Photo 5.9).

Stadsodling Stockholm The founder discovered a communal gardening project in a central park in Stockholm and also people to engage with. They wanted to know more, especially how to do farming in public places. Many plantations were of a temporary nature and it was much disorganized. The idea was that the different urban gardening/farming communities could help each other, so in 2013 they started off with a matching function – helping people with an interest or a project to find

Photo 5.9 Guided tour with all municipal politicians, officials and initiative representatives with the aim to create a shared understanding of their common interests – 2015 in Tyresta green wedge 10 years later, Stockholm, Sweden

each other. In a garden fair, a map over Stockholm was put up and people could pin their plantations to it. It later turned in to a digital map. The emerging network wanted to be a voice, influence and be a contrast to the city. To get permits they needed an organisation so they started an association, however it is mostly on paper, what they do is grow and cultivate. There are several reasons people get active and there are various styles between groups and within groups, e.g. to create community in the town, to produce food autonomously and safe, to protect the environment and make the city greener and more aesthetic. The leaders of the association are nowadays seen as nation leading experts in urban gardening, writing hand books, giving lectures and leading workshops as well as designing new types of urban gardens. In the spring 2014 Sundbyberg municipality was part of a smaller investigation regarding urban gardening and it was nearly nonexistent and the municipality showed very limited interest in the topic. Two years later, in spring 2016, the same municipality has decided to formulate an urban gardening strategy and invited the Stadsodling Stockholm association to arrange a workshop. This initiative is a clear example of restoring and regenerating nature-to-place dimension of NBS (Photo 5.10).

Photo 5.10 Sundbyberg
municipality in Stockholm
has decided to formulate a
strategy for urban
gardening and asks
Stadsodling Stockholm to
arrange a workshop to get
on the right track. (May
2016, Stockholm, Sweden)

5.3.3.2 Accelerating Stockholm's Transition with Nature-Based Solutions' Initiatives

The linkages to nature is very different in all the three examples of transition initiatives in Stockholm but also in a very direct way link nature to urban challenges such as social unrest, social segregation, food security and access to ecosystem services provided by functional urban green structures. Even if different, one commonality is the importance of collaboration in all three initiatives which is part of their success in accelerating transition to sustainable development in Stockholm city-region.

Miljöverkstan Flaten has created a solid local network as well as a network of significant partners throughout the region. Their well anchored activities both with residents and local authorities have given them legitimacy to represent the Flaten area in the pilot project which will be part of forming how Stockholm municipality will think of, design and maintain their protected areas in the future – potentially activating the replicating mechanism. The most important mechanisms in the case of Miljöverkstan have so far been coupling and embedding.

Stockholm green wedges collaboration started with the initiative that instrumentalised a policy report which triggered a long term, step-wise partnership process towards embedding into the municipal work. First different local initiatives and associations coupled to form a strong voice and consortium that invited the concerned municipalities. After 5 years this is now fully part of the municipal work and replicated to other similar situations in the city-region. One of the key success factors has been the carefully constructed partnership between different types of actors at different geographical scales. Stadsodling Stockholm aims to activate the mechanism coupling by linking urban gardening initiatives for exchange and power and perhaps also replicating. They are part of the strong emergence of urban gardening and agriculture in the city-region and forcefully seek to change the way these activities are discussed at regional and municipal level – embedding – also by bringing in examples from outside the region.

5.4   Implications for Accelerating Urban Sustainability Transitions Through Nature-Based Solutions

For accelerating urban sustainability transitions, social production and mediation of nature-based solutions requires collective agency – transition initiatives- and urban change agents to mediate and catalyse processes of transformation. More specifically, we found that transition initiatives have often depended on a small circle of urban change agents. Especially urban change agents have been central to promoting the activities of transition initiatives in the city-regions. They are a specific type of change agents who act as mediators, translators and networkers between both different sectors (i.e. local government, civil society) and different domains (i.e. food, energy). They can be members of a single or of multiple transition initiatives. They are capable of 'speaking the languages' of different sectors (i.e. public sector, private sector, civil society) and of developing a horizontal, integrated view of several domains, which enables them to identify and support synergies between these domains. They often also act as networkers between the city-region and the governance context and, in doing so, contribute to the diffusion of ideas, knowledge and experience. While on the one hand, these transition entrepreneurs can be important connectors and networkers between transition initiatives, on the other hand, the reliance on these few change agents makes the process of change very fragile. Activities of transition initiatives might decline again if these individuals leave the city-region or become overburdened by their voluntary commitment.

For accelerating urban sustainability transitions, investing in establishing resourceful and capacitated local governments is critical for ensuring productive collaborations and synergies with transition initiatives and ability to recognise the value of nature-based solutions as practices and outcomes of transition initiatives. Another common challenge across the three city-regions includes the insufficient capacities of local governments to establish synergies across domains, like food, education and biodiversity. The limited ability of local governments (local political and administrative authorities) to incorporate novelty due to compartmentalisation has been identified as a common obstacle across the city-regions. Local public administration tends to have separate departments each following a distinct administrative specialisation, which is associated with different objectives, legal frameworks and responsibilities. Due to this compartmentalisation, TIs often find it difficult to approach the public administration and find an appropriate contact person if the novel character of their activities does not fit the established specialisation of administrative departments. This in turn hampers their efforts in accelerating a sustainability transition in the city-region. For instance, urban gardening transition initiatives in Dresden and Stockholm reported that they did not fit into the established specialisations of the public administrations of their municipalities, which led to a situation of unclear mandates and specifications. The local governments have, therefore been struggling with integrating these progressive transition initiatives in their work or providing support for them. By contrast, the local government of Genk has taken a very proactive approach. It has recognised the problem and is seeking innovative ways of overcoming compartmentalisation and 'silo politics' to achieve better coordination and policy integration. It is exploring approaches of systems and transition thinking and has introduced a new, horizontal position for sustainability and a transversal transition team.

For accelerating sustainability transitions in cities with and through transition initiatives who experiment and innovate with nature-based solutions, a policy mix is required to support beyond 'seeding' for advancing and enabling transition initiatives to scale. Transition initiatives across the city-regions have reported that the resources available to them such as time, budget, space or political mandate are seen as insufficient, which impedes their activities and, thereby, change towards sustainability. The dependence of many TIs on external funding by the central government has been a major source of instability for the TIs, especially in the present era of austerity, where reduced funding for the voluntary sector generally has caused a loss of momentum of the TIs. In terms of governance, this implies that this projectification of funding should be overcome and more stable funding schemes established. For instance, local governments could create funds for NBS initiatives with a longer time frame. Conversely, the decentralised governance structure of the city-region of Stockholm gives much political autonomy to the municipalities within the city-region. Here, the political situation depends on the individual municipalities with some of them giving high priority to environmental sustainability and others pursuing different political priorities. TIs from both Dresden and Stockholm report that the trend of a projectification of funding, where most of the funding is short-term and project-based, poses a severe challenge to them because it creates high instability and uncertainty for their activities.

Last but not least, *for scaling nature-based solutions to contribute to accelerating sustainability transitions in cities, social and environmental agendas in cities need to connect or exploit synergies more strategically.* Nature-based solutions require a social process to be spatially integrated in a city, and produce social benefits in the form of sense of place, empowering communities and establishing ties between social groups. As such, even though on the outset nature-based solutions are 'environmental solutions', they produce multiple benefits and with our cases we show that they produce social benefits, addressing social challenges such as segregation and inclusion. Hence, it may be worth noticing that an urban agenda for nature-based solutions is intrinsically an integrated agenda for social and environmental issues.

Acknowledgements This book chapter is based on research carried out as part of the ARTS Project, Accelerating and Rescaling Sustainability Transitions Project funded by the European Union's Seventh Framework Programme (FP7) (grant agreement 603654). The views expressed in this article are the sole responsibility of the authors and do not necessarily reflect the views of the European Union.

References

Ambrey C, Fleming C (2014) Public greenspace and life satisfaction in urban Australia. Urban Stud 51(6):1290–1321. May 2014

Andersson E, Tengo M, McPhearson T, Peleg K (2015) Cultural ecosystem services as a gateway for improving urban sustainability. Ecosyst Serv 12:165–168

Avelino F, Rotmans J (2011) A dynamic conceptualization of power for sustainability research. J Clean Prod 19(8):796–804

Aylett A (2010) Conflict, collaboration, and climate change: participatory democracy and urban environmental struggles in Durban, South Africa. Int J Urban Reg Res 34:478–495

Aylett A (2013) Networked urban climate governance: neighborhood-scale residential solar energy systems and the example of Solarize Portland. Environ Plann C Gov Policy 31:858–875

Barr S, Shaw G, Coles T (2011) Sustainable lifestyles: sites, practices, and policy. Environ Plan A 43(12):3011–3029

Borzel TA, Risse T (2010) Governance without a state: can it work? Regul Gov 4:113–134. doi:10.1111/j.1748-5991.2010.01076.x

Boyer RHW (2015) Grassroots innovation for urban sustainability: comparing the diffusion pathways of three ecovillage projects. Environ Plan A 45:320–337

Bratman GN, Daily GC, Levy BJ, Gross JJ (2015) The benefits of nature experience: improved affect and cognition. Landsc Urban Plan 138:41–50

Buchel S, Frantzeskaki N (2015) Citizen's voice: a case study about perceived ecosystem services by urban park users in Rotterdam, The Netherlands. Ecosyst Serv 12:169–177

Burggraeve R (2015) Volunteering and ethical meaningfulness. Found Sci 20(2):1–4

Bussu S, Bartels KPR (2014) Facilitative leadership and the challenge of renewing local democracy in Italy. Int J Urban Reg Res 38(6):2256–2273. doi:10.1111/1468-2427.12070

Calhoun C (2012) The roots of radicalism: tradition, the public sphere, and early nineteenth-century social movements. The University of Chicago Press, Chicago

Carmin J, Hicks B, Beckmann A (2003) Leveraging local action: grassroots initiatives and transboundary collaboration in the formation of the white Carpathian Euroregion. Int Sociol 18(4):703. doi:10.1177/0268580903184004

Carrus G, Scopelliti M, Lafortezza R, Colangelo G, Ferrini F, Salbitano F, Agrimi M, Portoghesi L, Semenzato P, Sanesi G (2015) Go greener, feel better? The positive effects of biodiversity on the well-being of individuals visiting urban and peri-urban green areas. Landsc Urban Plan 134:221–228

Cerar A (2014) From reaction to initiative: potentials of contributive participation. Urbani izziv 25(1). doi:10.5379/urbani-izziv-en-2014-25-01-002

Chmutina K, Wiersma B, Goodier CI, Devine-Wright P (2014) Concern or compliance? Drivers of urban decentralized energy initiatives. Sustain Cities Soc 10:122–129

Christmann GB (2014) Investigating spatial transformation processes: an ethnographic discourse analysis in disadvantaged neighbourhoods. Hist Soc Res 39(2):235–256. doi:10.12759/hsr.39.2014.2.235-256

Coenen L, Truffer B (2012) Places and spaces of sustainability transitions: geographical contributions to an emerging research and policy field. Eur Plan Stud 20(3):367–374

Cote M, Nightingale AJ (2012) Resilience thinking meets social theory: situating social change in socio-ecological systems (SES) research. Prog Hum Geogr 36:475–489

Creamer E (2015) The double-edged sword of grant funding: a study of community-led climate change initiatives in remote rural Scotland. Local Environ 20(9):981–999. doi:10.1080/13549839.2014.885937

Desa G, Koch JL (2014) Scaling social impact: building sustainable social ventures at the base-of-the-pyramid. J Soc Entrep 5(2):146–174. doi:10.1080/19420676.2013.871325

Devolder S, Block T (2015) Transition thinking incorporated: towards a new discussion framework on sustainable urban projects. Sustainability 2015(7):3269–3289. doi:10.3390/su7033269

Eckerberg K, Bjarstig T, Zachrisson A (2015) Incentives for collaborative governance: top-down and bottom-up initiatives in the Swedish Mountain region. Mt Res Dev 35(3):289–298

Ernston H (2013) The social production of ecosystem services: a framework for studying environmental justice and ecological complexity in urbanized landscapes. Landsc Urban Plan 109:7–17

Foo K, Martin D, Wool C, Polsky C (2015) The production of urban vacant land: relational place-making in Boston, MA neighborhoods. Cities 40:175–182

European Commission (2015) Towards an EU research and innovation policy agenda for nature-based solutions & re-naturing cities. Final report of the Horizon 2020 expert group on "Nature-based solutions and re-naturing cities." Brussels.

Forrest N, Wiek A (2015) Success factors and strategies for sustainability transitions of small-scale communities – evidence from a cross-case analysis. Environ Innov Soc Trans 17:22–40

Frantzeskaki N, Kabisch N (2016) Designing a knowledge co-production operating space for urban environmental governance, lessons from Rotterdam, Netherlands and Berlin, Germany. Environ Sci Pol 62:90–98

Frantzeskaki N, Wittmayer J, Loorbach D (2014) The role of partnerships in 'realizing' urban sustainability in Rotterdam's City Ports Area, the Netherlands. J Clean Prod 65:406–417. doi:10.1016/j.jclepro.2013.09.023

Fraser JC, Kick EL (2014) Governing urban restructuring with city-building nonprofits. Environ Plan A 46:1445–1461

Fuchs G, Hinderer N (2014) Situative governance and energy transitions in a spatial context: case studies from Germany. Energy Sustain Soc 4:16

Garcia M, Eizaguirre S, Pradel M (2015) Social innovation and creativity in cities: a socially inclusive governance approach in two peripheral spaces of Barcelona. City Cult Soc 6:93–100

Gorissen L, Spira F, Meyers E, Velkering P, Frantzeskaki N (2017) Moving towards systemic change? Investigating acceleration dynamics of urban sustainability transitions in the Belgian City of Genk. J Clean Prod. Article in Press.

Green TL, Kronenberg L, Andersson E, Elmqvist T, Gomez-Baggethun E (2016) Insurance value of green infrastructure in and around cities. Ecosystems. doi:10.1007/s10021-016-9986-x

Haase D, Schwarz N, Strohbach M, Kroll F, Seppelt R (2012) Synergies, trade-offs, and losses of ecosystem services in urban regions: an integrated multiscale framework applied to the Leipzig-Halle region. Germany Ecol Soc 17(3). doi:10.5751/ES-04853-170322

Healey P (2015) Citizen-generated local development initiative: recent English experience. Int J Urban Sci 19(2):109–118. doi:10.1080/12265934.2014.989892

Healey P, Vigar G (2015) Creating a special place: the Ouseburn valley and trust in Newcastle. Plan Theory Pract 16(4):565–567. doi:10.1080/14649357.2015.1083153

Horsford SD, Sampson C (2014) Promise neighborhoods: the promise and politics of community capacity as urban school reform. Urban Educ 49(8):955–991. doi:10.1177/0042085914557645

Kabisch N, Frantzeskaki N, Artmann M, Davis M, Haase D, Knapp S, Korn H, Naumann S, Pauleit S, Stadler J, Zaunberger K, Bonn A et al (2016) Nature-based solutions to climate change mitigation and adaptation in urban areas – perspectives on indicators, knowledge gaps, opportunities and barriers for action. Ecol Soc 21(2):39. [online] URL: http://www.ecologyandsociety.org/vol21/iss2/art39/

Kazmierczak A (2013) The contribution of local parks to neighbourhood social ties. Landsc Urban Plan 109:31–44

Kothari A (2014) Radical ecological democracy: a path forward for India and beyond. Development 57(1):36–45. doi:10.1057/dev.2014.43

Magnani N, Osti G (2016) Does civil society matter? Challenges and strategies of grassroots initiatives in Italy's energy transition. Energy Res Soc Sci 13:148–157

Moss T, Becker S, Naumann M (2014) Whose energy transition is it, anyway? Organisation and ownership of the Energiewende in villages, cities and regions. Local Environ 20(12):1547–1563. doi:10.1080/13549839.2014.915799

Mullaney J, Lucke T, Trueman SJ (2015) A review of benefits and challenges in growing street trees in paved urban environments. Landsc Urban Plan 134:157–166

Olsson P, Galaz V, Boonstra WJ (2014) Sustainability transformations: a resilience perspective. Ecol Soc 19(4):1. doi:10.5751/ES-06799-190401

Reeves A, Lemon M, Cook D (2014) Jump-starting transition? Catalysing grassroots action on climate change. Energ Effic 7:115–132

Romero-Lankao P (2012) Governing carbon and climate in the cities: an overview of policy and planning challenges and options. Eur Plan Stud 20(1):7–26

Rotmans J, Kemp R, Asselt MV (2001) More evolution than revolution: transition management in public policy. Foresight 3:1–17

Sagaris L (2014) Citizen participation for sustainable transport: the case of "Living City" in Santiago, Chile (1997–2012). J Transp Geogr 41:74–83

Seyfang G, Longhurst N (2013) Desperately seeking niches: grassroots innovations and niche development in the community currency field. Glob Environ Chang 23(5):881–891. doi:10.1016/j.gloenvcha.2013.02.007

Seyfang G, Smith A (2007) Grassroots innovations for sustainable development: towards a new research and policy agenda. Environ Polit 16(4):584–603. doi:10.1080/09644010701419121

Seyfang G, Hielscher S, Hargreaves T, Martiskainen M, Smith A (2014) A grassroots sustainable energy niche? Reflections on community energy in the UK. Environ Innov Soc Trans 13:21–44. doi:10.1016/j.eist.2014.04.004

Staggenborg S, Ogrodnik C (2015) New environmentalism and Transition Pittsburgh. Environ Polit 24(5):723–741. doi:10.1080/09644016.2015.1027059

Tidball K, Stedman R (2012) Positive dependency and virtuous cycles: from resource dependence to resilience in urban social-ecological systems. Ecol Econ 86:292–299

Touchton M, Wampler B (2014) Improving social well-being through new democratic institutions. Comp Pol Stud 4(10):1442–1469. doi:10.1177/0010414013512601

Van der Brugge R, Van Raak R (2007) Facing the adaptive management challenge: insights from transition management. Ecol Soc 12:33

Verdini G (2015) Is the incipient Chinese civil society playing a role in regenerating historic urban areas? Evidence from Nanjing, Suzhou and Shanghai. Habitat Int 50:366–372

Wagenaar H, Healey P (2015) The transformative potential of civic enterprise. Plan Theory Pract 16(4):557–585. doi:10.1080/14649357.2015.1083153

Walker JS, Koroloff N, Mehess SJ (2014) Community and state systems change associated with the healthy transitions initiative. J Behav Health Serv Res 42:254–271. doi:10.1007/s11414-014-9452-5

Warshawsky DN (2014) Civil society and urban food insecurity: analyzing the roles of local food organization in Johannesburg. Urban Geogr 35(1):109–132. doi:10.1080/02723638.2013.860753

Westley FR, Tjornbo O, Schultz L, Olsson P, Folke C, Crona B, Bodin Ö (2013) A theory of transformative agency in linked social-ecological systems. Ecol Soc 18(3):27. doi:10.5751/ES-05072-180327

Wolfram M, Frantzeskaki N (2016) Cities and systemic change for sustainability: prevailing epistemologies and an emerging research agenda. Sustainability 8. doi:10.3390/su8020144

Zajontz T, Laysens A (2015) Civil society in Southern Africa – transformers from below? J South Afr Stud 41(4):887–904. doi:10.1080/03057070.2015.1060091

Part II
Evidence for Nature-Based Solutions to Adapt to Climate Change in Urban Areas

Chapter 6
Integrating the Grey, Green, and Blue in Cities: Nature-Based Solutions for Climate Change Adaptation and Risk Reduction

Yaella Depietri and Timon McPhearson

Abstract Cities are high emitters of greenhouse gases and are drivers of environmental modification, often leading to degradation and fragmentation of ecosystems at local and regional scales. Linked to these trends is a growing threat experienced by urban areas: the risk from hydro-meteorological and climatological hazards, further accentuated by climate change. Ecosystems and their services, though often overlooked or degraded, can provide multiple hazard regulating functions such as coastal and surface flood regulation, temperature regulation and erosion control. Engineering or grey approaches often do not tackle the root causes of risk and can increase the vulnerability of populations over the long-term. However, evidence of alternative approaches such as the role of healthy, functioning ecosystems in disaster risk reduction are still scarce, contentious, and with limited applicability in the urban context. This chapter explores the role of grey, green, and blue infrastructure and in particular hybrid approaches for disaster risk reduction and climate change adaptation to shed light on available sustainable adaptation opportunities in cities and urban areas. We highlight the dependence of cities on ecosystems as a key component of climate resilience building through case studies and literature review. At the same time, we highlight the limitation and drawbacks in the adoption of merely grey or merely green infrastructures. We suggest that an intermediate 'hybrid' approach, which combines both blue, green and grey approaches, may be the most effective strategy for reducing risk to hazards in the urban context.

Keywords Urban areas • Disaster risk reduction • Climate change adaptation • Green infrastructures • Hybrid approaches

Y. Depietri (✉) • T. McPhearson
Urban Ecology Lab, Environmental Studies Program, The New School, New York, USA
e-mail: depietry@newschool.edu; timon.mcphearson@newschool.edu

© The Author(s) 2017
N. Kabisch et al. (eds.), *Nature-based Solutions to Climate Change Adaptation in Urban Areas*, Theory and Practice of Urban Sustainability Transitions, DOI 10.1007/978-3-319-56091-5_6

6.1 Introduction

6.1.1 Challenges of Climate Change in Cities

Levels of greenhouse gases emissions per person are particularly high in cities in North America – often 25–50 times higher than in cities in low-income nations (Satterthwaite 2006). Cities are therefore drivers of climate change and at the same time increasingly at risk from its effects. At the global scale, climate change is expected to lead to significant sea level rise and to changes in frequency, intensity and spatial patterns of temperature, precipitation and other meteorological factors (IPCC 2015). Over the coming century, climate change scenarios project that urban regions will have to cope with and adapt to increasing extreme events (Rosenzweig et al. 2011a). Furthermore, cities already face aggravated impacts due to the higher presence of sealed surfaces which increase the magnitude of heat risk via the urban heat island (UHI) (Tan et al. 2010). Similarly, reduced water infiltration in highly paved urban areas generates increased risk of surface flooding at the local scale and regional scale (see Depietri et al. 2011 for a review), especially given that cities are often located in exposed coastal areas and floodplains. Negative impacts of climate extremes are likely to affect human health, energy and critical infrastructures, such as transportation, and water supply (McCarthy et al. 2010; Rosenzweig et al. 2011a). Many of these impacts are already being felt, especially by coastal communities (Spalding et al. 2014).

So far, most efforts by cities to respond to climate change have focused on mitigation (i.e. reduction of greenhouse gases emissions) and much less on adaptation (i.e. long term strategies to reduce exposure, susceptibility and improve coping capacity of communities to hazards) as these strategies imply taking a precautionary and anticipatory approach (Castán Broto and Bulkeley 2013). However, the implementation of adaptation plans is urgent. Changes in global climate are already underway and social, infrastructural, and economic costs of inaction are high (Bosello et al. 2012).

In this chapter, we explore the role of grey, green, blue and hybrid infrastructures for climate change adaptation (CCA) in cities in order to shed light on the different resilience and sustainability opportunities available and their pros and cons for urban areas. We highlight the dependence of cities on healthy ecosystems and support the case for 'hybrid' approaches as a key component of urban disaster risk reduction (DRR) and CCA through literature review and using New York City (NYC) as a case study. Natural capital (or the stock of biophysical resources), along with technological or infrastructural capital, are considered together in order to look closely at the interdependency and feedbacks between biophysical and technological domains of complex urban systems (McPhearson et al. 2016a,b) which challenge decision-makers faced with advancing CCA and DRR agendas. In the following sections we review the risk caused by to climate change in cities and introduce the social-ecological-technological systems (SETs) framework as a way for researchers and practitioners to explore adaptation strategies, particularly hybrid approaches, that work across interacting SET domains of urban systems.

6.1.2 Risk and Vulnerability to People, Ecosystems and Infrastructures in Cities

Cities, if exposed to hazards, are hotspots of vulnerability due to the concentration of people and infrastructures. It is increasingly acknowledged that the human vulnerability to natural hazards is the result of the socio-economic, physical and environmental processes that characterise a social-ecological system and is thus socially constructed (Oliver-Smith 1999). This view of hazards is even more relevant in urban areas where the environment is highly modified by physical infrastructures and socio-economic activities. Cities are centres of interchange of knowledge, cultures, innovations and goods. To facilitate exchanges, these are often located in the proximity of rivers and seas making them exposed to a number of hazards such as storms, flooding, cyclones, coastal erosion and sea level rise (Sherbinin et al. 2007). Urban sprawl can exacerbate impacts of hazards through "poor urban management, inadequate planning, high population density, inappropriate construction, ecological imbalances and infrastructure dependency" (Jacobs 2005). As a result, cities of developed countries may face the highest impacts in terms damages assets and economic losses (Dickson et al. 2012). In the US, catastrophic events have increased in the last 35 years according to MunichRe NatCatService.[1]

In healthy environments, ecosystems do not strictly experience disaster in the same way that we consider disaster in the human context. When discussing risks to ecosystems, ecologists tend to discuss this in terms of disturbance (e.g. Attiwill 1994; Swetnam and Betancourt 2010). In fact, variation and extremes in weather and climate and other disturbances have always been part of the functioning of natural ecosystems and provide a wide range of benefits such as soil fertilization in floodplains in the case of floods or groundwater recharge in the case of intense precipitation events associated, for instance, with typhoons. However, major impacts on the ecosystem might occur if hazards affect a degraded and less diverse ecosystems, as is often the case in and around cities (Alberti 2005). This could translate to a temporary or even permanent decline or impairment in supplying necessary ecosystem services to urban and peri-urban areas. Mitigating and adapting to climate change in and around cities thus needs to take into account the interacting effects of the built infrastructure and climate change on the ecological or biophysical components of local and regional ecosystems. If we are to utilize nature-based solutions (NBS) for CCA and DRR, then the health and function of urban ecosystems is of primary importance for providing effective climate regulating services.

---

[1] https://www.munichre.com/en/reinsurance/business/non-life/natcatservice/index.html (Retrieved on 13th of October 2016).

6.1.3   The SETS Framework

Due to the multiple factors of risk, management in cities and urban regions needs to be based on a multi-disciplinary and integrated approach. A social-ecological-technological systems (SETs) approach (illustrated in Fig. 6.1) can be a useful framework to understand the dynamic interactions between social, ecological, and technical-infrastructural domains of urban systems. The SETs approach aims at overcoming the limitation of a purely socio-technological approach which tends to exclude ecological functions, or of a social-ecological approach inclined to over-look critical roles of technology and infrastructure as fundamental constituents, and drivers of urban system dynamics (McPhearson et al. 2016a).

As the SETs approach, can broaden the spectrum of the options available for intervention (Grimm et al. 2016), it is therefore a suitable framework to explore the range of options available and needed to adapt to climate impacts in the urban context. Using this framework, we investigate the pros and cons of adapting to climate change through grey infrastructures (i.e. hard or engineering approaches), 'green' and 'blue' approaches (i.e. the restoration of ecosystems, various types of

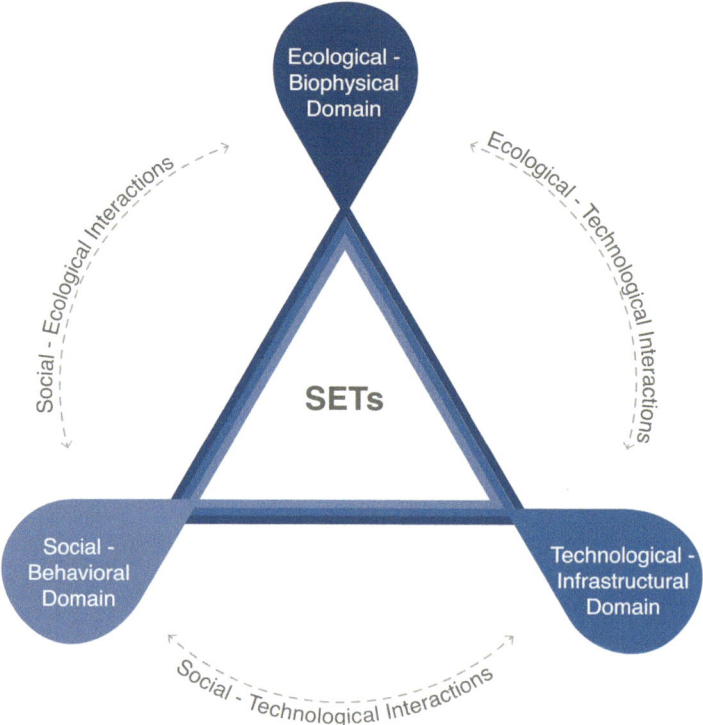

Fig. 6.1 Conceptualizing urban systems as social-ecological-technical systems (SETs) with emphasis on the *interactions* between social, ecological, and technical-infrastructural domains of cities and urban areas (Source: own elaboration)

Table 6.1 Flow of infrastructural adaptation options

Grey	Hybrid or mixed approaches	Green and blue
Hard, engineering structures	Blend of biological-physical and engineering structures	Biophysical, Ecosystems and their services
Very limited role of ecosystem functions	Allows for some ecosystem functions mediated by technological solutions	Mainly relying on existing or restored ecosystem functions and water bodies
e.g. canals, pipes and tunnels of the drainage system; dikes; wastewater treatment plants; water filtration plants	e.g. bioswales; porous pavement; green roofs; rain gardens; constructed wetlands; Sustainable Urban Drainage Systems (SUDS)	e.g. wetlands restoration; installation of grass and riparian buffers; urban trees; stream restoration; rivers, lakes, ponds, oceans and seas

Adapted from Grimm et al. (2016)

ecosystem-based adaptation -EbA strategies, or NBS) and more mixed or 'hybrid' approaches, based on ecosystem functions complemented by engineered infrastructures in urban areas. The contrast between these three different strategies is described in Table 6.1 and illustrated in Fig. 6.2. Table 6.1 defines grey, hybrid, green and blue infrastructures as a continuum from grey infrastructures, to hybrid, to green and blue where hybrid approaches make use of engineering and ecosystem functions together. In Fig. 6.2 we use an example illustrating a range from green, to grey, to hybrid options for managing challenges of precipitation and stormwater in the urban context.

Soft, organizational or institutional and economic approaches (such as early warning systems, insurance or risk transfer, evacuation plans or improvements in public health and insurance system) are of primary importance for DRR and CCA, though it is beyond the scope of this chapter. The social component of the SETS framework described below is thus not explored.

6.2 Approaches to Reducing Risk and Overall Effects of Urban Climate Change

Adaptation strategies to climate change can be evaluated in multiple ways, including: success in implementation of no-regret measures; in terms of favouring reversibility; flexibility; cost-effectiveness and feasibility; or long-term sustainability. Next, we review relevant literature to summarize key arguments for the three main (grey, green and blue as well as hybrid) approaches in DRR and CCA in cities. We conclude with a summary of the three main approaches across all evaluative factors in the discussion section (see also Table 6.2).

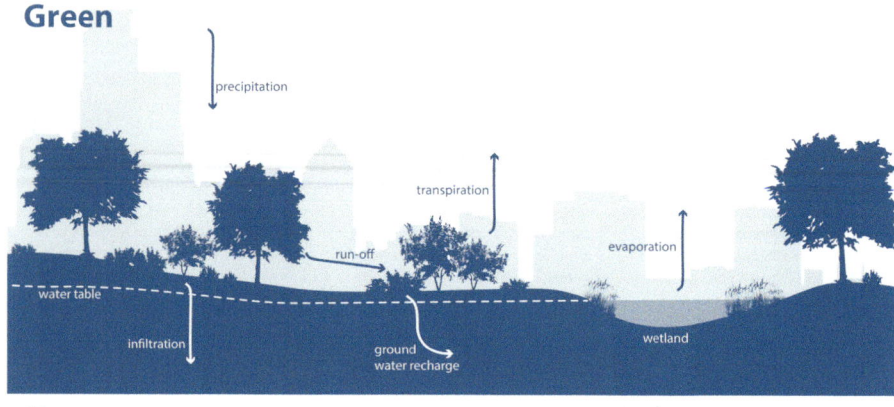

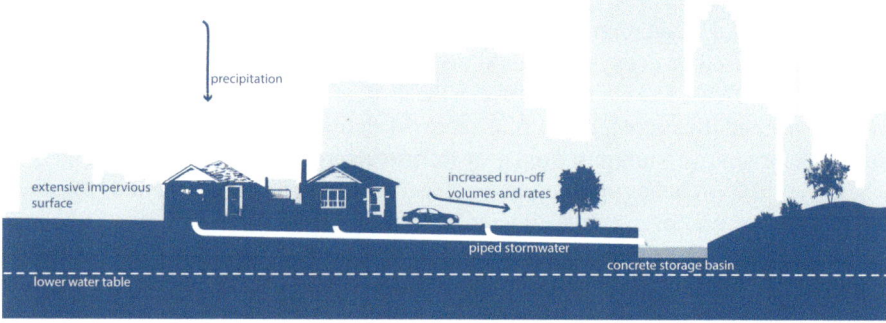

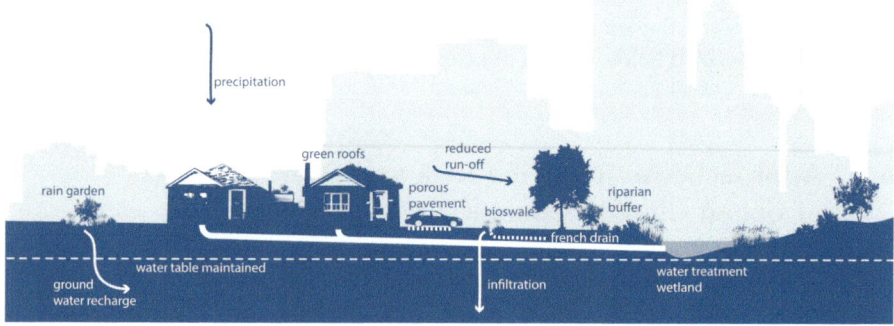

Fig. 6.2 Three contrasting approaches, green and blue only, grey only, and hybrid for dealing with urban water, in particular significant precipitation events and other stormwater challenges that cities face. Hybrid approaches illustrated in the *bottom panel* combine grey and green approaches to maximize water absorption and infiltration and limit costs of green infrastructure while providing potential co-benefits (Source: own elaboration)

Table 6.2 Summary table for comparison of the three approaches based on their suggested low-medium-high performance with respect to a list of factors identified in the literature

Aspect	Grey infrastructures	Green infrastructures	Hybrid approaches
Feasibility in the urban context	High (occupies a reduced area)	Low (But highly important and feasible in peri- and regional urban areas)	High
Reliability	Medium These measures do not completely eliminate risk Mixed success has been reported	Medium Role has been proven but some studies lead to contradictory results due to the multiple factors that play a role in determining the magnitude of a hazard Highly depends on the type of hazard	High
No-regret strategy	Often high regret measure	Low regret measure	Medium
Long-term durability or resilience	Durable, but can be maladaptive.	Medium Can be affected by hazards and ecosystems in and around cities are generally highly transformed and often degraded	Medium-high
Reversibility and flexibility	Little or not reversible	Medium Can be high or low reversibility depending on the type	Medium
Cost-effectiveness	Low. High building costs Depreciate in value over time	High Investments in green infrastructure can be much less expensive in short and long run than those in grey infrastructure	Medium to High
Biodiversity conservation	None	High Green infrastructures provide natural habitat for species	Medium
Other co-benefits	Low (but some examples of medium to high exist such as water and energy supply provided by riverine dikes initially designed for flood control)	High Vegetation provides local communities with critical ecosystem services such as those improving livelihoods, food security and recreation and that may enhance their resilience to extreme events in the long-term Broadly applicable	Medium Contributes to providing other services, such as pollution control and recreation, but will depend on the green infrastructure component of the hybrid approach

(Source: own elaboration)

6.2.1 Grey Strategies

Response to exposure of communities to natural hazards has traditionally relied on grey infrastructures (Jones et al. 2012). Grey infrastructures are built up, engineered and physical structure, often made of concrete or other long-lasting materials, that mediate between the human, built up system and the variability of the meteorological and climatic system. These include dikes, floodgates, levees, embankments, sea walls and breakwaters for riverine and coastal flood protection, drainage systems for storm water management such as storm sewers, pipes, detention basins, and air conditioning or cooling centers to cope with extreme heat. Engineering approaches largely ignore or supplant the functions of biophysical systems. Through the SETs lens these approaches tend to be located fully within the technological domain with little input from ecological domains. Grey infrastructures provide an important means of adapting to biophysical challenges including hazards and climate driven extreme events, but are often costly to install and maintain, have long-term effects on ecosystems, tend to have low flexibility, and when they fail can generate catastrophic impacts on social and ecological domains of urban SETs.

Despite this, in some cases grey infrastructures might still be needed. For instance, Khazai et al. (2007) reviewed the role and effectiveness of coastal structures in reducing damage to coastal communities from tsunamis and storm surges caused by cyclones, hurricanes and typhoons, finding that concrete seawalls are the most durable protection against various types of storm surges.

Still the he SETs framework helps elucidate the need to recognize that grey infrastructures are not isolated systems but embedded in and affect social and ecological components of the urban system. In urban systems, habitat loss is often a direct consequence of the hardening of coastal and inland water systems. Grey infrastructures for coastal defence have also been criticized for inhibiting normal coastal processes (Khazai et al. 2007). Social and ecological systems are co-evolutionary (Norgaard 1994; Kallis 2007) in the sense that they evolve while influencing eachother thus creating sharp contrast with fixed, long-lasting grey infrastructures that lack flexibility to be. For these and other reasons tied to the long term impacts engineered infrastructures, such as sea walls or wastewater treatment plants, have on neighbouring social communities, these projects increasingly encounter high resistance by residents especially those concerned with the environment, health and sustainability.

Though engineered systems can have enormous benefits (clean water, sanitation, etc.) they can also lead to undesirable system lock-ins and path dependency which make their negative impacts, and even the infrastructure itself, difficult to reverse (Dawson 2007). Expanding cities increasingly rely on grey infrastructures for their protection with sprawling areas and informal settlements are often located in hazard-prone areas. A false sense of security generated by these protective infrastructures can in fact lead populations to further expand in unsafe areas increasing their exposure to hazards and further aggravating risk (Mitchell 2003).

Grey infrastructures can fail, especially when confronted with climate driven extreme events. Ready examples include the devastating effects of Hurricane Katrina in New Orleans in 2005 and Hurricane Sandy in New York City in 2012. These types of weather-related hazards overtopped levees, sea walls, and storm barriers engineered to protect people and ecosystems from hurricanes and storm surges, and ultimately failed with disastrous consequences.

Climate change creates uncertainty and system non-stationarity. Thus future risk is not easily taken into account in planning and building of infrastructures (Hallegatte 2009). For example, a levee designed to accommodate a certain future level storm surge is useless if climate change causes extreme events that surpass the original infrastructure target. Grey infrastructures might not be able to respond and accommodate the uncertain future ahead of us. Huge flood defence works, in the Thames river in the UK or the MOSES project in Venice (Italy), have shown to have quite long time lags of implementation, about 30 years, which is also true in the case changes in urban planning (Hallegatte 2009). Maladaptation to climate change can also occur through the installation of energy-intensive machines or infrastructures (such as pumped drainage or desalination plants) (Dawson 2007). These usually do not meet climate mitigation objectives since they are often powered by climate polluting energy sources, and can also fail as was the case of overconsumption and power outages illustrated by the summer blackout in the U.S. Northeast in 2003 (Andersson et al. 2005). Especially in the case of CCA, construction costs can be extremely high (Bosello et al. 2012) while maintenance and restructuring (e.g. digging in sand for riverine dikes) can also be financially demanding.

There is however a wide variability among factors important to consider in the case of the implementation and maintenance of grey infrastructures. These factors also depend on the type of hazard under consideration. For some hazards, structural measures with highly sophisticated early warning and evacuation plans might be needed. Seawalls are for instance particularly effective in the case of tsunamis, even if these might be costly (Khazai et al. 2007). Additionally, to adapt to rising temperatures and heat-waves, air conditioning and cooling centers, will continue to be important for adaptation and reduction of risk. Grey infrastructures tend to require limited amounts of land, are replicable, can be monitored, and to some extent controlled (The Nature Conservancy 2013), all of which are characteristics that remain particularly suitable to the urban context in a fast changing climate.

6.2.2 Green and Blue Infrastructures

The vulnerability of social-ecological-technological systems can be expressed also through the type and the quality of the dependence of communities on ecosystems (Adger 2000; Renaud et al. 2010). Anthropogenic environmental change, by affecting the functioning of ecosystems and their services through land use and climate changes, is one of the main drivers of the increasing impacts of a number of natural hazards (Kaly et al. 2004). Healthy ecosystems play a significant role in buffering communities from climatological and hydro-meteorological hazards at different scales (McPhearson et al. forthcoming). However, despite the recognition of green approaches as "low regret" measures for DRR and CCA, also at the global level (UNISDR 2005, 2015; IPCC 2012), ecosystems approaches remain the most disregarded component of plans and strategies (Renaud et al. 2013; Matthews et al. 2015).

Green infrastructures are principally constituted by well-functioning biophysical systems to which some management and restoration may apply. They are represented, by healthy oyster reefs, coastal salt marshes, mangroves, coral reefs, sea grasses, sand beaches and dunes in the coast environment and mainly by forests, parks, street trees, and grasslands inland. Blue infrastructures include all bodies of waters, including ponds, wetlands, rivers, lakes and streams, as well as estuaries, seas and oceans. Since water and land come together in multiple ways, including riparian areas, beaches, wetlands, and more, combining green and blue infrastructure is gaining attention in both research and practice for CCA and DRR. Green and blue infrastructures, as we use the terms here, rely primarily on healthy, functioning ecosystems and allow for little or no technological/infrastructure intervention, thus situating them fully within the ecological domain of the SETs framework.

Initial research and practice has shown that well managed ecosystems and their regulating services can contribute to the reduction of risk and are very often cost-effective, multifunctional and win-win solutions especially in the long run (Renaud et al. 2013; Sudmeier-Rieux 2013). In addition, to be useful strategies for CCA and DRR, green and blue infrastructures provide multiple co-benefits such as recreation, psychological well-being and pollution-control opportunities (Gomez-Baggethun et al. 2013). These are also often flexible and applicable in a variety of settings (Jones et al. 2012).

Improvements in the well-being and security specifically of urban populations through green infrastructures have been reviewed by various authors (e.g. Gill et al. 2007; Foster et al. 2011; Depietri et al. 2011) and we do not attempt a comprehensive review here. An example of the benefits of the ecosystem-based approach to DRR is the case of flood regulation policies in The Netherlands. Investments in alternative flood control policies, such as land use changes and floodplain restoration, were found to be justified when additional ecological and socio-economic benefits in a long-term perspective were included (Brouwer and van Ek 2004). In the US, the case of the Boston's Charles River Basin is exemplary in this aspect too. The city was threatened by disastrous floods since urban expansion and industrial development converted land in large parts of the floodplain during the 1950s-1960s (Platt 2006). By 1983 the Army Corps of Engineers acquired the Charles River Natural Valley Storage areas, a total of about 32.8 km^2, and, after a decade of improvements and ecosystem restoration, the Charles River Water Association (CRWA) could measure significant benefits in terms of flood reduction, and improvements in water quality, and recreation opportunities (Platt 2006). Another example of the role of green and blue infrastructures is in Sheffield, UK where the temperature above the river crossing the city was found to be 1.5 °C lower compared to the neighbouring areas in the spring. Of course, the high heat capacity of water which helps maintain cooler temperatures during high heat events also has built in drawbacks due a capacity for thermal inertia, and so may have other consequences on urban SETs.

Ecosystem-based strategies can be cost-effective. In Portland, Oregon, USA, an increase in street trees has been estimated to be 3–6 times more effective in managing storm-water per US$1000 invested than conventional drainage systems. These estimates induced the city to invest US$8 million in green infrastructure in order to save US$250 million in hard infrastructure costs (Foster et al. 2011). Guadagno et al. (2013) gathered and reviewed a wide range of case studies worldwide demonstrating the effectiveness of promoting ecosystem management for DRR in urban areas, and not least for its reduced economic costs.

However, research on the role of ecosystems in mitigating hazards has led so far to contradictory results or has been overemphasized in some cases (Renaud et al. 2013). The available evidence is still scarce and in some cases contentious (Balmford et al. 2008). The lack of evidence of the direct role of ecosystems for human health and well-being may be one additional obstacle that helps to explain the lack of implementation of NBS for hazard mitigation in general, but also in Europe (Sudmeier-Rieux 2013) and in cities worldwide (Guadagno et al. 2013).

Few studies have analysed the way green infrastructures actually meet the demand for hazard related services in urban areas. A study in Cologne, Germany highlighted that ecosystem services might be effective in terms of microclimate regulation but much less in terms of air purification which has implications for risk to extreme heat (Depietri et al. 2013). Urban cooling by green spaces can be significant. In Singapore cooling by vegetation was estimated at 3.07 °C as a mean value by Wong and Yu (2005) while the urban heat island of the city reaches 7 °C (Chow and Roth 2006). The effectiveness of the removal of air pollutants by trees in NYC was estimated between 0.001% and 0.4% depending on the air pollutant (Nowak et al. 2006), which remains low. These cases express limitation in the possibility to rely merely on green infrastructures in the urban context for CCA and DRR. Also, green infrastructures generally require large amounts of land to deliver the service, which is often in short supply in many built up urban areas.

Another drawback is that trees and green areas in cities are generally distributed unevenly, are not always in locations where they are most needed (See Andersson et al., Chap. 4, this volume). Tree canopy cover is often concentrated in wealthier neighbourhoods as is the case in Phoenix, Arizona (Harlan et al. 2006; Jenerette et al. 2011). Plans for the implementation of green infrastructures across the urban fabric should then take into account concerns of social justice and equity lest new green infrastructure investment exacerbate existing inequalities in access to benefits of urban green space (see Chap. 13 by A. Haase in this issue). In short, more research is needed but literature so far shows that ecosystem-based approaches vary in their ability to mediate and mitigate climate threats, but can have a potentially much stronger role in DRR if appropriately managed, protected and better located where they are most needed (McPhearson et al. forthcoming).

6.2.3 Hybrid, Green-Grey Approaches

Hybrid, green-grey approaches utilize combined grey and green infrastructures. An example is when wetlands restoration is coupled with engineering measures such as small levees for coastal flood protection. Other examples are bioswales, rain gardens, green roofs, street trees installed in sidewalk tree pits, and other engineered ecosystem approaches to CCA and DRR. Hybrid approaches thus combine engineering and properly ecosystem functions and are situated at the intersection of the ecological and technological components of the SETs framework. It is important to note that in the literature, the term green infrastructure often tends to encompass what we defined here as hybrid approaches. However, we make a distinction between a system which relies merely on ecosystem functions (green or blue infra-structures) or where a technological or built infrastructure complement the service delivered by a green or blue infrastructure (= hybrid infrastructure).

There is increasing evidence that hybrid approaches provide cost-effective haz-ard protection solutions. Hamilton City, California, USA and in its surrounding rural areas are regularly exposed to floods. The option of setback levees, facilitating the natural functioning of the floodplain, was estimated to be a more cost effective strategy when compared to upgrading existing levees (The Nature Conservancy, n.d.). Biotechnologies or hybrid approaches like these are especially suitable in the urban context where relying solely on green infrastructures rarely meets demands in risk reduction but where urban planners have traditionally relied only on solely built structures. Hybrid approaches are intended to reduce reliance of the urban system on grey infrastructures and the drawbacks that these involve and improve sustain-ability of cities and the well-being of their inhabitants through co-benefits.

Although there is a wide array of emerging literature related to green infrastruc-tures for climate change adaptation in urban areas (e.g. review in McPhearson et al. forthcoming), literature remains thin on hybrid, green and grey approaches with an often confusing use of terminology. We suggest that hybrid approaches are of pri-mary importance in urban areas where purely green approaches may be insufficient to meet the rising impacts of climate change, where space is limited, and cost effec-tiveness is critical not only in a context of climate uncertainty, but also economic uncertainty.

Coastal and riverine urban areas, exploring how dunes, wetlands, and forest res-toration contribute to adaptation to climate change in and around cities are examples of how green infrastructure can be a first step in the planning process for CCA and DRR. However, complementary infrastructures such as small levees, embankments, bioswales, rain gardens, green roofs, porous pavements, and even more traditional grey infrastructures could be implemented simultaneously to utilize a more hybrid approach to maximize ability to provide safety during climate driven extreme events. Hybrid approaches can also benefit from a stronger local support as environ-mentalists generally generate opposition to engineering approaches sometimes forcing abandonment of grey infrastructure projects (e.g. the case of Napa River also in California, The Nature Conservancy, n.d.).

6.3   Focusing on Key Urban Climate Challenges

6.3.1   New York City and Climate Change

Most adaptation strategies in the US are still at the initial stage of drafting and implementation (Bierbaum et al. 2012). In this section we briefly highlight New York City as a case study and examine the planned and potential opportunities for the implementation of integrated green and grey approaches to climate change adapta-tion, focusing on surface and coastal floods.

NYC is the largest city in the USA with about 8.3 million people in 2010 according to U.S. census bureau and the largest also in terms of economic activity. In the city live approximately 1.4 million elderly (age 65 and older), which constitute 17% of the population and an example of one of the many vulnerable populations to climate extremes along with low-income, minority, and children among other indicators of climate vulnerability. The elderly proportion is projected to increase in the next 20 years (Goldman et al. 2014) creating significant challenges for the city to prepare for and build resilience to predicted climate extremes including heat waves, coastal flooding, and risk of major storms. Additionally, rising sea level poses increasing risk to city infrastructure and residents. NYC is built around a networks of rivers, estuaries and islands with much of the Metropolitan region less than 5 m above mean sea level (MSL) (Colle et al. 2008). New York City climate is already changing with higher temperatures and heavy downpours increasingly frequent (Rosenzweig and Solecki 2015). With these changes hazards such as urban flooding and coastal storms are also projected to increase.

6.3.2 Surface and Coastal Flooding in NYC

Combined sewer overflows, occurring when sewage and storm water are discharged from sewer pipes without treatment, are frequent in NYC and are a significant source of environmental pollution (Rosenzweig et al. 2006; McPhearson et al. 2014). Precipitation has increased at a rate of approximately 20.3 mm per decade from 1900 to 2013 in Central Park and this trend is likely to continue according to climate projections (Horton et al. 2015). Even relatively small precipitation events (over 4.4 cm) can overwhelm the combined sewage system causing raw sewage to be discharged into adjacent waterways.

NYC is low-lying with nearly 15% of the its area within the 100-year flood zone (Maantay and Maroko 2009). It is one of the top ten cities in the world in terms of assets exposed to coastal floods aggravated by climate change (Nicholls et al. 2008). The most frequent coastal storms affecting NYC are tropical storms and Nor'easters (cyclones occurring along the upper East Coast of the United States and Atlantic Canada). In NYC, even moderate Nor'eastern events can cause significant flooding (Colle et al. 2008) and are often associated with extended periods of high winds and high water (Rosenzweig et al. 2011b). Hurricanes affect the city infrequently. Five major hurricanes of category 3 have affected the New York area between 1851 and 2010, most in the month of September (Blake et al. 2011), but generally leading to large damages (Rosenzweig et al. 2011b). Hurricane Sandy which made landfall in 2012 caused 43 deaths in New York City of which nearly half were adults ages 65 or older (Kinney et al. 2015). Yet infrastructural and other damage resulted in US$67 billion of total economic losses in the country (NOAA 2015).

In 2010 the city committed to a hybrid infrastructure plan for storm water management, investing US$ 5.3 billion over 20 years to absorb 10% of the first inch (25.4 mm) of rainfall to reduce unwanted storm water run-off (NYC 2010). Of this,

US$2.4 billion is targeted for green infrastructure investments which were shown in cost-benefit analysis to have significant savings compared to a scenario of traditional pipe and tanks improvements (NYC 2010). The city Green Infrastructure Plan (NYC 2010) is a clear example of how the SETs approach can be implemented for DRR and CCA in cities (note: In this plan green infrastructure means both ecosystem-based and hybrid approaches). Overall, NYC's 2010 Green Infrastructure Plan aims to reduce the city's sewer management costs by US$2.4 billion over 20 years (Foster et al. 2011). The plan estimates that every approx. 4000 m² of green infrastructure would provide total annual benefits of US$8522 in reduced energy demand, US$166 in reduced CO_2 emissions, US$1044 in improved air quality, and US$4725 in increased property value. It also estimates that the city can reduce combined sewage overflow volumes by 2 billion gallons by 2030, using vegetated areas at a total cost of US$1.5 billion less than traditional methods (Foster et al. 2011).

6.4 Discussion

6.4.1 Embrace Both Green and Grey Approaches

Research is beginning to demonstrate the importance of preserving well-functioning ecosystems in and around urban areas for DRR and to protect and enhance human well-being (see Depietri et al. 2011 for a review; Depietri et al. 2013; Andersson et al., Chap. 4, this volume). Green, blue and grey protection systems in combination may provide some of the most effective and broadly beneficial solutions against hurricane, cyclone, typhoon and storm surges in urban areas. Hybrid approaches, like all approaches, have pros and cons. Based on our literature review, we derived the main factors through which the three approaches (grey, green and blue, and hybrid) have been so far described and evaluated. These factors are analysed and listed in the first column of Table 6.2 as hypotheses, which need to be examined empirically to better understand the effectiveness of hybrid approaches for DRR and CCA. We assigned classes of low to medium to high performance to the three main strategies with respect to the factors identified in the literature, with particular considerations of the urban context.

Grey infrastructures for DRR provide a wide array of drawbacks under most factors we consider (Table 6.2), but are, on the other hand, easily adaptable to the urban context. Green infrastructures, on the other hand, are flexible, no-regret measures and provide a wide range of benefits and co-benefits, which go beyond the mere protective or buffering functions. However, in the urban context these are often difficult to implement. Hybrid approaches fit well to the already hybrid nature of urban areas while providing solid solutions including many, if not all, of the co-benefits that more traditional green or blue approaches. Thus, at the very local scale, hybrid approaches are suggested as the way forward for DRR and CCA solutions in cities and urbanized regions.

6.4.2   Urban SETS and Importance of Bringing Together Engineering and Ecological Approaches

Built and technical infrastructures continue to be viewed by local policy makers as the most important line of defense against hazards and disasters in cities. In much of the developed world, however, urban infrastructure is aging and proving inadequate for protecting city populations (for the US see ASCE 2013). And in much of the developing, rapidly urbanizing world, new infrastructure is being constructed at breath-taking pace, often without the benefit of ecologically based design (McHale et al. 2015; McPhearson et al. 2016b). Urban infrastructure mediates the relationships between human activities and ecosystem processes and may exacerbate or reduce human impact depending on its approach (McPhearson et al. 2016c). We suggest that a fundamental rethinking is urgently needed of what makes both grey and green built infrastructures – as well as human communities with their social, ecological, and technological couplings – resilient to environmental hazards and climate extremes. We argue that urban decision makers need to move beyond traditional engineering approaches and compliment stand-alone ecological interventions to consider how to utilize combined green-grey or hybrid approaches to advance CCA and DRR. Hybrid approaches are fundamentally ecosystem-based and take advantage of ecosystem functions together with the efficacy of more engineered systems to deliver the needed level of service. Examples such as using vegetation, porous surfaces, and temporary water storage in a combined hybrid approach to limit combined sewage overflows in New York City is a useful benchmark on how cities can transmit less water through the grey infrastructure drainage system to often overloaded wastewater treatment plants.

Additionally, we suggest that viewing cities as interactive urban SETs can help to keep in mind the need for more combined approaches to dealing with climate driven hazards and improve urban sustainability. The SETs framework offers the fundamental concept that urban systems and all urban services have a combination of all three domains (social, ecological, and technical-infrastructural) as part of their production, dynamics, and efficacy.

6.5   Conclusion

6.5.1   Critical Opportunities for Working with Hybrid Approaches in Cities for CCA and DRR

Local, state and national governments are developing a range of adaptation plans to climate change. We reviewed grey, green and blue, and hybrid infrastructures approaches to CCA and DRR as a way to suggest avenues for future research and for guidance on urban development strategies. The future is ultimately uncertain with inherent challenges, in part due to climate change, and therefore difficult to make clear predictions to guide safe, secure, and urban sustainable development practices. Development and implementation strategies that are flexible, adaptive and can accommodate change are important in an era of non-stationarity and uncertainty. In this context, grey infrastructures tend to have problematic risks, and are often not cost-effective, nor fit easily into long-term sustainability goals. On the other hand, the implementation of purely green infrastructures at the urban level for CCA, though offering short and long-term benefits, might not be sufficient to meet the scale of predicted future climate hazards. Additionally, they often encounter resistance in city planning departments due to institutional path dependency form a history of utilizing grey infrastructures to meet city needs for hazard mitigation.

We suggest that cities should rely on a mix of grey, green and blue infrastructure solutions, which balance traditional built infrastructures with more nature-based solutions, especially to improve the management of urban water, heat, and other climate driven threats. Instead of turning to grey infrastructures as the default solution, town and regional planners should assess and investigate opportunities for restoring and expanding ecosystems to provide hybrid, more flexible and sustainable approaches to CCA and DRR.

Acknowledgments Timon McPhearson is supported by the Urban Resilience to Extreme Weather-Related Events Sustainability Research Network (URExSRN; NSF grant no. SES 1444755) and the Urban Sustainability Research Coordination Network (NSF grant no. RCN 1140070).

References

Adger WN (2000) Social and ecological resilience: are they related? Prog Hum Geogr 24:347–364. doi:10.1191/030913200701540465
Alberti M (2005) The effects of urban patterns on ecosystem function. Int Reg Sci Rev 28:168–192. doi:10.1177/0160017605275160
Andersson G, Donalek P, Farmer R et al (2005) Causes of the 2003 major grid blackouts in North America and Europe, and recommended means to improve system dynamic performance. IEEE Trans Power Syst 20:1922–1928. doi:10.1109/TPWRS.2005.857942
ASCE (2013) 2013 Report Card for America's Infrastructure. American Society of Civil Engineers (ASCE), Reston
Attiwill PM (1994) The disturbance of forest ecosystems: the ecological basis for conservative management. For Ecol Manag 63:247–300. doi:10.1016/0378-1127(94)90114-7
Balmford A, Rodriguez ASL, Walpole M et al (2008) The economics of ecosystems and biodiversity: scoping the science. European Commission, Cambridge
Bierbaum R, Smith JB, Lee A et al (2012) A comprehensive review of climate adaptation in the United States: more than before, but less than needed. Mitig Adapt Strateg Glob Chang 18:361–406. doi:10.1007/s11027-012-9423-1
Blake ES, Lansea CW, Gibney EJ (2011) The deadliest, costliest, and most intense United States Tropical Cyclones from 1851 to 2100 (and other frequently requested Hurricane facts). National Weather Service, National Hurricane Center, Miami
Bosello F, Nicholls RJ, Richards J et al (2012) Economic impacts of climate change in Europe: sea-level rise. Clim Chang 112:63–81. doi:10.1007/s10584-011-0340-1

Brouwer R, van Ek R (2004) Integrated ecological, economic and social impact assessment of alternative flood control policies in the Netherlands. Ecol Econ 50:1–21. doi:10.1016/j.ecolecon.2004.01.020

Castán Broto V, Bulkeley H (2013) A survey of urban climate change experiments in 100 cities. Glob Environ Chang 23:92–102. doi:10.1016/j.gloenvcha.2012.07.005

Chow WTL, Roth M (2006) Temporal dynamics of the urban heat island of Singapore. Int J Climatol 26:2243–2260. doi:10.1002/joc.1364

Colle BA, Buonaiuto F, Bowman MJ et al (2008) New York City's vulnerability to coastal flooding: storm surge modeling of past cyclones. Bull Am Meteorol Soc 89:829–841. doi:10.1175/2007BAMS2401.1

Dawson R (2007) Re-engineering cities: a framework for adaptation to global change. Philos Trans R Soc Math Phys Eng Sci 365:3085–3098. doi:10.1098/rsta.2007.0008

Depietri Y, Renaud FG, Kallis G (2011) Heat waves and floods in urban areas: a policy-oriented review of ecosystem services. Sustain Sci 7:95–107. doi:10.1007/s11625-011-0142-4

Depietri Y, Welle T, Renaud FG (2013) Social vulnerability assessment of the cologne urban area (Germany) to heat waves: links to ecosystem services. Int J Disaster Risk Reduct 6:98–117. doi:10.1016/j.ijdrr.2013.10.001

Dickson E, Baker JL, Hoornweg D, Asmita T (2012) Urban risk assessments: an approach for understanding disaster and climate risk in cities. The World Bank, Washington, DC

Foster J, Lowe A, Winkelman S (2011) The value of green infrastructures for urban climate adaptation. The Center for Clean Air Policy. Washington, DC

Gill S, Handley J, Ennos A, Pauleit S (2007) Adapting cities for climate change: the role of the green infrastructure. Built Environ 33:115–133. doi:10.2148/benv.33.1.115

Goldman L, Finkelstein R, Schafer P, Pugh T (2014) Resilient communities: empowering older adults in disasters and daily life. The New York Academy of Medicine, New York

Gómez-Baggethun E, Gren Å, Barton DN et al (2013) Urban ecosystem services. In: Elmqvist T, Fragkias M, Goodness J et al (eds) Urbanization, biodiversity and ecosystem services: challenges and opportunities. Springer Netherlands, Dordrecht, pp 175–251

Grimm N, Cook EM, Hale RL, Iwaniec DM (2016) A broader framing of ecosystem services in cities: benefits and challenges of built, natural, or hybrid system function. In: Seto KC-Y, Solecki W, Griffith C (eds) The Routledge handbook of urbanization and global environmental change, 1st edn. Routledge/Taylor & Francis Group, New York

Guadagno L, Depietri Y, Fra Paleo U (2013) Urban disaster risk reduction and ecosystem services. In: Renaud FG, Sudmeier-Rieux K, Estrella M (eds) The role of ecosystems in disaster risk reduction. United Nations University Press, Shibuya-ku, pp 389–415

Hallegatte S (2009) Strategies to adapt to an uncertain climate change. Glob Environ Chang 19:240–247. doi:10.1016/j.gloenvcha.2008.12.003

Harlan SL, Brazel AJ, Prashad L et al (2006) Neighborhood microclimates and vulnerability to heat stress. Soc Sci Med 63:2847–2863. doi:10.1016/j.socscimed.2006.07.030

Horton R, Bader D, Kushnir Y et al (2015) New York City Panel on Climate Change 2015 Report. Chapter 1: Climate observations and projections: NPCC 2015 Report Chapter 1. Ann N Y Acad Sci 1336:18–35. doi:10.1111/nyas.12586

IPCC (ed) (2012) Managing the risks of extreme events and disasters to advance climate change adaption: special report of the Intergovernmental Panel on Climate Change. Cambridge University Press, New York

IPCC (2015) Climate change 2014: synthesis report. Intergovernmental Panel on Climate Change (IPCC), Geneva

Jacobs B (2005) Urban vulnerability: public management in a changing world. J Conting Crisis Manag 13:39–43. doi:10.1111/j.1468-5973.2005.00454.x

Jenerette GD, Harlan SL, Stefanov WL, Martin CA (2011) Ecosystem services and urban heat riskscape moderation: water, green spaces, and social inequality in Phoenix, USA. Ecol Appl 21:2637–2651. doi:10.1890/10-1493.1

Jones HP, Hole DG, Zavaleta ES (2012) Harnessing nature to help people adapt to climate change. Nat Clim Chang 2:504–509. doi:10.1038/nclimate1463

Kallis G (2007) When is it coevolution? Ecol Econ 62:1–6. doi:10.1016/j.ecolecon.2006.12.016

Kaly U, Pratt C, Mitchell J (2004) The environmental vulnerability index (EVI). South Pacific Applied Geoscience Commission, Suva

Khazai B, Ingram JC, Saah D (2007) The protective role of natural and engineered defence systems in coastal hazards. State of Hawaii and the Kaulunani Urban and Community Forestry Program of the Department of Land and Natural Resources by the Spatial Informatics Group, LLC, San Leandro

Kinney PL, Matte T, Knowlton K, et al (2015) New York City Panel on Climate Change 2015 Report Chapter 5: Public health impacts and resiliency: NPCC 2015 Report Chapter 5. Ann N Y Acad Sci 1336:67–88. doi: 10.1111/nyas.12588

Maantay J, Maroko A (2009) Mapping urban risk: flood hazards, race, & environmental justice in New York. Appl Geogr 29:111–124. doi:10.1016/j.apgeog.2008.08.002

Matthews T, Lo AY, Byrne JA (2015) Reconceptualizing green infrastructure for climate change adaptation: barriers to adoption and drivers for uptake by spatial planners. Landsc Urban Plan 138:155–163. doi:10.1016/j.landurbplan.2015.02.010

McCarthy MP, Best MJ, Betts RA (2010) Climate change in cities due to global warming and urban effects. Geophys Res Lett 37:n/a-n/a. doi: 10.1029/2010GL042845

McHale M, Pickett S, Barbosa O et al (2015) The new global urban realm: complex, connected, diffuse, and diverse social-ecological systems. Sustainability 7:5211–5240. doi:10.3390/su7055211

McPhearson T, Hamstead ZA, Kremer P (2014) Urban ecosystem services for resilience planning and management in New York City. Ambio 43:502–515. doi:10.1007/s13280-014-0509-8

McPhearson T, Haase D, Kabisch N, Gren Å (2016a) Advancing understanding of the complex nature of urban systems. Ecol Indic. doi:10.1016/j.ecolind.2016.03.054

McPhearson T, Parnell S, Simon D et al (2016b) Scientists must have a say in the future of cities. Nature 538:165–166. doi:10.1038/538165a

McPhearson T, Pickett STA, Grimm NB et al (2016c) Advancing urban ecology toward a science of cities. BioScience 66:198–212. doi:10.1093/biosci/biw002

McPhearson T, Madhav K, Herzog C, et al (forthcoming) Urban ecosystems and biodiversity. In: Rosenzweig C, Solecki B (eds) Urban Climate Change Research Network second assessment report on climate change in cities (ARC3-2). Cambridge University Press, Cambridge

Mitchell JK (2003) European river floods in a changing world. Risk Anal 23:567–574. doi:10.1111/1539-6924.00337

Nicholls RJ, Hanson S, Herweijer C et al (2008) Ranking port cities with high exposure and vulnerability to climate extremes. Organisation for Economic Co-operation and Development, Paris

NOAA (2015) U.S. billion-dollar weather & climate disasters 1980–2015.

Norgaard RB (1994) Development betrayed: the end of progress and a coevolutionary revisioning of the future. Routledge, London/New York

Nowak DJ, Crane DE, Stevens JC (2006) Air pollution removal by urban trees and shrubs in the United States. Urban For Urban Green 4:115–123. doi:10.1016/j.ufug.2006.01.007

NYC (2010) NYC green infrastructure plan: a sustainable strategy for clean waterways. City of New York, New York

Oliver-Smith A (1999) "What is a disaster?": anthropological perspectives on a persistent question. In: Oliver-Smith A, Hoffman SM (eds) The angry earth: disaster in anthropological perspective. Routledge, New York, pp 18–34

Platt RH (2006) Urban watershed management. Sustainability, one stream at a time. Environment 48:26–42

Renaud FG, Birkmann J, Damm M, Gallopín GC (2010) Understanding multiple thresholds of coupled social–ecological systems exposed to natural hazards as external shocks. Nat Hazards 55:749–763. doi:10.1007/s11069-010-9505-x

Renaud FG, Sudmeier-Rieux K, Estrella M (eds) (2013) The role of ecosystems in disaster risk reduction. United Nations University Press, Shibuya-ku, Tokyo

Rosenzweig C, Solecki W (2015) New York City Panel on Climate Change 2015 Report Introduction: NPCC 2015 Report Introduction. Ann N Y Acad Sci 1336:3–5. doi:10.1111/nyas.12625

Rosenzweig C, Gaffin S, Parshall L (eds) (2006) Green roofs in the New York Metropolitan region. Research report. Columbia University Centre for Climate Systems Research and NASA Goddard Institute for Space Studies, New York

Rosenzweig C, Solecki WD, Hammer SA, Mehrotra S (eds) (2011a) Climate change and cities: first assessment report of the Urban Climate Change Research Network. Cambridge University Press, Cambridge/New York

Rosenzweig C, Solecki WD, Blake R et al (2011b) Developing coastal adaptation to climate change in the New York City infrastructure-shed: process, approach, tools, and strategies. Clim Chang 106:93–127. doi:10.1007/s10584-010-0002-8

Satterthwaite D (2006) Climate change and cities. International Institute for Environment and Development (IIED), London

Sherbinin AD, Schiller A, Pulsipher A (2007) The vulnerability of global cities to climate hazards. Environ Urban 19:39–64. doi:10.1177/0956247807076725

Smith A, Katz R (2013) US billion-dollar weather and climate disasters: data sources, trends, accuracy and biases. Nat Hazards 67:387. doi:10.1007/s11069-013-0566-5

Spalding MD, Ruffo S, Lacambra C et al (2014) The role of ecosystems in coastal protection: adapting to climate change and coastal hazards. Ocean Coast Manag 90:50–57. doi:10.1016/j.ocecoaman.2013.09.007

Sudmeier-Rieux K (2013) Ecosystem approach to DRR. Basic concepts and recommendations to governments, with a special focus on Europe

Swetnam TW, Betancourt JL (2010) Mesoscale disturbance and ecological response to decadal climatic variability in the American southwest. In: Stoffel M, Bollschweiler M, Butler DR, Luckman BH (eds) Tree rings and natural hazards. Springer, Dordrecht, pp 329–359

Tan J, Zheng Y, Tang X et al (2010) The urban heat island and its impact on heat waves and human health in Shanghai. Int J Biometeorol 54:75–84. doi:10.1007/s00484-009-0256-x

The Nature Conservancy (2013) The case for green infrastructure. Joint Industry White Paper. Dow Chemical Company, Shell, Swiss RE, Unilever, The Nature Conservancy

The Nature Conservancy (n.d.) Reducing climate risk with natural infrastructures

UNISDR (2005) Hyogo framework for action 2005–2015: building the resilience of nations and communities to disasters

UNISDR (2015) Sendai Framework for Disaster Risk Reduction 2015–2030. United Nation Office for Disaster Risk Reduction, Geneva

Wong NH, Yu C (2005) Study of green areas and urban heat island in a tropical city. Habitat Int 29:547–558. doi:10.1016/j.habitatint.2004.04.008

Chapter 7
Urban Wetlands and Riparian Forests as a Nature-Based Solution for Climate Change Adaptation in Cities and Their Surroundings

Dagmar Haase

Abstract Wetlands and riparian forests belong to the most productive, but also the most vulnerable, ecosystems in urban regions and cities due to their complex watershed system, often very high biodiversity and the pressure from urban land use and surface sealing. Wetlands and floodplain forests are often highly valued recreational areas, providing many benefits for urban dwellers, such as fresh air, moisture, oxygen and biogenic essentials as well as many cultural and place-based values. Wetlands and riparian forests are very efficient spaces for water and matter regulation, pollutants fixation and flood water retention. Thus, particularly for dense urban areas, they represent almost perfect nature-based solutions for risk mitigation and adaptation concerning both climate extremes: flood and drought. Moreover, they can serve as a buffer against high air temperatures and provide wetness during heat waves. However, urban wetlands and riparian forests are often endangered by urbanisation pressure, land take for construction purposes and pollution. This chapter provides arguments that urban wetlands being a nature-based solution for cities facing climate change and presents design options to expand and even create such wetlands in cities where remnants are no longer available.

Keywords Urban wetlands • Riparian forests • Nature-based solution • Climate change adaptation • Cities

D. Haase (✉)
Department of Geography, Humboldt University Berlin, Berlin, Germany

Department of Computational Landscape Ecology, Helmholtz Centre for Environmental Research – UFZ, Leipzig, Germany
e-mail: dagmar.haase@geo.hu-berlin.de

© The Author(s) 2017   111
N. Kabisch et al. (eds.), *Nature-based Solutions to Climate Change Adaptation in Urban Areas*, Theory and Practice of Urban Sustainability Transitions,
DOI 10.1007/978-3-319-56091-5_7

7.1 Introduction: What Is the Value of Wetlands and Riparian Forests in Cities?

Wetlands and riparian forests belong to the most complex ecosystems on the planet. Their water balance and watershed systems are highly complex in terms of the ongoing interactions between ground-, interflow and surface waters (Haase and Nuissl 2007). Wetlands are able to store an enormous amount of water in their sediments, soils and vegetation. They can buffer extreme air temperatures due to the flows of evapotranspiration they create. Their typically loamy, clay and silt-rich floodplain soils are fantastic buffers and transmitters of all kinds of organic and inorganic pollutants emitted; they represent a kind of 'translocation area' for matter and are a sink and source of pollutants at the same time. Mature floodplain trees sequester and store considerable amounts of CO_2 contributing to climate change mitigation. In terms of flora and fauna, wetlands and riparian forests provide multiple wet and perennial habitats and, consequently, are home to thousands of species (Fig. 7.1).

Since the industrial revolution and in particular since WW II, cities represent the main 'habitat' for humans, with about 55% of the world population and current estimates assume that some 75% of the global population will live in cities by 2030 (UN 2016). Cities are places of close contact between humans and nature, where society and biodiversity are privy to complex interactions and co-evolution in terms of typical urban niches adapted gene mutation and species creation (Alberti 2015).

Many cities have been established along rivers due to the economically and defense-related strategic location and the availability of water resources and fertile soils (Kühn and Klotz 2006). Thus, wetlands, riparian forests and cities already share a long history punctuated with synergistic interactions and risk creation (Schanze 2006). More specifically, river floods endanger humans and their properties and humans in turn destroy wetlands and floodplain forests by means of construction, surface sealing, groundwater regulation and tree cuttings. However, when society supports the proper functioning of wetlands, these habitats and their forests can produce significant ecosystem services for urban dwellers, including the aforementioned regulation functions as well as being recreational areas and places for all age groups to experience and

Leipzig, Weiße-Elster-floodplains   Vienna, Danube floodplains

Fig. 7.1 Typical urban wetland and associated floodplain forest in Leipzig, Germany, and Vienna, Austria (Photos: Dagmar Haase)

enjoy nature and, what is more, providing cool places during hot summers and thus to support physical health of urban dwellers (Haase 2003; Andersson et al. 2015).

As outlined previously, wetlands and riparian forests offer a multitude of services which benefit both society and biodiversity. They are ideal nature-based solutions for regulating the effects of climate stress and climate change that cities are increasingly faced with, including heat waves, long dry spells, floods and related accumulation of polluted sediments. Furthermore, and in addition to their importance in water flow regulation, they are stepping stones for species of different taxa and place of the creation of novel or altered habitats and the survival of endangered species (Elmqvist et al. 2015).

However, and not least because of the aforementioned properties, river regulation and drainage of wetlands have been common practice in most areas of Europe for centuries, with interventions increasing over the past 50–100 years. More than 50% of the European wetlands that existed 100 years ago have been lost mainly to urban developments (conversion into other land types) (European Commission 1995), leading to a substantial decrease in the number, size and natural habitat of large bogs and marshes and small or shallow lakes. As a result, both European legislation (e.g. the EU Birds Directive (1979) and the EU Flora, Fauna and Habitats Directive (1992; European Commission) and international agreements (i.e. the Convention on Wetlands of International Importance especially as Waterflow Habitat (Ramsar, Iran, 1971)) have implemented protective measures. A considerable part of European wetlands (>40,000 km^2 or about 0.5%) has been designated as 'Ramsar sites' under the international convention.

Next to cities themselves, roads have a major impact on wetlands and riparian areas in central parts of Europe characterised by a dense infrastructure, such as Austria, Belgium, Denmark, Germany, Luxembourg and the Netherlands. And, what is more, this pressure along with peri-urbanisation is expected to increase as settlement areas and transport networks expand (Fig. 7.2).

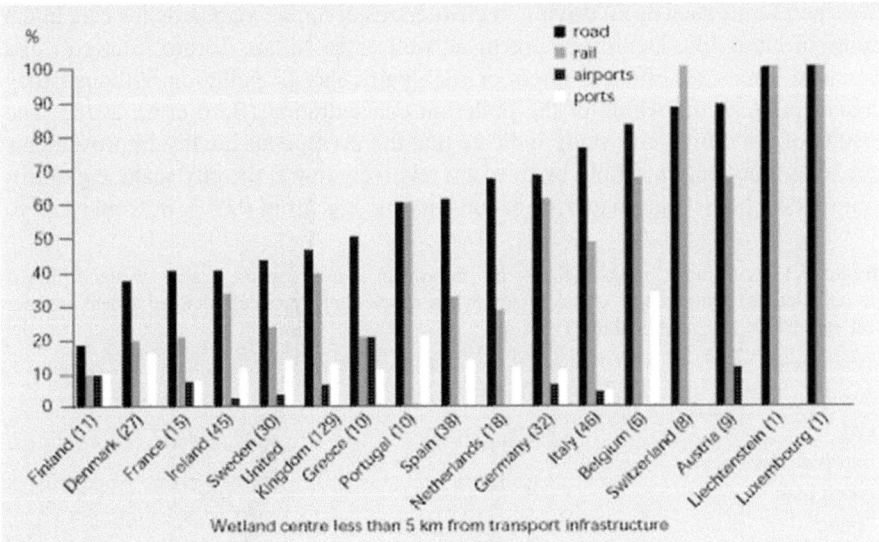

Fig. 7.2 Proximity of major transport infrastructure to Ramsar areas in selected European countries (European Commission 1995)

7.2 Ecosystem Services Relevant for Climate Change Adaptation Provided by Wetlands and Riparian Forests and Trade-Offs

Wetlands and riparian forests provide a multitude of ecosystem services (Haase 2014; Kroll et al. 2012). They are extremely efficient and important cooling elements in urbanised areas due to their continuous high groundwater levels, their wet soils and, in the event that they include a forest stand, the leaf-based shading and transpiration effects (Breuste et al. 2013). Concerning climate change adaptation, the suitability of wetlands is manifold ranging from their organic soils as carbon stores to its alluvial parts as areas for local climate cooling, what is more, the role of floodplains as means to slow down flood peaks, as space for water and allowing for floods to happen and for water storage to counteract droughts.

In Europe, almost all riparian or floodplain forests are broadleaf forests with originally unique but heavily modified, except a few situations, tree stands and forest habitats. Natural floodplain tree stands with large, green leaves which help to cool the surroundings more efficiently than typical park trees (Table 7.1; Haase and Gläser 2009; Elmqvist et al. 2015). Data from Manchester (UK) show, for example, that a 10% increase in tree canopy cover may result in a 3–4 °C decrease in ambient temperature (Gill et al. 2007) and save large amounts of energy used in air conditioning (Nicolson-Lord 2003; Akbari 2002).

Due to increasing car traffic and travel activities in most of the large and megacities worldwide, air pollution with particular matter (PM_{10}, $PM_{2.5}$ and $PM_{<1}$), NO_x and ozone, has been considerably increasing in the last three decades since 1990 (Weber et al. 2014a, b). This pollution heavily and considerably impacts the respiratory and cardiovascular systems of urban populations. Consequently, the health thresholds are exceeded in many cases, which have led to transport regulations and limits such as no driving in city centres or higher standards for cars in the cities in cities like Delhi or Karachi as well as in Milan, Torino, Stuttgart and Sarajevo. Trees are efficient filters of such particular air pollution, although they cannot mitigate the whole of the pollution concentration (Baró et al. 2015). The results of a multiple-city study indicate that the average air quality improvements due to air purification supply by trees are relatively low at the city scale, e.g. nitrogen oxides, particulate matter or ozone varying, e.g. from 0.07% in Rotterdam to

Table 7.1 Leaf area index (LAI) of floodplain forest species and wetland lawns (in italic letters) compared to 'classical' urban tree species and grassland (defined as leaf area per soil unit, BFI = A(leaf)/A(soil unit)

Tree species	BFI
Beech	*6–8*
Oak	*5–7*
European larch	2–4
Forest pine	3–4
Spruce	5–10
Lawn	*7*
Grassland	1–2

0.81% in Stockholm for NO_2), although positive effects are likely to be more relevant in highly tree-covered areas such as urban forests; for example, the expected air improvements are higher than 6% for PM_{10} in Stockholm (Sweden) and Salzburg (Austria), assuming a 100% tree cover (Baró et al. 2015). For the city of Leipzig (Germany), the whole tree cover including the floodplain forests and the street and park trees is able to balance the CO_2 emissions of 6000 residents, which is about 1.5% of the urban population (Strohbach and Haase 2012).

Wetlands are also a sink for matter and could become, depending on the geochemical milieu and flooding, a source of matter (Haase and Neumeister 2001). Fluvisols, the typical soil type of wetlands, are able to bind many organic and inorganic compounds due to their high contents of fine grain material (clay, silt), Fe oxides and, in their topsoils, up to 10% organic carbon (Table 7.2; Haase and Neumeister 2001). Thus, a closed coverage layer of wetland soils is the best protection against the pollution of the local and regional aquifers that provide fresh water for urban dwellers (Haase 2009). Thus, the soils and tree cover of wetlands and floodplains serve as a highly efficient metabolic body for nutrient fluxes and pollutant, benefiting the humans settling around these areas.

Next to the many abiotic properties and processes that are beneficiary for humans, wetlands and riparian forests provide excellent and unique habitats for not only humans but also many other species (Elmqvist et al. 2015). Floodplain forest trees such as *Quercus* (oak) or *Fraxinus* (ash) as well as *Salix* (willow) or *Carpinus* (beech) are habitat to up to 1000 animal species and thus critical elements of local and regional food chains. Whole food webs/chains are linked through particularly old floodplain forest trees with mighty stems and thick bark, and wetlands are larger landscape elements that link habitats at the regional scale as fauna migration and nesting areas (Alberti 2015). Due to the high variety of species that inhabit wetlands, these landscapes are specifically robust against single pests and thus act as natural pest control (Alberti 2015) as the survival rate of the ecosystem is higher when not all species are affected. At the same time, being wet habitats and charac-

Table 7.2 Matter sink and matter flow balancing functions of fluvisols compared to other urban substrates

Soil/substrate property	Fluvisol	Regosol (sand)	Cambiosol	Gleysol (clay)	Regosol (sand)
Infiltration capacity	+	+ +	±	− −	+ +
Nutrient accumulation	+	− −	−	+ +	− −
Nutrient cycling	+ +	−	+	+	−
Pollutant accumulation	+ +	−	+	+ +	−
Water capacity	+ +	− −	+	+ +	− −
Water flow balancing	+	−	+ +	−	−
Mechanical filtering	+	+	+ +	−	+
Chemical filtering	+	− −	−	+ +	− −
Drainage	±	+ +	− −	−	+ +
Erosion susceptibility	−	±	+	− −	±

with ++ is very good/high, + is good (high), ± is medium, − is poor/low and − − is very poor/low terised by closeness to water, they provide ideal breeding grounds for the spread of invasive species such as the tsetse fly, *Aedes aegypti* or *Amenophelis* that inflow from south and are carrier of so far unknown illnesses such as yellow or dengue fever or types of ticks bringing *meningitis* and *encephalitis* as well as bringing endemic malaria possibly back to Europe (Medlock and Leach 2015). Thus, while using wetlands as a nature-based solution to flood alleviation, air cooling and pollutant fixers to address those threats, they can also create trade-offs and disservices (Döhren and Haase 2015).

Potentially the most important and widely known property of wetlands and floodplain forests are their role as a flood moderator; their function as drought moderator is far less known (Haase 2011). Wetlands and floodplains are spaces to naturally store and 'save' water in times of high rainfall and long periods of precipitation. Historical wetland management has made use of the high water storage and spatial inundation capacities of floodplains and thus created the most fertile and productive landscapes in former centuries, such as the Hungarian Tisza floodplains (Haase 2011), the Oder valley in Germany (Dalchow and Bork 1998) or the area of Kaliningrad in Northwestern Russia. Only in case they were object of an artificial embankment and river regulation, wetlands can change from places of saving water to places that let water easily migrate into surrounding spaces creating loss of life, loss of property and loss of assets being a clear trade-off (Scheuer et al. 2012).

7.3   Urban Wetlands as a Nature-Based Solution and Options for Their Design

As discussed in the previous sections, wetlands and their riparian or floodplain forests as such are solutions created by nature to store, distribute and hold water in ecosystems and whole landscapes. This function has been largely impacted by humans, mainly in cities and urban areas to use the water as resource and the river as transportation means, on the one hand, and to protect human lives and assets from floods on the other hand (see Haase 2011, Kubal et al. 2009 and Meyer et al. 2009). However, the last big floods in Europe (Elbe, Oder, Mulde, Tisza, Rhine to list a few) drew attention to the fact that technical or built protection against water will not provide 100% protection to people in cities against floods, particularly when taking climate change and longer heavy precipitation events into consideration. Moreover, events such as the hurricane Katrina in the USA in 2005 or the last Elbe flood in Germany in 2013 with hazardous bursting of large dams made very clear that technological solutions in case they fail can produce enormous damage and casualties as their capacities are enormous in terms of how much water they can collect/hinder to flow, actually much more than any nature-based solution can (see also a global map by Scheuer et al. 2012).

But, which efforts could instead be reinvested in restoring and maintaining the functionality of wetlands/floodplains? The revitalisation of large and smaller rivers and thus a continuous and more natural inflow and distribution of rainfall water in a larger area have been shown to improve protection against horrible flood events along rivers and coasts (Fig. 7.3; Middelkoop et al. 2004; Krysanova et al. 2008). A restoration of the (remnant) riparian vegetation would assist in reconnecting rivers with floodplains and to provide greater instream ecosystem complexity, particularly in urban areas (see examples in Fig. 7.3). In addition to restoration, wetlands could also be created, including so-called balancing ponds, and thus natural functions created by a kind of nature mimicry in the surroundings of cities. Also bioswales as their smaller version could be created along streets in, e.g. housing areas to take up and hold rainfall water and, what is more, moisturize the dry urban air after the rainfall event for a while (Fig. 7.3; Haase 2016). Also permeable surfaces can support this moistening of urban areas. A high channel diversity in the wetlands themselves but also in form of bioswales and integrated into the street network would reduce the speed of (superficial) flood transmission. All these measures are known in hydrology and hydrological construction; however, in urban planning they need better introduction and a cooperation between landscape planners, ecologists, green infrastructure specialists on the one hand and hydraulic engineers on the other to

Recreation areas to promote health and happiness	Tree planting to clean the air and to filter pollutants	(Re-)Greening waters to buffer noise and to filter pollutants

Revitalization of wetlands/rivers to balance the urban water cycle	Bioswales to store rainwater in streets and to lower the flood risk	Constructed wetlands to regulate ground- and surface water flows
	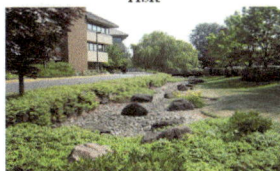	
Restoring fluvisols to bind and immobilize pollutants	Room for rivers to alleviate floods and to safe cities from hazards	Nutrient-cycling for food production in cities and prevent food import

Fig. 7.3 Nature-based solutions linked to wetlands, riparian forests and floodplains (Photos: Dagmar Haase)

Table 7.3 Nature-based and artificial/built water retention and climate change adaptation measures for relevant spatial response units

Spatial response unit	Grey measures	Green measures
Cities	Infiltration devices	Filter strips and bioswales
	Dams, dikes	Permeable surfaces and filter drains
	Walls	
	Underground drainage	Green roofs and green walls
Urban regions	Basins and ponds	Wetland restoration/construction
	Dams, dikes	Floodplain restoration
	Reservoirs	Re-meandering
		Restoration of lakes
		Natural bank stabilisation (also at catchment level)
Larger river catchments	Dam and dike systems	Afforestation and forest protection
	Reservoirs	Maintaining and developing riparian forests
		Afforestation of agricultural land
		Afforestation of unused land and pastures
		Buffer stripes of agricultural fields

According to Vanneuville and Werner (2012), modified

make the suggested nature-based solutions effective and successful (as already mentioned by Pahl-Wostl 2007).

The restoration and construction of wetlands would increase and enhance all of the aforementioned ecosystem services provided by wetlands and riparian forests and provide additional health benefits for urban dwellers such as cooling by tree shade, aesthetic improvement of biodiversity and healthy forest products to be foraged (e.g. fruits, mushrooms, herbs). In terms of the spatial configuration of wetlands as parts of larger river catchments, the aforementioned measures might be carried out either within cities or also upstream to be most efficient for flood protection, for example. This is a typical case for regional planning. Table 7.3 summarises again the various options of grey and green solutions in wetlands to face the effects of climate change.

7.4 Discussion and Conclusions

Several options have been outlined which utilise wetland and riparian forests' functionality as nature-based solutions to better face the consequences of climate change in cities and urban regions. The following are the key observations:

1. Wetlands, floodplains and riparian forests are very precious and in many respects unique ecosystems of high complexity that are able to 'deal' with large amounts of water in a spatio-temporal distribution scheme (uptake, storage, release) that

saves humans, their assets and whole cities from extreme flooding. They have to be kept, and in case of their destruction, they have to be restored whenever economic and social trade-offs permit it. Moreover, rivers should be—as far as possible and at locations where possible—reconnected with floodplains to enhance natural water storage in the whole catchment system.

2. Wetlands can provide inspiration for how water storage could work also for non-lowland situations such as top hill situations and inner city areas. Wetlands could be created/constructed as a kind of nature mimicry in the form of constructed wetlands in the peri-urban landscape and smaller bioswales along streets in densely built inner city locations as 'smart and space saving' accepting the basic structure of the grey environment. Together, natural and newly created/constructed wetlands could support the regulation of the magnitude and timing of water runoff and flooding as well as the recharging of aquifers.

3. We need to better understand and make use of the enormous balancing, buffering and metabolic capacities of wetlands and their forests in cities as well as the recreational opportunities and air pollution filtering capacities, as they cannot simply be replaced by technology or grey infrastructure solutions.

4. As cities have been built and transformed over centuries and as humans are the dominating creative factor in cities and urban regions, natural remnants of wetlands and floodplain forests have to be combined with technological/technical solution such as discussed under (2). Implementing wetland functionality, be it restoration or creation, can be a way and a new job creator to realise 'mini-wetlands' and 'riparian trenches' in areas disconnected from the river. The complementarity of both nature and technology under the supervision of society might be a clever, pragmatic but at the same time innovative solution to complex problems created by man and nature response.

5. A combination of 'C&C'—conservation and construction—might be a coupled solution for most of our cities when it is about climate change adaptation. Along with this, a new way of urban governance including co-development and co-production to create real societal benefits and acceptance is needed to give this new 'C&C' a real chance.

However, as also discussed in this short contribution, wetlands as natural systems create trade-offs when it comes to human health and infectious diseases, particularly vector-borne diseases and regarding the loss of land for new developments. Knowledge about this fact or 'disservice' is a first step to create awareness. Strategies for public protection and, if needed, intervention are following steps to balance these trade-offs. Climate change itself can increase the 'danger' of new vectors and new infection risks, no doubt, but such trade-offs do not diminish the high value of wetlands and their riparian forests as natural solutions for adapting to climate change in cities and urban regions. For many cities across Europe, climate change and a gradual warming are already a reality, and based on recent temperature prognoses (IPCC 2016), the situation gets worse in terms of an increase in mean summer day and night temperatures and more heavy and less effective rainfall events. Thus, and despite a potential increase of vectors, there is a need to 'switch tactics' replacing grey for green.

Investing in restoring, protecting and enhancing ecological functionality and ecosystem services in urban wetlands is not only ecologically and socially desirable (cf. Elmqvist et al. 2015), but as the contribution has shown, wetlands are socio-ecologically viable recognising their multiple services and all their associated benefits for the large number of beneficiaries in cities.

References

Akbari H (2002) Shade trees reduce building energy use and CO_2 emissions from power plants. Environ Pollut 116(1):119–126

Alberti M (2015) The effects of urban patterns on ecosystem function. Int Reg Sci Rev 28(2):168–192. doi:10.1177/0160017605275160

Andersson E, McPherson T, Kremer P, Frantzeskaki N, Gomez-Baggethun E, Haase D, Tuvendal M, Wurster D (2015) Scale and context dependence of ecosystem service providing units. Ecosyst Serv 12:157–164

Baró F, Haase D, Gómez-Baggethun E, Frantzeskaki N (2015) Mismatches between ecosystem services supply and demand in urban areas: A quantitative assessment in five European cities. Ecol Indic 55:146–158

Breuste J, Haase D, Elmquist T (2013) Urban landscapes and ecosystem services. In: Sandhu H, Wratten S, Cullen R, Costanza R (eds) Ecosystem services in agricultural amd urban landscapes. Wiley, Chichester, pp 83–104

Dalchow C, Bork HR (1998) Das Oderbruch – Einführung in die Entstehungsgeschichte. In: Darkow G, Bork HR (eds) Die Bewirtschaftung von Niederungsgebieten in Vergangenheit und Gegenwart, ZALF-Bericht 34. ZALF, Müncheberg, pp 13–18

Döhren P, Haase D (2015) Ecosystem disservices research: a review of the state of the art with a focus on cities. Ecol Indic 52:490–497

Elmqvist T, Setälä H, Handel SN, van der Ploeg S, Aronson J, Blignaut JN, Gómez-Baggethun E, Nowak DJ, Kronenberg J, de Groot R (2015) Benefits of restoring ecosystem services in urban areas. Curr Opin Environ Sustain 14:101–108

European Commission (1995) Commission's communication to the Council and the Parliament: wise use and conservation of wetlands. European Commission, Brussels

Gill SE, Handley JF, Ennos AR, Pauleit S (2007) Adapting cities for climate change: the role of the green infrastructure. Built Environ 33:115–133

Haase D (2003) Holocene floodplains and their distribution in urban areas – functionality indicators for their retention potentials. Landsc Urban Plan 66:5–18

Haase D (2009) Effects of urbanisation on the water balance – a long-term trajectory. Environ Impact Assess Rev 29:211–219

Haase D (2011) Participatory modelling of vulnerability and adaptive capacity in flood risk management. Nat Hazards 67:77. doi:10.1007/s11069-010-9704-5

Haase D (2014) The nature of urban land use and why it is a special case. In: Seto K, Reenberg A (eds) Rethinking global land use in an urban era. Strüngmann Forum Reports, vol 14, Julia Lupp, series editor. MIT Press, Cambridge, MA

Haase D (2015) Reflections about blue ecosystem services in cities. Sustainability Water Qual Ecol 5:77. Available online http://www.sciencedirect.com/science/article/pii/S2212613915000112

Haase D, Gläser J (2009) Determinants of floodplain forest development illustrated by the example of the floodplain forest in the District of Leipzig. For Ecol Manag 258:887–894. doi:10.1016/j.foreco.2009.03.025

Haase D, Neumeister H (2001) Anthropogenic impact on fluvisols in German floodplains. Ecological processes in soils and methods of investigation. Int Agrophys 15(1):19–26

Haase D, Nuissl H (2007) Does urban sprawl drive changes in the water balance and policy? The case of Leipzig (Germany) 1870–2003. Landsc Urban Plan 80:1–13

IPCC (2016) http://www.ipcc.ch

Kroll F, Müller F, Haase D, Fohrer N (2012) Rural-urban gradient analysis of ecosystem services supply and demand dynamics. Land Use Policy 29(3):521–535. doi:10.1016/j.landusepol.2011.07.008

Krysanova V, Buiteveld H, Haase D, Hattermann FF, Van Niekerk K, Roest K, Martínez-Santos P, Schlüter M (2008) Practices and lessons learned in coping with climatic hazards at the river-basin scale: floods and droughts. Ecol Soc 13(2): 32. [online] URL: http://www.ecologyandsociety.org/vol13/iss2/art32/

Kubal T, Haase D, Meyer V, Scheuer S (2009) Integrated urban flood risk assessment – transplanting a multicriteria approach developed for a river basin to a city. Nat Hazards Earth Syst Sci 9:1881–1895

Kühn I, Klotz S (2006) Urbanisation and homogenization – comparing the floras of urban and rural areas in Germany. Biol Conserv 127:292–300

Medlock JM, Leach SA (2015) Effect of climate change on vector-borne disease risk in the UK. Lancet Infect Dis 15(6):721–730

Meyer V, Scheuer S, Haase D (2009) A multi-criteria approach for flood risk mapping exemplified at the Mulde river, Germany. Nat Hazards 48:17–39. doi:10.1007/s11069-008-9244-4

Middelkoop H, van Asselt MBA, van't Klooster SA, van Deursen WPA, Kwadijk JCJ, Buiteveld H (2004) Perspectives on flood management in the Rhine and Meuse Rivers. River Res Appl 20:327–342

Nicholson-Lord D (2003) Green cities: and why we need them. New Economics Foundation, London

Pahl-Wostl C (2007) Transition towards adaptive management of water facing climate and global change. Water Resour Manag 21(1):49–62

Schanze J (2006) Flood risk management—a basic framework. In: Schanze J, Zeman E, Marsalek J (eds) Flood risk management—hazards, vulnerability and mitigation measures. Springer, New York, pp 149–167

Scheuer S, Haase D, Meyer V (2012) Spatial explicit multi-criteria flood risk – fundamentals and semantics of multicriteria flood risk assessment. In: Wong TSW (ed) Flood risk and flood management. Nova Science Publishers Inc, Hauppauge

Strohbach MW, Haase D (2012) Estimating the carbon stock of a city: a study from Leipzig, Germany. Landsc Urban Plan 104:95–104. doi:10.1016/j.landurbplan.2011.10.001

United Nations (2016) http://www.un.org/en/development/desa/population/

Vanneuville W, Werner B (2012) Water resources in Europe in the context of vulnerability. EEA 2012 state of water assessment, EEA Report No 11/2012. Office for Official Publications of the European Communities, Copenhagen

Weber N, Haase D, Franck U (2014a) Assessing traffic-induced noise and air pollution in urban structures using the concept of landscape metrics. Landsc Urban Plan 125:105–116

Weber N, Haase D, Franck U (2014b) Zooming into the urban heat island: how do urban built and green structures influence earth surface temperatures in the city? Sci Total Environ 496:289–298

Chapter 8
Making the Case for Sustainable Urban Drainage Systems as a Nature-Based Solution to Urban Flooding

McKenna Davis and Sandra Naumann

Abstract European cities continue to experience a steady increase in the intensity and frequency of floods, largely due to high urban densities and resultant soil sealing. In the last decade, flooding as a natural hazard has produced the highest economic losses in Europe and storm water management has become a serious urban challenge.

The traditional solution to cope with excess rainwater in western cities has been piped drainage systems. These are mainly single-objective oriented designs that often no longer have the capacity to keep pace with on-going urbanisation and the impacts of climate change, and frequently involve high construction, maintenance, and repair costs. While such approaches have certainly reduced the damages incurred from flooding events during the past two centuries and are arguably still necessary for extreme flood events in the future, alternative approaches that accomplish these aims and offer additional benefits are progressively being pursued. Given these conditions, one increasingly utilised solution for managing flood risk by dealing with water at the source is sustainable urban drainage systems (SUDS). Other terms which also aim to minimise potential impacts on the neighbouring environment, people and development include *inter alia* BMP (Best Management Practices); LID (Low Impact Development); WSUD (Water Sensitive Urban Design) (see: Fletcher et al., Urban Water J 12(7):525–542, 2015 for a complete taxonomy).

SUDS as a promising nature-based solution are the focus of this chapter, utilizing a range of case studies and evidence from across Europe to underline the arguments presented. Besides reducing the negative effects of urban flooding and interlinked water pollution, the many supplementary benefits and potential cost-effectiveness of SUDS as compared to grey infrastructure solutions are also presented. In addition to highlighting relative advantages, the chapter also outlines current challenges facing a wider uptake of SUDS and presents approaches to help overcome existing social and political barriers.

The promise of ongoing research, targeted collaboration and partnerships and an ever-growing evidence base on the effectiveness and associated costs and benefits of SUDS serve as strong tools to improve the confidence and competence associated

M. Davis (✉) • S. Naumann
Ecologic Institute, Berlin, Germany
e-mail: mckenna.davis@ecologic.eu; sandra.naumann@ecologic.eu

© The Author(s) 2017   123
N. Kabisch et al. (eds.), *Nature-based Solutions to Climate Change Adaptation in Urban Areas*, Theory and Practice of Urban Sustainability Transitions,
DOI 10.1007/978-3-319-56091-5_8

with their design and implementation. Such data will help to refute existing public and political hesitation as compares to traditional grey infrastructure approaches to water management. However, the significant potential for more widespread uptake remains largely untapped. Further targeted actions are necessary for increasing the acceptance and application of this nature-based solution and realizing its full potential.

Keywords SUDS • Sustainable urban drainage system • nature based solution • flood management • Europe • cost-benefit analysis

8.1 Introduction

European cities continue to experience an increase in the intensity and frequency of floods, with further escalations projected as a result of climate change and rapid urbanization (Santato et al. 2013). Particularly in urban areas, the management and drainage of storm water presents a serious challenge. The high urban density within cities and resultant soil sealing has lead to a reduction in the potential of water infiltration in the ground, which increases run-off water and flood risk (EEA 2012).

The traditional solution to these challenges in western cities has been 'grey' infrastructure – such as piped drainage systems – which are mainly single-objective oriented designs to cope with rainwater within the urban landscape. However, these drainage infrastructures often no longer have the capacity to keep pace with on-going urbanisation and the increasing rate of storm water due to climate change and soil sealing, and can lead to increased run-off and a higher risk of urban flooding (EEA 2012; Perales-Momparler et al. 2016; Zhou 2014). Additional indirect consequences are an insufficient discharge of excess water to the regional water system, and an increase of pollutants in the water caused by run-off (e.g., oil, organic matter and toxic metals), leading to increases in algal blooms, harm to wildlife and reductions in amenity value (Sharma 2008). Furthermore, managing storm water runoff through grey infrastructure approaches typically entails high construction, maintenance, and repair costs (Hair et al. 2014).

While 'grey' approaches have certainly reduced the damages incurred from flooding events during the past two centuries and are arguably still necessary for extreme flood events in the future, alternative approaches that accomplish these aims while offering additional benefits are progressively being pursued (Jones and Macdonald 2007; Perales-Momparler et al. 2016). Sustainable urban drainage systems (SUDS),[1] which are outlined in detail in the subsequent section, represent one such promising alternative flood risk management tool in the transition towards achieving regenerative urban built environments.

SUDS as a type of nature-based solution are the focus of this chapter, particularly concentrating on the existing evidence base regarding the potential for delivering a wide range of benefits as compares to purely grey solutions. The chapter also highlights challenges that are currently limiting more widespread uptake of these greener approaches

---

[1] Other terms are used elsewhere: *inter alia* BMP (Best management practices); LID (Low Impact Development); WSUD (Water Sensitive Urban Design) (see: Fletcher et al. 2015 for a complete taxonomy)

and identifies potential solutions and needs to improve the confidence and competence associated with designing and implementing SUDS. The research results presented were gathered by the authors via an analysis of literature, expert interviews and EU level stakeholder workshops in the context of the EU research project RECREATE.[2]

8.2   Using Green Alongside Grey as an Alternative Approach to Flood Protection

Instead of focusing on 'end-of-pipe' or 'at the point of the problem' solutions as is the case with many purely 'grey' infrastructure solutions, sustainable urban drainage systems aim to slow down and reduce the quantity of surface water runoff in an area in order to minimize downstream flood risk and reduce the risk of resultant diffuse pollution to urban water bodies (Rose and Lamond 2013; Woods Ballard et al. 2015; Zhou 2014). As a nature-based solution, SUDS achieves these aims by utilizing a mix of natural processes[3] and green/grey components[4] to harvest, infiltrate, slow, store, convey and treat runoff onsite; examples include the following (from Woods Ballard et al. 2015):

- **Rainwater harvesting systems** – collect and store rainwater from roofs and other paved surfaces (such as car parks) for re-use
- **Green roofs** – involve constructing a soil layer on a roof to create a living surface that reduces surface runoff
- **Permeable pavements** – act as a hard surface for walking or driving, while enabling rainwater to infiltrate to the soil or underground storage
- **Bioretention systems** (such as rain gardens) – collect runoff in a temporary surface pond before it filters through vegetation and underlying soils
- **Trees** – capture rainwater while also providing evapotranspiration, biodiversity and shade
- **Swales, detention basins, retention ponds and wetlands** – slow the flow of water, store and treat runoff while draining it through the site and encouraging biodiversity
- **Soakways and infiltration basins** – promote infiltration as an effective means of controlling runoff and supporting groundwater recharge

These solutions are diverse in nature and can take many different forms both above and below ground, depending on the state and characteristics of the drainage system in place and the components utilized (State of Green 2015). Table 8.1 provides illustrative examples of different forms of SUDS which have been implemented across Europe.

SUDS can be implemented either as a new development or as a retrofit of existing structures. Regardless of the type, the central objective of all SUDS is to fully

---

[2] REsearch network for forward looking activities and assessment of research and innovation prospects in the fields of Climate, Resource Efficiency and raw mATErials (RECREATE): URL: http://www.recreate-net.eu/

[3] These could include, for example, evaporation, infiltration, re-use and plant transpiration.

[4] Including, for example, permeable surfaces, filter strips, filter and infiltration trenches, green roofs, swales, detention basins, underground storage, wetlands and/or retention ponds.

Table 8.1 European examples of SUDS and their components

Lamb Drove, Cambourne, United Kingdom[a]	
A demonstrative SUDS scheme was implemented in a residential development area in Cambourne to highlight innovative sustainable water management techniques within new developments. A variety of SUDS elements were implemented across the one hectare site, including permeable paving, green roofs, swales, filter strips, wetlands, and a retention pond. Results indicate e.g., improvements in biodiversity and water quality leaving the site, increased amenity and social values and cost savings to residents by avoiding stormwater disposal charges. Furthermore, the project concludes that many aspects of SUDS can be installed and maintained at lower costs than more traditional forms of drainage.	
Valencia, Spain	
Within the framework of the EU-funded AQUAVAL project, SUDS (infiltration basin, green roof, swales, etc.) were implemented in six sites across the Valencian region. The measures came as a response to shortcomings of the existing urban sewer system, which insufficiently abated frequent rainfall and caused pluvial flooding and the discharge of combined sewage into the receiving water courses. Monitoring results showed that SUDS performed well hydraulically under Mediterranean climate conditions and improve the water quality.	
Monnikenhuizen, Arnhem, Netherlands	
The Monnikenhuizen study site was selected for its unique and challenging topographic and contextual conditions for utilizing SUDS, as it is located on a hill and surrounded by forests and contains 204 residences. On a small scale, greenroofs and permeable parking lots were created; on a larger scale, water from the road is led via gutters to an infiltration and storage pond.	

Text/photo sources: Pledger n.d. (UK); Perales-Momparler et al. 2016/ Perales-Momparler 2012 (Spain); 2BG 2008 (Netherlands)
[a]The project was part of a European funded programme (FLOWS), which featured 40 projects throughout Germany, the Netherlands, Norway, Sweden and the UK

exploit the opportunities and benefits that can be obtained from surface water management (Woods Ballard et al. 2015).

8.3   Making the Investment Case for SUDS

In the past decade, more than 165 major floods evoking significant economic damages have taken place across Europe, making flooding (including from rivers, the sea and direct rainfall) the most widespread natural hazard on the continent in terms of economic loss (CRED 2009). In 2002, for example, flooding events occurred in six EU Member States and created infrastructure damages amounting to more than €18.7

billion (Santato et al. 2013). These damages highlight the shortcomings of piped systems as a stand-alone solution to address flooding, and imply the vast potential for replacing and complementing these systems with alternative nature-based approaches.

By integrating natural elements into their design, SUDS can serve to not only address the primary objective of improved water quality and quantity management, but can also offer a wide range of additional benefits which are supplementary to those of purely "grey" solutions (Charlesworth et al. 2016). SUDS can, for example, improve public health, create amenity values in the targeted areas, provide recreational opportunities, support the local ecology and biodiversity, and capture carbon (e.g., Burns et al. 2012; Charlesworth 2010; Norton et al. 2015; Novotny et al. 2010). Table 8.2 presents a more comprehensive overview of these and other potential benefits.

Table 8.2 SUDS benefit types, descriptions and provisioning details

Benefit category	Description	Aspects of the SUDS design that provide the benefit
Air quality	Reduced damage to health from improved air quality	Air particulate filtering via vegetation (e.g., trees and green roofs)
Air and building temperature	Cooling or insulation; thermal comfort and energy savings	Green and blue spaces, green roofs
Biodiversity and ecology	Sites of ecological value	Habitat creation and enhancement, connecting habitats
Carbon reduction and sequestration	Reduced energy/water use and planting	Low energy needs (materials, construction and maintenance); sequestration (e.g., trees and wetlands)
Climate change adaptation	Ability to make incremental changes to systems	Designing for exceedance, adaptability of scheme
Community cohesion and crime reduction	Crimes against property or people	See visual character, economic growth/ inward investment and education
Economic growth and inward investment	Business, jobs, productivity, tourism, property prices	See visual character, recreation and air and building temperature
Education opportunities	Enhanced access to and existence of educational possibilities	Community engagement (before and after construction), information boards, education programmes, play features
Flood risk reduction	Damage to property and people	Peak flow attenuation, volume control
Groundwater and soil moisture recharge	Improved water availability or quantity	Interception, infiltration, runoff treatment
Health and well-being	Physical, emotional and mental health benefits	See air quality and building temperature, recreation, crime reduction, reduced flood risk
Recreation	Involvement in specific recreational activities	Green and blue spaces and play features
Security of water supply	Reduced flows and reduced pollution	Rainwater harvesting; also see groundwater and soil moisture recharge
Sewerage systems and sewage treatment	Reduced flows and volume to treat in combined systems	Interception and further runoff volume reduction
Visual character	Attractiveness and desirability of area	Visual enhancement (as part of surface SUDS)
Water quality	Surface water quality improvements	Pollution prevention strategies, interception, runoff treatment

Source: Ashley et al. 2015 and Woods Ballard et al. 2015

While the extent and nature of SUDS benefits are site-specific and depend on the attributes of the build or retrofit, several resources exist to support the quantification and monetisation of benefits offered in a given context. Such data serves to support decision-making processes and ultimately mainstream SUDS by providing comparative information on purely "grey" versus "green" or "mixed" solutions. The BeST tool ('Benefits of SUDS Tool'), for example, was developed within the project 'Demonstrating the multiple benefits of SUDS' to enable practitioners to evaluate the wider benefits of SUDS in cases where surface water management is a key driver[5] (Digman et al. 2015). The UK's SUDS Manual (see Woods Ballard et al. 2015) also outlines key concepts in estimating the costs and benefits of SUDS schemes and provides tools and further resources for assessments and comparisons to purely 'grey' infrastructure. A case study from this manual is provided in Box 8.1, illustrating the potential benefits arising from a SUDS scheme versus a conventional drainage solution.

Box 8.1 Application of the BeST Tool to Compare the Benefits of Different Drainage Options: Roundhay Park, Leeds (UK)

Yorkshire Water utilized the BeST tool in order to compare the potential of different options to reduce combined sewer overflow (CSO) spills in Roundhay Park in Leeds (UK). As an additional decision-making criterion, the benefits that could be delivered by each option were also assessed (see Table 8.3). The four options considered used a range of conventional drainage and/or SUDS approaches, namely:

- **Option 1**: a conventional solution to store water in tanks at CSOs to limit the volume spilling
- **Option 2**: a conventional option also solving predicted flooding in the catchment, giving similar hydraulic performance in the combined sewer network as in options 3 and 4
- **Option 3**: a SUDS approach in public areas to disconnect surface water from the combined system and pass it through the conveyance and storage SUDS
- **Option 4**: as option 3, with measures added in residential private locations

Ultimately, **Option 1** lowered the CSO spills but failed to generate other benefits. **Option 2** would offer similar drainage benefits to the sewer network as **Options 3 and 4**, but created less benefits due to underground infrastructure and was less resilient to climate change. As **Options 3 and 4** created wider additional benefits to the community and environment with similar costs and benefits, the final selection was to pursue the public SUDS scheme as it had the best net present value (**Option 3**). The associated costs and benefits are illustrated in Fig. 8.1.

---

[5] See http://www.susdrain.org/resources/best.html

Table 8.3 Summary of options and potential benefits

| Option summary | Cultural | | Regulating | | | | Provisioning | Supporting |
	Recreation	Amenity	Climate resilience	Carbon retention	Water quality	Flooding	Treating wastewater	Biodiversity & ecology
1: Conventional			X	X	✔		X	
2: Conventional +			X	X	✔	✔		
3: Public SUDS	✔	✔	✔	✔	✔	✔	✔	✔
4: Public-private SUDS	✔	✔	✔	✔	✔	✔	✔	✔

Key: X indicates a negative impact; ✔ indicates a positive impact.

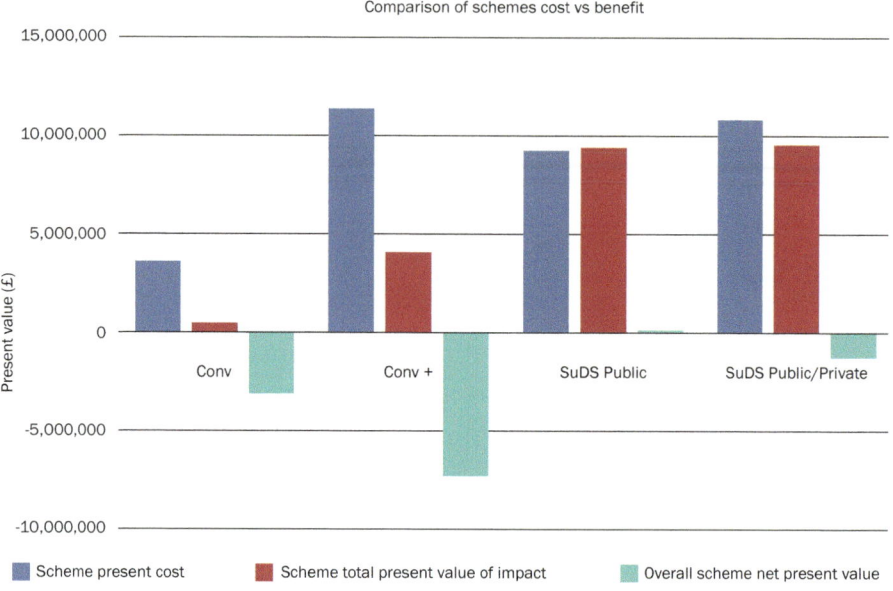

Fig. 8.1 Comparison of options: costs vs benefits (Source: Woods Ballard et al. 2015)

As shown in the above example, research also indicates that where SUDS "are designed to make efficient use of the space available, they can often cost less to implement than underground piped systems" (Woods Ballard et al. 2015: 8) as well as less to maintain. Further comparative studies on the capital (and sometimes the maintenance) costs and benefits of traditional drainage and SUDS have been conducted by Defra as part of their work on the Flood and Water Management Act (see e.g., Defra 2011, 2015). SUDS were found to offer cost savings of between approximately 10% and 85% as compared to traditional drainage approaches, with variations due to site and installation differences. Significant cost savings can be incurred inter alia due to the storage provided within landscape features and resultant reductions in the need for expensive boxed storage and creating low maintenance and monitoring costs (Defra 2011).

It should be emphasized, however, that further long-term research is necessary on the delivery and valuation of benefits as compares to piped solutions and particularly on the aspect of cost-effectiveness in different scenarios, contexts and combinations. Additional data is needed here to improve the targeted deployment of particular aspects and combinations of these technologies and design an optimal framework integrating technical, social, environmental, economic, legal and institutional aspects (Zhou 2014). These gaps as well as additional challenges to be addressed to foster a wider uptake of SUDS are outlined in the subsequent section.

8.4   Fostering a Wider Uptake and Implementation of SUDS

Despite the strong drivers and manifold benefits, SUDS have not been exploited or implemented to their full potential. Uncertainty about long-term maintenance, performance and (cost-) effectiveness both independently and as compared to purely grey infrastructure solutions serve as limitations to wider uptake. These informational limitations are particularly challenging to address as the restricted implementation in turn prohibits new data and evidence from being generated. Furthermore, the data and quantification of these aspects that do already exist are not widely known by the necessary actors, and therefore are commonly not considered alongside "grey" infrastructure. Given that SUDS are a rapidly evolving technology and are very site-specific in nature, another complication is that the levels of effectiveness, fulfilment of associated land requirements, costs and benefits vary greatly from case to case (Green Nylen and Kiparsky 2015). Technical, institutional/political, financial and social barriers relating to the above considerations are further impediments. Key challenges include obtaining the revenue to undertake maintenance, the potential land take and physical requirements involved in new developments, and the role of regulation (Ashley et al. 2015).

Strategies, regulatory frameworks and national level targets which exist to support SUDS implementation are currently scattered, with the majority of information on implementation and case studies limited to only a few countries (e.g. the UK, United States and Australia). For example, as a frontrunner in the field, the UK has SUDS legislation in place as part of the National Planning Policy Framework, which requires local authorities to include SUDS on new developments of 10 or more homes and all major new commercial and mixed use developments, unless demonstrated to be inappropriate. In consultation with the Lead Local Flood Authority, the Local Planning Authority then needs to approve drainage schemes (in line with non-statutory standards[6]). Also in the UK, CIRIA[7] has published an extensive guidance manual addressing the planning, design, construction and maintenance of SUDS as well as tools for maximising amenity and biodiversity benefits alongside flood risk reduction and water quality improvement (see Woods Ballard et al. 2015).

However, even in the limited contexts where SUDS are comparatively more widely implemented, such as in the UK, the challenge remains to overcome silo thinking. SUDS are often raised as a possible approach only when targeting surface water and flooding issues, despite their potential to also address water quality challenges and deliver wider benefits in parallel. This stems in part from dispersed responsibilities amongst agencies for these topics as well as a problematic disconnect between research/development activities and implementation in many cases, leading to limited knowledge of available data and potential scepticism regarding its validity.

---

[6] See   https://www.gov.uk/government/uploads/system/uploads/attachment_data/file/415773/sustainable-drainage-technical-standards.pdf

[7] CIRIA is a neutral, independent and not-for-profit British construction industry research and information association (see http://www.ciria.org/).

8.5   Addressing Silos and Informational Gaps

In order to address the outlined issues and improve the confidence and competence associated with designing and implementing SUDS, further coordinated research and targeted implementation initiatives are crucial. In addition to monitoring the performance, implementation and effectiveness of SUDS in cities, there is a need to disseminate this information highlighting the proven utility of SUDS in a targeted format to key stakeholder groups and decision makers. Such evidence could help appease existing hesitation and scepticism in choosing such a nature-based solution over the traditional grey alternative by providing evidence to questions of performance uncertainty. By providing a wealth of good practice experiences and accompanying monitoring data, a gradual change in stakeholder perception could be facilitated and therewith increased uptake and ecologic, socio-economic and monetary gains (Castro-Fresno et al. 2013; Perales-Momparler et al. 2016). Other research needs are on adequate institutional arrangements, human resource requirements, and performance indicators for urban drainage, which include the range of technical, economical, social and environmental aspects of SUDS (Ashley et al. 2013) as well as the improved quantification of benefits in order to capitalize on the potential future market.

Several research projects are aiming to fill these existing gaps by utilizing sound science to develop tools and guidance materials and implement demonstrative or pilot projects. The Danish 2BG "Black, Blue & Green" project, for example, commits to integrated infrastructure planning for sustainable urban water systems (DTU 2011), Ireland has several regional drainage assessment projects on integrated constructed wetlands, and the Swedish "Sustainable Urban Water Management" project focuses on protecting valuable water resources in urban areas.[8] The EU LIFE+ funded AQUAVAL project ("The efficient management of rain water in urban environments") was highlighted earlier. It aims to find, implement and promote innovative solutions to decrease the impacts of developments on quantity and quality of urban runoff in Valencia, Spain, and implements SUDS as an important step in a paradigm shift (AQUAVAL 2010). Thames Water in the UK has also launched the 'Twenty 4 Twenty' initiative, a ca. 26 million Euro campaign aiming to transform at least 20 acres of grey impermeable concrete into sustainable drainage projects by 2020 (Thames Water 2015). Finally, a newly published study by Allitt et al. (2015) identifies the wider benefits of SUDS and provides guidance to water and sewerage companies on approaches to maximise the potential for benefits to be realised. The UK's SUDS Manual (see Woods Ballard et al. 2015) also provides numerous good practice examples, tools and approaches for successful SUDS design, implementation and maintenance.

Due to the inherent need for cross-sectoral cooperation in designing, implementing and maintaining SUDS, efforts could also be placed on involving local communities in decision-making processes, instead of only presenting these actors

---

[8] See http://www.urbanwater.se/en

with end results. This is underlined by Hair et al. (2014) and Ashley et al. (2013), who suggest to encourage stakeholder involvement and education at all levels of decision-making processes to improve transparency and foster trust and therewith increase acceptance and engagement by addressing citizen, business and political concerns (see Box 8.2). In this context, also employing collaborative governance approaches as suggested by Kabisch et al. (2016) to foster collaboration between decision makers and citizens, businesses and civil society connecting demands for action with responsible actors or partnerships for action could be a promising instrument to reduce barriers for adopting and implementing SUDS. Investments in social-cultural research and the development of cross-disciplinary language could be valuable venues by which to increase public acceptability and support, particularly given that many of the decisions on SUDS retrofits are the responsibility of property owners.

Box 8.2 Herne Hill and Dulwich Flood Alleviation Scheme: A Model for Citizen Engagement and Public-Private Cooperation in SUDS Implementation

Several linked SUDS were installed in a public and two private parks in Southwark, London (UK) in order to stop the recurrent flooding of homes and businesses along the River Effra. The award-winning scheme involved a public-private partnership and shared costs between Southwark Council and Thames Water, with support from the UK Environment Agency (EA). The scheme was designed and delivered in close collaboration and represents one of the first multi-agency SUDS schemes to be implemented in London. Of the total costs, the Council contributed 5%, Thames Water 54% and the EA mediated flood defence grant 41% in aid. Furthermore, Thames Water have provided funding to the Council for long-term maintenance, which is an important aspect for continued delivery given the 100-year design life of the project.

As a result of the project, 447 properties are at reduced risk of surface water flooding and over 80 properties have a reduced risk of sewer flooding. In addition to the direct economic benefits (valued at ca. 12 million pounds), the SUDS scheme has been praised for the extensive stakeholder involvement with the local community interest groups, businesses and residents. The invested outreach efforts were central to gaining support for the scheme and ensuring the continued delivery of amenity and environmental benefits. After receiving the ICE Engineering Award 2015, the EU Project Excellency Award 2015 (for partnership) and being shortlisted for the British Construction Industry Awards 2015, the SUDS project serves as a strong example for future approaches to reduceing surface water flooding risk in urban regions.

Source: Woods Ballard et al. (2015)

Some initiatives have already been set in motion that support a more integrated approach or novel partnerships. For example, the Cooperative Research Centre for Water Sensitive Cities in Australia joins over 70 inter-disciplinary partners together to deliver sustainable water strategies that facilitate a city-wide transformation into a more liveable and resilient environment (CRC 2016). Further innovative approaches encouraging the implementation of SUDS and relevant green infrastructure elements by citizens were developed by the city of Hamburg, Germany. The RISA-project,[9] a cooperation between the city council and a private water company, aims to identify sustainable responses to avoid flooding of basements, streets and properties as well as water pollution from combined sewer overflow and urban run-off. It also seeks to integrate water management measures into urban and regional planning and develop a plan and guidance for rainwater management in the future. In addition, the city of Hamburg launched a green roof strategy[10] in 2016 providing financial support to citizens to install green roofs. In result not only water retention capacity in the city can be increased, but fees for sewage water and rainwater can also be reduced.

Efforts in the UK are also frontline in this regard, recognizing the value of partnerships to secure multiple sources of capital funding and share responsibility for long-term costs. The 'Herne Hill and Dulwich flood alleviation scheme', for instance, is an award-winning example of a successful public-private partnership for delivering SUDS (Southwark Council 2016; see Box 8.2). The UK Water Industry Research (UKWIR)[11] also finances proposals for collaborative research with joint funding and leverage, welcoming new partnerships and innovative associations for common research on SUDS.

8.6   Ways Forward for Increased SUDS Deployment

As urban populations grow alongside projected threats from climate change, demand is mounting for resilient future cities that can both protect the population from climatic events and offer further benefits in parallel. This chapter has thus presented a nature-based alternative to the historically pursued purely piped systems for addressing urban flooding which has the potential to support sustainable urban developments and provide recreational, aesthetic, environmental and socio-economic benefits. SUDS offer significant potential in this regard as evidence indicates a high potential for being sustainable, cost-effective approaches which can complement pure grey infrastructure, and can be applied within new developments or used to retrofit existing systems.

While a range of challenges have been outlined which threaten the wider uptake of SUDS, the promise of ongoing research, targeted collaborative and dissemination

---

[9] http://www.risa-hamburg.de/english.html

[10] http://www.hamburg.de/gruendach/4364756/gruendachfoerderung/

[11] See https://www.ukwir.org/

initiatives and an ever-growing evidence base of the effectiveness and associated costs/benefits of SUDS serve as strong countermeasures. Here, it is important to make lessons learned and data gathered from existing cases more widely available. New pilot or demonstration projects should also be promoted and invested in which are collaborative in nature and strengthen the links between researchers, practitioners and relevant community stakeholders. Finally, the targeted involvement of groups that are perhaps not traditionally interested in drainage matters, such as those in the health or transport sectors, and encouragement of exchanges between companies having implemented SUDS and those pursuing purely grey solutions can also benefit the mainstreaming of SUDS.

These efforts can in turn strengthen the 'business case' for SUDS by instilling more confidence in and drawing attention to their wider benefits produced, low comparative associated costs, and climate change compatible nature. Such evidence will help to refute public and political hesitation as compares to traditional grey infrastructure approaches to water management. Highlighting the delivery of the multiple benefits produced in addition to flood protection which traditional engineered flood protection schemes cannot deliver is a central element. New business-models for public-private partnerships are a further aspect of this process, combining blue/green spaces, human well-being, water management and climate change adaptation interests (see Box 8.2 for an example). Establishing such 'business case' arguments will serve as the foundation for increased investment, public and political support and ultimately SUDS deployment.

Once confidence exists that SUDS are effective and affordable as a nature-based solution technology, governments can increasingly support wider implementation. Means to do so include establishing an adequate legal framework that builds upon the evidence gathered and – alongside financial agreements/investment banks – helping to bridge the gap between short-term thinking and long-term investments via intentional regulatory design. At the EU level and in other industrialized countries, potential actions could experiment with and adjust institutional settings, considering alternative local capacities and site-specific cultural aspects. By requiring use of the technology and establishing duties for adoption and maintenance, governments can ensure the implementation of SUDS and facilitate a transition to becoming a 'business as usual' option and highlight the importance of such an approach as a national priority. More specifically, national regulators can use their authority to more actively accelerate and improve SUDS development by adopting standardized monitoring and reporting protocols and guidance and by incentivising and highlighting the importance of voluntary monitoring.

A strong evidence base exists which demonstrates the effectiveness of SUDS and highlights their promise as a sustainable solution to reduce urban flooding. Yet, the significant potential for more widespread uptake remains largely untapped. Further targeted actions are necessary for increasing the acceptance and application of this nature-based solution and realizing its full potential.

References

2BG (2008) Sustainable urban drainage systems – 8 case studies from the Netherlands. Working Paper, University of Copenhagen & Technical University of Denmark, Black, Blue and Green PhD course, pp 20–22

Allitt R, Tanzir C, Massie A, Sherrington CA (2015) Realising the wider benefits of sustainable drainage. ISBN: 184057 769 X

AQUAVAL (2010) Project Objectives. Accessed on 30 Sept 2016 at http://www.aquavalproject.eu/adaptingContenidos/muestrausuario.asp?accADesplegar=113&IdNodo=14

Ashley R, Shaffer P, Walker L (2013) Sustainable drainage systems: research roadmap. On behalf of UKWIR. Final report

Ashley RM, Walker L, D'Arcy B, Wilson S, Illman S, Shaffer P, Woods-Ballard B, Chatfield P (2015) UK sustainable drainage systems: past, present and future. Civil Engineering – Proceedings of the Institution of Civil Engineers. ICE Publishing

Burns et al (2012) Hydrologic shortcomings of conventional urban stormwater management and opportunities for reform. Landsc Urban Plan 105:230–240

Castro-Fresno et al (2013) Sustainable drainage practicies in Spain, specially focused on pervious pavements. Water 5:67–93

Charlesworth (2010) A review of the adaptation and mitigation of global climate change using sustainable drainage in cities. J Water Clim Change 1(3):165–180

Charlesworth et al (2016) Renewable energy combined with sustainable drainage: ground source heat and pervious paving. Renewable and Sustainable Energy Reviews, Articile in Press, pp 1–8

CRC (2016) Global partnerships. Available online at: https://watersensitivecities.org.au/collaborate/global-partnerships/

CRED (2009) Disaster data: a balanced perspective. CRED Crunch, Centre for Research on the Epidemiology of Disasters (CRED), Issue No. 17, Brussels

Defra (2011) Comparative costings for surface water sewers and SUDS. Cases: Daniels Cross, Newport, Shropshire; Red Hill C.of E. Primary School, Worcester; Caledonian Road Housing, Islington, London; Railfreight Terminal, Telford, Shropshire; Marlborough Road, Telford, Shropshire; URL: http://www.susdrain.org/delivering-suds/using-suds/the-costs-and-benefits-of-suds/comparison-of-costs-and-benefits.html

Defra (2015) Cost estimation for SUDS – summary of evidence. Report –SC080039/R9

Digman CJ, Horton B, Ashley RM et al (2015) Getting the BeST from SUDS. In: Proceedings of the Chartered Institution of water and environmental management urban drainage group spring conference, Birmingham, UK

DTU (2011). 2BG: black, blue and green – Integrated infrastructure planning as key to sustainable urban water systems. Accessed 30 Sept 2016 at http://orbit.dtu.dk/en/projects/2bg-black-blue-and-green--integrated-infrastructure-planning-as-key-to-sustainable-urban-water-systems(36195120-7c7e-4efe-bc00-78d01bb7a678).html

EEA (2012) Urban adaptation to climate change in Europe – challenges and opportunities for cities together with supportive national and European policies

Fletcher T, Shuster W, Hunt WF, Ashley R, Butler D, Arthur S, Trowsdale S, Barraud S, Semadeni-Davies A, Bertrand-Krajewski J, Mikkelsen PS, Rivard G, Uhl M, Dagenais D, Viklander M (2015) SUDS, LID, BMPs, WSUD and more – the evolution and application of terminology surrounding urban drainage. Urban Water J 12(7):525–542

Green Nylen N, Kiparsky M (2015) Accelerating cost-effective green stormwater infrastructure: learning from local implementation. Center for Law, Energy & the Environment, U.C. Bekeley School of Law. http://law.berkeley.edu/cost-effective-GSI.htm

Hair L, Clements J, Pratt J (2014) Insights on the economics of green infrastructure: a case study approach. Water Economics Federation, WEFTEC 2014, pp 5556–5585

Jones P, Macdonald N (2007) Making space for unruly water: sustainable drainage systems and the disciplining of surface runoff. Geoforum 38:534–544

Kabisch N, Frantzeskaki N, Pauleit S, Naumann S, Davis M et al (2016) Nature-based solutions to climate change mitigation and adaptation in urban areas –perspectives on indicators, knowledge gaps, barriers and opportunities for action. Ecol Soc 21(2):39

Norton et al (2015) Planning for cooler cities: a framework to prioritise green infrastructure to mitigate high temperatures in urban landscapes. Landsc Urban Plan 135:127–138

Novotny et al (2010) Water centric sustainable communities: planning, retrofitting, and building the next Urban environment. Wiley, New Jersey. ISBN 978-0-470-47608-6

Perales-Momparler S (2012) Demonstration SUDS in the Mediterranean region of Valencia. SUDSnet International Conference. 5th September 2012, Conventry University, UK. Available for download at: http://sudsnet.abertay.ac.uk/presentations/National%20Conf%202012/Session7_AQUAVALDemonstrationSUDSInTheMediterraneanRegionOfValencia_Perales.pdf

Perales-Momparler S, Andrés-Doménech I, Hernández-Crespo C, Vallés-Morán F, Martín M et al (2016) The role of monitoring sustainable drainage systems for promoting transition towards regenerative urban built environments: a case study in the Valencian region, Spain. J Clean Prod, Article in Press, pp 1–12

Pledger S (n.d.) Lamb Drove, Residential SUDS scheme, Cambourne. Accessed at http://www.susdrain.org/case-studies/case_studies/lamb_drove_residential_suds_scheme_cambourne.html, on 18 Sept 2016

Rose C, Lamond J (2013) Performance of sustainable drainage for urban flood control, lessons from Europe and Asia. In: International conference on flood resilience, experiences in Asia and Europe, Exeter, United Kingdom, 5–7 Sept 2013

Santato S, Bender S, Schaller M (2013)The European floods directive and opportunities offered by land use planning. CSC Report 12, Climate Service Center, Germany. Available for download at: http://www.climate-service-center.de/imperia/md/content/csc/csc-report_12.pdf

Sharma D (2008). Sustainable drainage system (SUDS) for stormwater management: a technological and policy intervention to combat diffuse pollution. 11th International conference on urban drainage, Edinburgh, Scotland, UK, 2008. Available for download at: http://web.sbe.hw.ac.uk/staffprofiles/bdgsa/11th_International_Conference_on_Urban_Drainage_CD/ICUD08/pdfs/753.pdf

Southwark Council (2016) Herne Hill and Dulwich Flood Alleviation Scheme. Accessed 30 Sept 2016 at http://www.southwark.gov.uk/info/200448/flood_risk_management/3889/herne_hill_and_dulwich_flood_alleviation_scheme

State of Green (2015) Sustainable urban drainage systems: using rainwater as a resource to create resilient and liveable cities. Available online at: https://stateofgreen.com/files/download/8247

Thames Water (2015) Twenty 4 twenty: sustianable drainage. Accessed on 30 Sept 2016 at http://www.thameswater.co.uk/about-us/19122.htm

Woods Ballard W, Wilson S, Udale-Clarke H, Illman S, Scott T, Ashley R, Kellagher R (2015) The SUDS Manual. CIRIA C697, London, 2015

Zhou Q (2014) A review of sustainable urban drainage systems considering the climate change and urbanization impacts. Water 6:976–992

Chapter 9
Assessing the Potential of Regulating Ecosystem Services as Nature-Based Solutions in Urban Areas

Francesc Baró and Erik Gómez-Baggethun

Abstract Mounting research assesses the provision of regulating ecosystem services by green infrastructure in urban areas, but the extent to which these services can offer effective nature-based solutions for addressing urban climate change-related challenges is rarely considered. In this chapter, we synthesize knowledge from assessments of urban green infrastructure carried out in Europe and beyond to evaluate the potential contribution of regulating ecosystem services to offset carbon emissions, reduce heat stress and abate air pollution at the metropolitan, city and site scales. Results from this review indicate that the potential of regulating ecosystem services provided by urban green infrastructure to counteract these three climate change-related pressures is often limited and/or uncertain, especially at the city and metropolitan levels. However, their contribution can have a substantially higher impact at site scales such as in street canyons and around green spaces. We note that if regulating ecosystem services are to offer effective nature-based solutions in urban areas, it is critically important that green infrastructure policies target the relevant implementation scale. This calls for a coordination between authorities dealing with urban and environmental policy and for the harmonization of planning and management instruments in a multilevel governance approach.

F. Baró (✉)
Institute of Environmental Science and Technology (ICTA), Universitat Autònoma de Barcelona (UAB), Edifici Z (ICTA-ICP), Carrer de les Columnes s/n, Campus de la UAB, 08193 Cerdanyola del Vallès, Barcelona, Spain

Hospital del Mar Medical Research Institute (IMIM),
Edifici PRBB, Carrer Doctor Aiguader 88, 08003 Barcelona, Spain
e-mail: francesc.baro@uab.cat

E. Gómez-Baggethun
Department of International Environment and Development Studies (Noragric), Norwegian University of Life Sciences (NMBU), P.O. Box 5003, N-1432 Ås, Norway

Norwegian Institute for Nature Research (NINA), Gaustadalléen 21, 0349 Oslo, Norway
e-mail: erik.gomez@nmbu.no

© The Author(s) 2017   139
N. Kabisch et al. (eds.), *Nature-based Solutions to Climate Change Adaptation in Urban Areas*, Theory and Practice of Urban Sustainability Transitions, DOI 10.1007/978-3-319-56091-5_9

Keywords Regulating ecosystem services • Urban green infrastructure • Global climate regulation • Local climate regulation • Air quality regulation • Multi-scale assessment

9.1 Introduction

In an increasingly urban planet, cities and metropolitan areas are facing multiple climate change-related challenges, including heat stress, inland and coastal flooding, drought, increased aridity, and air pollution (Revi et al. 2014; UN 2015). Making cities and human settlements resilient, sustainable and safe should be thus a major priority on any government's agenda, as reflected in one of the seventeen United Nations (UN) Sustainable Development Goals (SDGs[1]). In this context, policy-makers, practitioners and scientists are paying growing attention to the sustainable planning and management of urban and periurban green spaces as a way to cope with threats affecting urban areas (McDonnell and MacGregor-Fors 2016). In the European Union (EU), strategies relying on ecosystems and their processes are mostly built on the concepts of 'green infrastructure' (GI, see EC 2013) and, more recently, 'nature-based solutions' (NBS, see EC 2015). Both terms are very much related as reflected in the EU GI strategy, which defines GI as "a successfully tested tool for providing ecological, economic and social benefits through natural solutions" and states that GI is based on the principle that "the many benefits human society gets from nature, are consciously integrated into spatial planning and territorial development" (EC 2013:2).

GI and NBS are useful notions for the operationalisation of the ecosystem services (ESS) framework, a powerful way of examining the interaction between ecosystems and human well-being (see also Pauleit et al., this volume). Since the seminal paper by Bolund and Hunhammar (1999), a growing body of literature has advanced our understanding of urban ESS in their spatial, temporal, value or practical dimensions (Gómez-Baggethun et al. 2013; Haase et al. 2014). Gómez-Baggethun and Barton (2013) synthesized knowledge and methods to classify and value urban ESS for planning, management and decision-making. Regulating ESS such as air purification, noise reduction, urban temperature regulation or runoff mitigation, not explicitly considered in MEA (2005) and TEEB (2010) classifications, were highlighted in that work due to their expected relevance for the quality-of-life of the urban population. Further, NBS examples in cities are often referred to air quality improvements, local temperature regulation, or increased energy savings provided by green roofs, urban parks or street trees (see Kabisch et al. 2016 and Enzi et al., this volume).

Although regulating ESS are the most frequently assessed ESS group in urban areas (Haase et al. 2014; Luederitz et al. 2015), the actual and potential contribution of regulating ESS to climate change mitigation and adaptation policies is often

---

[1] See http://www.un.org/sustainabledevelopment/sustainable-development-goals/

overlooked in these evaluations, and therefore unknown to local authorities (see Baró et al. 2014). Considering both the potential magnitude of regulating ESS and the scope of the associated pressures to be addressed (e.g., greenhouse gas emissions, heat stress, air pollution) is essential to understand the extent to which regulating ESS can offer effective NBS at different spatial scales (Pataki et al. 2011). According to the framework developed by Villamagna et al. (2013), the flow of regulating ESS contributes to maintain or improve environmental quality within socially acceptable ranges (defined by standards or policy targets) up to a certain level of pressure. Once this threshold of pressure is exceeded, regulating ESS flow will no longer sustain a good environmental quality and therefore its impact as NBS will cease (see Fig. 9.1 and Baró et al. 2015).

In this chapter, we synthesize knowledge and findings of urban GI assessments carried out in Europe and beyond to evaluate the potential contribution of regulating ESS to cope with climate change-related challenges across metropolitan, city and site scales. Improving our understanding on the scale at which regulating ESS can offer most effective NBS is essential to link greening strategies to appropriate levels of planning and decision-making (Scholes et al. 2013; Demuzere et al. 2014). Here we focus on the role of regulating ESS in climate change mitigation (carbon sequestration and avoided emissions), climate change adaptation (urban temperature regulation) and air quality regulation (indirectly related to climate change adaptation). Following a sample of studies assessing the potential of regulating ESS as NBS in urban areas (Sect. 9.2), the case study of Barcelona, Spain, is described in more

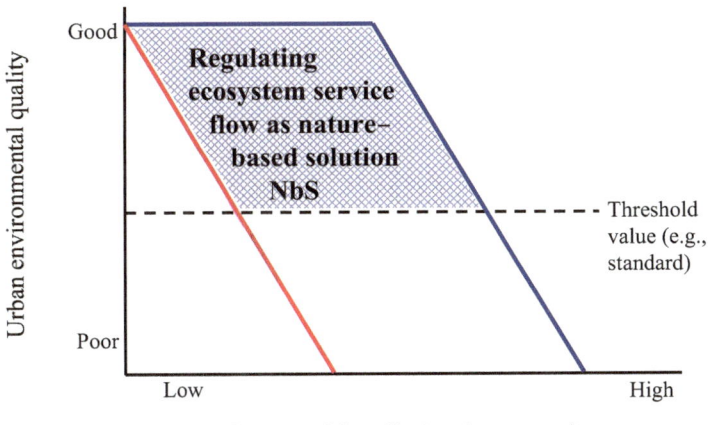

Fig. 9.1 Effects of climate change-related pressures (e.g., air pollution, GHG emissions, heat stress) on urban environmental quality within a system with low to no regulating capacity (*red line*) and a system with high regulating capacity (*blue line*). In the latter system, the flow of regulating ecosystem service contributes to maintain environmental quality within socially acceptable ranges (defined by standards or policy targets) up to a certain level of pressure. Once this threshold of pressure is exceeded, regulating ecosystem service flow will no longer sustain a good environmental quality and therefore its impact as nature-based solution will cease (Source: own elaboration building on Villamagna et al. (2013))

detail (Sect. 9.3). Section 9.4 synthesizes our main findings and points out the main policy implications as well as the priorities for the research agenda on the role of regulating ESS as NBS in urban areas.

9.2 Regulating Ecosystem Services as Nature-Based Solutions in Urban Areas

9.2.1 Global Climate Regulation (Carbon Sequestration and Avoided Emissions)

According to Satterthwaite (2008), 60–70% of total anthropogenic greenhouse gas (GHG) emissions could be assigned to urban activities. Urban climate change-related risks, such as droughts, flash floods and heatwaves, have increasing impacts on urban population (Revi et al. 2014). In response to this trend, a mounting number of cities worldwide are committing themselves to reduce their local GHG emissions by implementing climate change mitigation policies within their territories (see Bulkeley 2010).

Urban vegetation, in particular trees, can directly offset GHG emissions by sequestering carbon dioxide (CO_2) through photosynthesis and biomass storage (Nowak et al. 2013b). Further, urban trees can avoid GHG emissions associated to energy use in buildings due to their micro-climate regulation effects related to shading and evapotranspiration (McPherson et al. 2013, see also next subsection). Some studies suggest that urban green spaces can play an important role as carbon sinks (e.g., Nowak et al. 2013b) and that carbon sequestration rates are comparable to other local mitigation strategies based on energy savings (Escobedo et al. 2010). However, some authors argue that global climate regulation does not stand amongst the most relevant regulating ESS in urban areas because cities can benefit from carbon offsets performed by ecosystems located elsewhere (Bolund and Hunhammar 1999).

Most studies quantifying carbon storage and sequestration by urban vegetation use methods based on tree biomass and growth equations (e.g., i-Tree Eco model; Nowak et al. 2008). Main data inputs include field survey data on urban vegetation structure and remote sensing imagery (e.g., Liu and Li 2012). Recent meta-analyses in USA and China showed that urban GI can sequester and store substantial amounts of carbon. Nowak et al. (2013b) estimated total tree carbon storage and annual gross sequestration in USA urban areas at 643 and 25.6 million tonnes respectively (year 2005). Chen (2015) estimated carbon storage and yearly sequestration by the urban vegetation in 35 major Chinese cities at 18.7 and 1.9 million tonnes respectively (year 2010). However, the latter study also revealed that the offsetting impact by this regulating ESS represented only 0.33% of the carbon emissions from fossil fuel consumption in the case study cities. Generally, studies estimating carbon budgets in urban areas show very modest or marginal impacts in terms of carbon offsetting by urban vegetation (e.g., Escobedo et al. 2010; Liu and Li 2012; Vaccari et al. 2013; Zhao and Sander 2015; see also Table 9.1). Besides, Baró et al. (2015) showed

Table 9.1 Selected sample of modelling and empirical studies on carbon offsetting by vegetation in urban areas at different spatial scales

Study sites	Scale and green infrastructure considered	Methods and data	Indirect energy effects considered?	Annual % offset of total CO_2 emissions	References
35 Chinese cities	City (green space in general)	Meta-analysis of various empirical studies	No	From 0.01 (Hohhot) to 22.45 (Haikou). 0.33 (overall)	Chen (2015)
Shenyang (China)	Metropolitan (urban trees)	Biomass equations, field survey data and satellite images	No	0.26	Liu and Li (2012)
Beijing (China)	City (street trees)	Field surveys, tree growth measurements and statistical data	No	0.2	Tang et al. (2016)
Urbanized portion of Miami-Dade County and city of Gainesville (USA)	Metropolitan and city (Urban trees and palms)	UFORE model (allometric equations), field data	Yes	3.4 (Gainesville) 1.8 (Miami-Dade)	Escobedo et al. (2010)
Municipality of Florence (Italy)	City (urban green space in general)	Eddy covariance technique, GIS data	No	6.2 (total) 1,1 (urban green) 5.1 (periurban green)	Vaccari et al. (2013)
Urbanized areas of Dakota and Ramsey County (USA)	Metropolitan (urban trees)	Allometric models and LiDAR data	No	1.08	Zhao and Sander (2015)
5 EU cities (Barcelona, Berlin, Rotterdam, Stockholm, Salzburg)	City (urban trees)	i-Tree Eco model, tree cover data	No	From 0.12 (Rotterdam) to 2.75 (Salzburg)	Baró et al. (2015)
Residential neighbour-hoods in Singapore and Mexico City	District (Trees and other vegetation, soils)	Eddy covariance technique, biomass and growth equations, tree survey	No	1.4 (Mexico City)-4.4 (Singapore)	Velasco et al. (2016)
Salt Lake Valley (USA)	Metropolitan (urban trees)	Forest growth model and satellite imagery	No	0.2 (relative to a scenario of doubling the tree-planting density after 50 years)	Pataki et al. (2009)

Note: Annual % offset of total CO_2 emissions are based on different baseline years and considering different carbon inventories (see corresponding references)

that contribution is also minor in relation to local GHG reduction targets, suggesting that greening strategies are not likely to be an effective carbon mitigation strategy in cities. For example, Pataki et al. (2009) found that doubling the tree-planting density in the urban region of Salt Lake Valley (USA) would offset only 0.2% of total annual CO_2 emissions over the period 1980–2030. Most of these studies, however, only consider direct carbon sequestration, omitting the indirect effects of urban vegetation that can lead to reduced energy use in cities (see Table 9.1). Yet, the assessments considering emissions avoided due to micro-climate regulation by urban GI show that the related offsets are lower than those related to direct sequestration (Escobedo et al. 2010; McPherson et al. 2013).

Generally, estimates of direct carbon sequestration and indirect energy effects provided by urban GI face multiple uncertainties and limitations. Urban vegetation is usually exposed to unique environmental conditions (e.g., restricted rooting volumes, higher temperature and CO_2 concentration than in rural areas) and maintenance characteristics (e.g., intensity of pruning and irrigation) which can positively or negatively impact their total carbon offsetting capacity (Pataki et al. 2011; Tang et al. 2016). Allometric and growth equations used to quantify carbon storage and sequestration are mostly based on non-urban conditions, yet adjustment factors are often considered in the modelling to minimize error (Nowak et al. 2008). In addition, fossil fuel emissions associated to urban green space maintenance (e.g., pruning) and decomposition rates of removed trees can eventually compensate sequestration gains or even generate negative carbon balances (Nowak et al. 2002). Using a life cycle approach, Strohbach et al. (2012) predicted positive carbon balances of an urban green space project in Leipzig (Germany) over a lifetime of 50 years considering different design and maintenance scenarios. However, the study revealed that small increases in tree mortality can lead to substantial sequestration reductions, thus adequate tree species selection and management can play a key role in carbon offsetting potential.

Most part of the above-mentioned studies only consider the CO_2 flux associated to urban trees and other vegetation, omitting the contribution related to soils. Urban soils can act as relevant carbon sinks (Pouyat et al. 2006), especially those primarily composed by organic materials (e.g., histosols or peat soils). However, soil respiration can constitute an important emission source too (Velasco et al. 2016), thereby adding a new layer of complexity in urban carbon budget estimates.

9.2.2   Local Climate Regulation (Urban Temperature Regulation)

The negative impacts of heat stress on human health, particularly during heatwaves, are singularly strong in cities due to the exacerbating effect of the urban heat island (UHI) (EEA 2012). Human health vulnerability to temperature extremes depends on a complex interaction between different factors such as age, health status and socio-economic variables such as housing (Kovats and Hajat 2008; Fischer and Schär 2010). However, general critical temperature thresholds for health impacts in Europe have been estimated based on the spatial and temporal variance in excess mortality during recent heatwaves episodes (Fischer and Schär 2010). For example, more than 70,000 excess deaths were attributed to the European heatwave occurred during the summer of 2003 (Robine et al. 2008). Consequently, there is a pressing need to develop effective adaptation strategies against mounting heat stress associated to more frequent and intense extreme heat events in cities expected from human-induced climate change (Revi et al. 2014).

Urban greening has been proposed as an effective strategy to mitigate the human health impacts from increased temperatures in urban areas (e.g., EEA 2012). Basically, urban vegetation can reduce local temperatures through evapotranspiration and shading. Obviously, urban trees have a major role in both processes compared to other types of vegetation such as shrubs or grass (Bowler et al. 2010).

The extensive review by Bowler et al. (2010) of the empirical evidence for the cooling effect of urban GI revealed that this impact can be especially relevant at the site scale. The main findings of this meta-analysis were: (1) urban parks are, on average, around 1 °C cooler than non-green sites in the day, with maximum difference values around 2 °C or even higher (e.g., Jansson et al. 2007); (2) street trees have a cooling effect at the urban canyon level, but its magnitude depends on a number of factors such as tree species, canyon orientation or canyon width (see also Norton et al. 2015); (3) studies show that other types of urban GI elements such as green roofs and green walls can also regulate urban temperature at the site scale (see Alexandri and Jones 2008 and Enzi et al., this volume); and (4) the extension of the cooling effect of green space beyond its boundaries is likely, but uncertain, especially at the wider city and metropolitan scales. By using a modelling approach, Chen et al. (2014) predicted substantial reductions in heat stress-related mortality in the city of Melbourne (Australia) associated to the urban cooling effects generated by city-scale greening strategies.

9.2.3 Air Quality Regulation (Air Pollution Removal)

Abatement of air pollution is a pressing challenge in most major urban areas worldwide, either in low-, middle or high-income countries (World Health Organization – WHO – Global Urban Ambient Air Pollution Database, update 2016[2]). For example, the 2015 annual report on air quality in Europe (EEA 2015) estimated that, during the period 2011–2013, 17–30% of the European urban population was exposed to

---

[2] See http://www.who.int/phe/health_topics/outdoorair/databases/cities/en/

PM$_{10}$ (particulate matter with a diameter of 10 μm or less) concentrations above the limit value set by the EU Air Quality Directive (50 μg m^{-3}, 24-h mean value; EU 2008). This percentage of people exposed to problematic pollution levels increases to 61–83% if the more stringent WHO standard (WHO 2005) is applied (20 μg m^{-3}, annual mean value). The harmful impacts of ambient air pollution on human health are consistently and increasingly supported by scientific evidence (Brunekreef and Holgate 2002; EEA 2015) and its global burden of disease was estimated to be 3.7 million deaths during 2012 (WHO 2014). Urban air quality in most cities is compromised by local air pollution emissions from transport, industry and other sources, but it is also sensitive to climate change (Revi et al. 2014). Recent literature shows evidence that climate change will generally increase ground-level ozone in the USA and Europe, but the impacts on air quality in particular urban areas are highly uncertain, as are the effects on other pollutants' concentrations such as particulate matter (Jacob and Winner 2009).

Vegetation in urban landscapes, in particular trees, can remove pollutants from the atmosphere, mainly through leaf stomata uptake and interception of airborne particles (Irga et al. 2015). Further, urban vegetation can act as physical barrier that prevents the penetration of pollutants into specific areas (Salmond et al. 2013). Thus, urban greening strategies have been proposed as a means to reduce air pollution levels (e.g., Nowak et al. 2006). However, the potential for vegetation to improve urban air quality (and consequently population health) in meaningful ways is contested due to uncertainties associated to the modelled estimations and the scarcity of empirical studies (Pataki et al. 2011). Further, urban vegetation can emit biogenic volatile compounds (BVOCs) which eventually contribute to the formation of ground-level ozone and CO (carbon monoxide) air pollutants (Kesselmeier and Staudt 1999).

Most studies estimating air pollution removal by urban vegetation are based on dry deposition models such as i-Tree Eco[3] (e.g., Yang et al. 2005; Nowak et al. 2006; Escobedo and Nowak 2009; Nowak et al. 2013a; Selmi et al. 2016). Generally, these models are applied at a city or metropolitan scale considering green space attributes (such as leaf area index, LAI), pollution concentration data (from available monitoring stations) and meteorological data (Nowak et al. 2008). Results from these modelling studies show that urban vegetation can remove substantial amounts of air pollution. For example, Nowak et al. (2006) estimated that total annual air pollution removal (considering five different pollutants) by urban trees and shrubs in conterminous US amounted to 711.300 t during 1994. Nevertheless, estimated average percent air quality improvements in the 55 selected USA cities attributable to air pollution removal by vegetation were very low (from 0.1 to 0.6% for nitrogen dioxide, NO$_2$, and 0.2 to 1.0% for PM$_{10}$). Modelling studies in urban areas of South America (Escobedo and Nowak 2009), Asia (Yang et al. 2005) or Europe (Selmi et al. 2016) showed similar marginal impacts on air quality

---

[3] Formerly known as UFORE (Urban Forest Effects), see http://www.itreetools.org/

at the city or metropolitan scale (see also Table 9.2). These results suggest that greening strategies (e.g., implementing tree-planting programs) might have a limited effectiveness to address air pollution problems (e.g., if pollutant concentrations are surpassing air quality standards) at the city scale (Baró et al. 2015). Still, modelling studies also show that air quality improvements by vegetation are likely to be more relevant at the site scale. For example, Escobedo and Nowak (2009) and Baró et al. (2015) estimated average percent air quality improvements higher than 6% for PM_{10} in urban areas with an hypothetical 100% tree cover (e.g., contiguous forest stands). In street canyons, however, some modelling studies (e.g., Wania et al. 2012; Vos et al. 2013; Jin et al. 2014) reveal that most part of green street designs (such as double tree row) have a negative effect on local air quality because they reduce ventilation and hence dispersion of traffic emitted pollutants such as particulate matter (PM) and NO_2. Jin et al. (2014) suggested intense pruning of street tree canopies (optimal canopy density was estimated at 50–60%) in order to minimize their negative trapping effect on particles. In contrast, Pugh et al. (2012) argued that GI elements such as green roofs and especially green walls can substantially reduce street-level concentrations (as much as 43% for NO_2 and 62% for PM_{10}) because they increase pollutant deposition without the negative aerodynamic effects on ventilation.

As for other models attempting to simulate complex biophysical processes, there are many uncertainties and limitations in dry deposition models which prevent a more accurate determination of air pollution uptake by urban vegetation. For instance, some sources of uncertainty include non-homogeneity in spatial distribution of air pollutants, particle re-suspension rates, soil moisture status, transpiration rates or leaf boundary resistance (Manning 2008). Local fine-scale input data for these variables are not usually available and empirical data on the actual uptake of pollutants by urban vegetation is still limited (Pataki et al. 2011; Setälä et al. 2013). In general, available experimental studies show that green space is quantifiably associated with reduced air pollution levels at the site scale, especially in regard to particulate matter (Irga et al. 2015; see also Table 9.2). For example, urban parks in Shanghai, China, could remove pollution at ground-level by a maximum of 35% for TSP (total suspended particles) and 21% for NO_2 (Yin et al. 2011); an approximate average removal of 50% for TSP was attributed to greenbelts in Khulna City, Bangladesh (Islam et al. 2012); and the average reduction of air pollutants under tree canopy in two Finnish cities was as much as 40.1% for airborne particles and 7.1% for NO_2 relative to pollutant concentrations in open areas (Setälä et al. 2013). However, this last study found no significant associations between the variation in pollution concentrations and vegetation structure attributes such as canopy closure or number and size of trees. Janhäll's review (2015) concluded that design and selection of urban vegetation is critical for air quality improvements at the site level. Low, dense and porous vegetation close to pollution sources was suggested as the most effective design because it increases pollutants deposition and at the same time does not hinder dilution of emissions with the higher clean atmospheric layer.

Table 9.2 Selected sample of modelling and empirical studies assessing the role of air purification by vegetation in urban areas at different spatial scales

Study site(s)	Scale and green infrastructure considered	Air pollutants assessed	Method	Estimated % air quality improvement	References
55 USA cities	City (urban trees and shrubs)	CO, NO$_2$, O$_3$, PM$_{10}$, SO$_2$	Dry deposition model (i-Tree Eco)	0.2–1.0 (PM$_{10}$) 0.1–0.6 (NO$_2$)	Nowak et al. (2006)
Santiago Metropolitan Region (Chile)	Metropolitan (urban trees)	CO, NO$_2$, O$_3$, PM$_{10}$, SO$_2$	Dry deposition model (i-Tree Eco)	0.6–1.6 (PM$_{10}$) 0.2–0.4 (NO$_2$)	Escobedo and Nowak (2009)
10 USA cities	City (urban trees)	PM$_{2.5}$	Dry deposition model (i-Tree Eco)	0.05–0.24	Nowak et al. (2013a)
5 EU cities (Barcelona, Berlin, Rotterdam, Stockholm, Salzburg)	City (urban trees and shrubs)	PM$_{10}$, NO$_2$, O$_3$	Dry deposition model (i-Tree Eco)	0.20–2.42 (PM$_{10}$) 0.07–0.81 (NO$_2$) 0.10–1.16 (O$_3$)	Baró et al. (2015)
Central London (UK)	Site - Street Canyon (Green roofs and walls scenarios)	PM$_{10}$, NO$_2$	Street-canyon chemistry and deposition model (CiTTy-Street)	6.4–42.9 (NO$_2$) 10.8–61.9 (PM$_{10}$)	Pugh et al. (2012)
19 different real-life urban vegetation designs (Belgium and Netherlands)	Site- Street Canyon (Trees and other green barriers)	PM$_{10}$, NO$_2$ and EC	Computational fluid dynamics (CFD) model (ENVI-met)	Most part of roadside urban vegetation designs have a negative effect on air quality	Vos et al. (2013)
Pudong District, Shanghai (China)	Site (six urban parks)	TSP, NO$_2$ and SO$_2$	Empirical data (mid-flux air and passive samplers)	2–35 (TSP) 2–27 (SO$_2$) 1–21 (NO$_2$)	Yin et al. (2011)
Khulna City, Bangladesh	Site (two greenbelts)	TSP	Empirical data (active monitors)	Approx. 50–65	Islam et al. (2012)
Two Finnish cities (Helsinki and Lahti)	Site (tree-covered park areas and treeless open areas, twenty sites in total)	NO$_2$, VOC and TSP	Empirical data (passive samplers)	2.0–7.1 (NO$_2$) 36.1–40.1 (TSP)	Setälä et al. (2013)
Sydney (Australia)	Site (eleven sites in central Sydney with various green space conditions)	CO$_2$, CO, VOC, NO, NO$_2$, SO$_2$, TSP, PM$_{10}$, PM$_{2.5}$	Empirical data (active monitors)	Green space is quantifiable associated with reduced PM levels	Irga et al. (2015)

Notes: CO$_2$ (carbon dioxide), CO (carbon monoxide), VOC (volatile organic compounds), NO (nitric monoxide), NO$_2$ (nitrogen dioxide), SO$_2$ (sulphur dioxide), TSP (total suspended particulate matter), PM$_{10}$ (suspended particles <10 μm in diameter), PM$_{2.5}$ (suspended particles <2.5 μm in diameter), O$_3$ (ground-level ozone). Estimated % air quality improvements indicate the minimum-maximum average value range (if available). In empirical studies it refers to average removal of air pollutants in green areas relative to treeless areas (see corresponding references)

9.3   The Case Study of Barcelona

9.3.1   Case Study Area

For the urban area of Barcelona, located northeast of Spain on the Mediterranean Sea, regulating ESS have been assessed both at the city (Barcelona municipality) and regional (Barcelona metropolitan region, BMR) scales. See Baró et al. (2014, 2016) for a complete assessment description. The BMR hosts 5.03 million inhabitants living in a total area of 3243 km^2 (Statistical Institute of Catalonia 2015). It embeds 164 municipalities, but its urban core is mainly constituted by the municipality of Barcelona (1.61 million inhabitants and 101.4 km^2) and several adjacent middle-size cities. The BMR still contains a rich variety of natural habitats of high ecological value, including Mediterranean forests (1185 km^2; 36.5%) and scrubland (449 km^2; 13.8%), extensive agro-systems (655 km^2; 20.2%) with a substantial share of vineyard, and various inland water bodies (24 km^2; 0.7%). Currently, almost 70% of the land is protected from urbanisation including, totally or partially, 14 Natura 2000 sites. In contrast, green space in the municipality of Barcelona is scarce. The total green space within the municipality of Barcelona (including urban parks, periurban forests and other green land covers) amounts to 27.2 km^2 representing 26.8% of the municipal area and a ratio of 16.9 m^2 of green space per inhabitant (based on Land Cover Map of Catalonia 4th edition[4], year 2009).

The multi-scale assessment covers two relevant regulating ESS for the case study area: air quality regulation and carbon sequestration. The city of Barcelona and other urban areas in the BMR have repeatedly exceeded the EU limit values for average annual concentrations of NO_2 and PM_{10} (both set at 40 μg/m^3) in the last decade (ASPB 2011). The City Council of Barcelona signed the 'Covenant of Mayors' initiative[5], committing to reduce 23% municipal GHG emissions until 2020 (baseline year 2008). Other municipalities in the BMR have also set similar reduction targets.

9.3.2   Data and Main Results

The multi-scale assessment was based on the definition and quantification of indicators of regulating ESS provision and pressure, building on different models and data sources described in Baró et al. (2014, 2016) and Baró (2015). See also an overview in Table 9.3. Pressure indicators (i.e., NO_2 pollution and carbon emissions) can be considered a proxy of regulating ESS demand since the higher the pressure magnitude, the higher the policy demand for regulating processes by ecosystems (see Burkhard et al. 2014; Baró et al. 2015; Wolff et al. 2015).

---

[4] Available from http://www.creaf.uab.es/mcsc/

[5] See http://www.covenantofmayors.eu/index_en.html

Table 9.3 Overview of regulating ecosystem service indicators and related pressures calculated for the Barcelona multi-scale assessment

Regulating ecosystem service	Scale/study area	Provision and pressure indicators	Unit	Main input data	Methods	Main references
Carbon sequestration	Regional (BMR)	Carbon sequestration (Provision)	kg/ha year	National forest inventory data (IFN2 & IFN3). Land use data and other spatial predictors.	Land use regression modelling	Pino (2007) Baró (2015)
	Local (Barcelona municipality)	Carbon sequestration (Provision)	t/year	Field data. Allometric equations from literature	i-Tree Eco Model	Nowak et al. (2008) Baró et al. (2014)
	Regional (BMR)	Carbon emissions (Pressure, demand proxy)	kg/ha year	Carbon emissions per sector and municipality (year 2012)	Available carbon inventories	Sustainable Energy Action Plans (Barcelona Regional Council data)
	Local (Barcelona municipality)	Carbon emissions (Pressure, demand proxy)	t/year	Carbon emissions per sector in Barcelona (year 2008)	Available carbon inventories	Baró et al. (2014) PECQ (2011)
Air quality regulation	Regional (BMR)	NO_2 removal flux (Provision)	kg/ha year	Air quality data from BMR monitoring (year 2013). Various spatial predictors (see main references)	Land use regression modelling (ESTIMAP)	Zulian et al. (2014) Baró et al. (2016)
	Local (Barcelona municipality)	NO_2 removal flux (Provision)	t/year	Field data. Air quality data from Barcelona monitoring stations (year 2008). Meteorological data (year 2008)	i-Tree Eco Model	Nowak et al. (2008) Baró et al. (2014)
	Regional (BMR)	Annual mean NO_2 concentration (Pressure, demand proxy)	$\mu g/m^3$	Air quality data from BMR monitoring (year 2013). Various spatial predictors (see main references)	Land use regression modelling (ESTIMAP)	Zulian et al. (2014) Baró et al. (2016)
	Local (Barcelona municipality)	NO_2 emissions (Pressure, demand proxy)	t/year	NO_2 emissions in Barcelona and background pollution impact (year 2008)	Available emissions data	Baró et al. (2014) PECQ (2011)

At the level of Barcelona municipality, results show that the contribution of urban GI to climate change mitigation is very low (5187 t carbon sequestered in 2008), accounting for 0.47% of the overall city-based GHG emissions in that year. Similarly, NO_2 removal by urban GI in the municipality of Barcelona (55 t/year) only represented 0.53% of the total city-based emissions in 2008, indicating a marginal air quality improvement.

At the regional level (BMR), the provision of both regulating ESS shows similar spatial patterns (see Figs. 9.2a and 9.3a). Regulating ESS fluxes are especially relevant in periurban forest areas such as the mountain range of Collserola and other tree-covered sites located in the hinterland. However, NO_2 removal in some of these areas (e.g., Montseny massif) is relatively low because pressure (pollutant concentrations) is also moderate (see Fig. 9.2b). The lowest provision values for both regulating ESS are located in urban and agricultural land. As expected, the highest pressure values are mostly located in the municipality of Barcelona and adjacent middle-size cities (see Figs. 9.2b and 9.3b). As observed in the local scale assessment, the urban core is characterized by a compact urban form, very high population density and a relative small share of inner green areas. The other middle-size municipalities, located both along the coastline and hinterland, show mostly middle to low pressure values. The higher spatial resolution of NO_2 concentration compared to carbon emissions also reveals that high capacity roads are major sources of NO_2 pollution. The spatial indicator of pressure related to air purification (annual mean NO_2 concentration) expresses the remaining air pollution after regulating ESS provision (Guerra et al. 2014 refer to it as 'ESS mitigated impact'). Thus, the resulting map (Fig. 9.2b) indirectly shows where regulating ESS provision cannot sustain a good air quality level according to the NO_2 limit value set by the EU Air Quality Directive (40 µg/m³). The carbon offsetting impact of urban GI is small on average (less than 5%) across BMR municipalities (see Fig. 9.3c). Only in 5 out of 164 BMR municipalities, the estimated carbon emissions are completely offset by carbon sequestration by the local vegetation. These municipalities are characterized by very low population density (less than 500 inhabitants) and predominance of forest land cover.

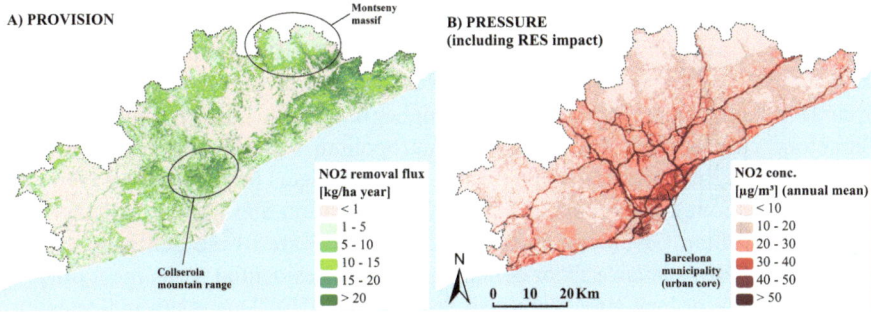

Fig. 9.2 (**a, b**) Provision and pressure maps related to the air purification in the Barcelona metropolitan region (Source: own elaboration building on Baró et al. (2016). Map '2a' is reused from Baró et al. (2016) with kind permission from Elsevier Ltd. See Table 9.3 for data sources)

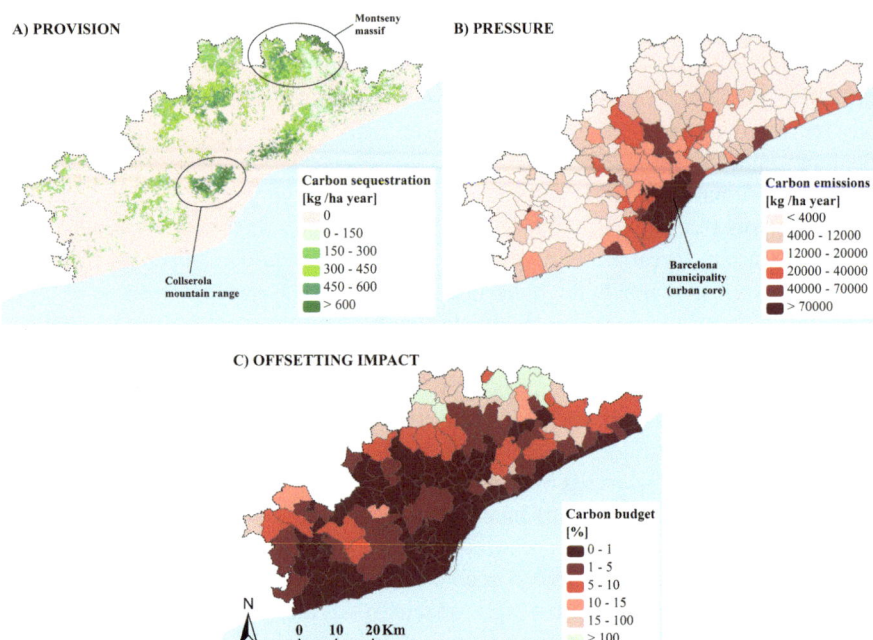

Fig. 9.3 (**a**, **b**, **c**) Provision and pressure maps related to carbon sequestration in the Barcelona metropolitan region (Source: own elaboration building on Baró (2015). See Table 9.3 for data sources)

9.4 Synthesis and Concluding Remarks

Our review indicates that the potential of regulating ESS provided by urban GI to counteract carbon emissions, air pollution and heat stress is often limited and/or uncertain, especially at the city level. In other words, most studies suggest that the magnitude of these environmental problems is usually too high at the city scale relative to the actual or potential contribution of urban ecosystems in mitigating their impacts. At the metropolitan scale, the proportion of urban GI versus built-up or urbanized land is generally substantially higher than at the core city level (e.g., see Barcelona case described above). Yet, metropolitan regulating ESS assessments also show marginal impacts in the overall carbon budgets (e.g., less than 1% in the case of Barcelona). The estimated high air purification and cooling capacities of large metropolitan GI blocks (e.g., protected natural areas) are generally 'underused' due to their distance from demand sites (i.e., residential areas most affected by air pollution or heat stress; see also Baró et al. 2016). This result indicates that the relevant spatial scales for NBS with respect to air pollution and cooling are probably confined to the city or site level. Results from empirical and modelling studies are largely supportive that urban GI, especially urban trees, can improve air quality, offset carbon emissions and reduce heat stress at the site level (especially within and around green spaces). Yet, factors such as species selection, design and management practices of NBS can have a critical impact on the performance of regulating ESS provision. Table 9.4 summarizes the evidence associated with the potential of the three regulating ESS considered here as NBS at three different spatial scales: metropolitan, city and site. Our findings are consistent with previous similar assessments (Pataki et al. 2011; Demuzere et al. 2014).

On the basis of current knowledge and associated uncertainties regarding the potential of regulating ESS as NBS for air quality improvement, carbon offsetting and reduction of heat stress in urban areas, we advance the following policy and research implications:

- More empirical research is needed in order to decrease the levels of uncertainty associated to the impact of regulating ESS provision on urban environmental quality, especially at the city and metropolitan scales, which mostly rely on modelling studies.

Table 9.4 Potential magnitude of the assessed regulating ESS as NBS relative to the scope of the associated urban pressure on three spatial scales

Regulating ecosystem service	Potential as NBS			
	Metropolitan (regional scale)	City (local scale)	Green space (site scale)	Street canyon (site scale)
Air quality regulation	Low to moderate	Low	Moderate	Depending on vegetation design and composition
Carbon sequestration and avoided carbon emissions	Low	Low	Moderate	Not defined
Local temperature regulation	Not defined	Low to moderate	Moderate to high	Moderate

Source: own elaboration based on the evidence discussed above (Tables 9.1 and 9.2) and Pataki et al. (2011)

Notes: The potential was considered high on a specific scale when the evidence from the reviewed studies showed that urban GI can substantially contribute to environmental quality (i.e., air quality, local temperature, carbon offsets). The regulating ESS potential was considered low when most part of studies show that urban GI has a marginal impact on environmental quality at the corresponding spatial scale. In some cases, this qualitative assessment could not be defined due to unclear, conflicting or even lacking evidence. Additionally, grid colours correspond to the current level of uncertainty (considering both empirical and modelling analyses) associated to the potential magnitude of regulating ESS at the different spatial scales: low (*green*); moderate (*orange*) and high (*red*)

- Urban climate change and air pollution mitigation policies should primarily focus on the sources of pollution (built infrastructure and transport systems), not on the sinks (urban GI absorbing carbon and pollutants). Our assessment clearly shows that air pollution problems and local GHG reduction targets are to be dealt with emission reduction policies (e.g., road traffic management, energy efficiency measures). The role of urban GI strategies can be complementary to these policies, but not alternative. Additionally, carbon offsets associated to GI can be fostered by local and metropolitan authorities beyond urban boundaries (see Seitzinger et al. 2012).
- Urban GI can contribute to site-scale strategies related to air quality and heat stress. For example, urban parks, street trees or green roofs/walls can act as clean air/cool zones and corridors within cities. The potential of green roofs and walls can be particularly relevant due to lack of available land in urban cores (see Enzi et al., this volume).
- Trade-offs and disservices related to NBS should be considered in planning and management in order to estimate 'net' contributions to environmental quality. Even if most urban GI elements, such as urban trees, are multi-functional in relation to the three regulating ESS considered in this analysis, some trade-offs have been identified in the literature. For example, dense tree canopies provide a high shading effect, but they are also associated to lower dispersion rates of air pollution in street canyons (e.g., Jin et al. 2014).

The scope of this analysis is limited to three tested regulating ESS (air quality regulation, local climate regulation and global climate regulation through carbon sequestration and avoided emissions) in urban areas, while obviously urban GI can also provide additional ESS and benefits to the urban population, such as water regulation, health and social benefits (see chapters in this issue and other synthesis reviews, e.g., Pataki et al. 2011; Demuzere et al. 2014). Unlike standard 'grey' or technological infrastructures that are normally designed as single-purpose, an added value of urban GI resides on its multi-functionality (see Demuzere et al. 2014 for a comprehensive analysis of synergies or co-benefits associated to different types of urban GI). Therefore, we contend that planning and managing urban GI in the context of NBS for climate change mitigation and adaptation requires an holistic approach, considering the whole range of ESS potentially provided by different types of urban GI and the interactions between them, together with the different spatial scales at which these ESS can be relevant for the resilience, sustainability and safety of urban areas. This calls for a strong multi-scale institutional coordination between all the authorities dealing with urban and environmental policy and for the harmonization of planning and management instruments at different levels.

Acknowledgments This chapter builds on various research projects focusing on urban ecosystem services. We thank the following people for contributing directly or indirectly to this work: Johannes Langemeyer; Lydia Chaparro; David J. Nowak; Jaume Terradas; Ignacio Palomo; Grazia Zulian; Pilar Vizcaino; Dagmar Haase; Coloma Rull; Margarita Parès; Montserrat Rivero; and Carles Castell. This research was partially funded by the following organizations or programs: ERA-Net BiodivERsA network through the Spanish Ministry of Economy and Competitiveness projects 'URBES' (ref. PRI- PIMBDV-2011-1179) and 'ENABLE' (ref. PCIN-2016-002); 7th Framework Program of the European Commission project 'OpenNESS' (code 308428); Horizon 2020 Program of the European Commission project 'NATURVATION' (code 730243); Fundación Iberdrola España (*Ayudas a la Investigación en Energía y Medio Ambiente 2015*); Barcelona City Council (*Ajuntament de Barcelona*); and Barcelona Regional Council (*Diputació de Barcelona*).

References

Alexandri E, Jones P (2008) Temperature decreases in an urban canyon due to green walls and green roofs in diverse climates. Build Environ 43:480–493

ASPB (Agency for Public Health of Barcelona) (2011) Report on evaluation of the air quality in the city of Barcelona, year 2011. Agency for Public Health of Barcelona (ASPB), Barcelona, Spain, 75pp (In Catalan)

Baró F (2015) A multi-scale assessment of regulating ecosystem services in Barcelona. In: Nuss-Girona S, Castañer M (eds) Ecosystem services: concepts, methodologies and instruments for research and applied use, Quaderns de medi ambient; 6. Documenta Universitaria, Girona. ISBN 978-84-9984-308-7

Baró F, Chaparro L, Gómez-Baggethun E, Langemeyer J, Nowak D, Terradas J (2014) Contribution of ecosystem services to air quality and climate change mitigation policies: the case of urban forests in Barcelona, Spain. Ambio 43:466–479

Baró F, Haase D, Gómez-Baggethun E, Frantzeskaki N (2015) Mismatches between ecosystem services supply and demand in urban areas: a quantitative assessment in five European cities. Ecol Indic 55:146–158

Baró F, Palomo I, Zulian G, Vizcaino P, Haase D, Gómez-Baggethun E (2016) Mapping ecosystem service capacity, flow and demand for landscape and urban planning: a case study in the Barcelona metropolitan region. Land Use Policy 57:405–417

Bolund P, Hunhammar S (1999) Ecosystem services in urban areas. Ecol Econ 29:293–301

Bowler DE, Buyung-Ali L, Knight TM, Pullin AS (2010) Urban greening to cool towns and cities: a systematic review of the empirical evidence. Landsc Urban Plan 97:147–155

Brunekreef B, Holgate ST (2002) Air pollution and health. Lancet 360:1233–1242

Bulkeley H (2010) Cities and the governing of climate change. Annu Rev Environ Resour 35:229–253

Burkhard B, Kandziora M, Hou Y, Müller F (2014) Ecosystem service potentials, flows and demands – concepts for spatial localisation, indication and quantification. Landsc Online 32:1–32

Chen WY (2015) The role of urban green infrastructure in offsetting carbon emissions in 35 major Chinese cities: a nationwide estimate. Cities 44:112–120

Chen D, Wang X, Thatcher M, Barnett G, Kachenko A, Prince R (2014) Urban vegetation for reducing heat related mortality. Environ Pollut 192:275–284

Demuzere M, Orru K, Heidrich O, Olazabal E, Geneletti D, Orru H, Bhave AG, Mittal N, Feliu E, Faehnle M (2014) Mitigating and adapting to climate change: multi-functional and multi-scale assessment of green urban infrastructure. J Environ Manag 146:107–115

EC (European Commission) (2013) Green infrastructure (GI) — enhancing Europe's natural capital. European Commission. COM(2013) 249 final, Brussels, 6 May 2013

EC (European Commission) (2015) Nature-based solutions & re-naturing cities. Final report of the horizon 2020 expert group on 'Nature-based solutions and re-naturing cities' (full version). European Commission. Publications Office of the European Union, Luxembourg. ISBN 978-92-79-46051-7. doi: 10.2777/765301

EEA (European Environment Agency) (2012) Urban adaptation to climate change in Europe. Challenges and opportunities for cities together with supportive national and European policies, EEA Report 2/2012. Publication office of the European Union, Luxembourg, p 143

EEA (European Environment Agency) (2015) Air quality in Europe – 2015 report, EEA Report 5/2015. Publication Office, Luxembourg, p 57. doi:10.2800/62459

Escobedo FJ, Nowak DJ (2009) Spatial heterogeneity and air pollution removal by an urban forest. Landsc Urban Plan 90:102–110

Escobedo F, Varela S, Zhao M, Wagner JE, Zipperer W (2010) Analyzing the efficacy of subtropical urban forests in offsetting carbon emissions from cities. Environ Sci Pol 13:362–372

EU (European Union) (2008) Directive 2008/50/EC of the European Parliament and of the Council of 21 May 2008 on ambient air quality and cleaner air for Europe, OJ L 152, 11 June 2008., pp. 1–44. Retrieved from: http://eurlex.europa.eu/LexUriServ/LexUriServ.do?uri=OJ:L:2008:152:0001:0044:EN:PDF

Fischer EM, Schär C (2010) Consistent geographical patterns of changes in high-impact European heatwaves. Nat Geosci 3:398–403

Gómez-Baggethun E, Barton DN (2013) Classifying and valuing ecosystem services for urban planning. Ecol Econ 86:235–245

Gómez-Baggethun E, Gren Å, Barton D, Langemeyer J, McPhearson T, O'Farrell P, Andersson E, Hamstead Z, Kremer P (2013) Urban ecosystem services. In: Elmqvist T et al (eds) Urbanization, biodiversity and ecosystem services: challenges and opportunities. Springer, Dordrecht, pp 175–251

Guerra C, Pinto-Correia T, Metzger M (2014) Mapping soil erosion prevention using an ecosystem service modeling framework for integrated land management and policy. Ecosystems 17:878–889

Haase D, Larondelle N, Andersson E, Artmann M, Borgström S, Breuste J, Gomez-Baggethun E, Gren Å, Hamstead Z, Hansen R, Kabisch N, Kremer P, Langemeyer J, Rall E, McPhearson T, Pauleit S, Qureshi S, Schwarz N, Voigt A, Wurster D, Elmqvist T (2014) A quantitative review of urban ecosystem service assessments: concepts, models, and implementation. Ambio 43:413–433

Irga PJ, Burchett MD, Torpy FR (2015) Does urban forestry have a quantitative effect on ambient air quality in an urban environment? Atmos Environ 120:173–181

Islam MN, Rahman K-S, Bahar MM, Habib MA, Ando K, Hattori N (2012) Pollution attenuation by roadside greenbelt in and around urban areas. Urban For Urban Green 11:460–464

Jacob DJ, Winner DA (2009) Effect of climate change on air quality. Atmos Environ 43:51–63

Janhäll S (2015) Review on urban vegetation and particle air pollution – deposition and dispersion. Atmos Environ 105:130–137

Jansson C, Jansson P-E, Gustafsson D (2007) Near surface climate in an urban vegetated park and its surroundings. Theor Appl Climatol 89:185–193

Jin S, Guo J, Wheeler S, Kan L, Che S (2014) Evaluation of impacts of trees on $PM_{2.5}$ dispersion in urban streets. Atmos Environ 99:277–287

Kabisch N, Frantzeskaki N, Pauleit S, Naumann S, Davis M, Artmann M, Haase D, Knapp S, Korn H, Stadler J, Zaunberger K, Bonn A (2016) Nature-based solutions to climate change mitigation and adaptation in urban areas: perspectives on indicators, knowledge gaps, barriers, and opportunities for action. Ecol Soc 21. doi:10.5751/ES-08373-210239

Kesselmeier J, Staudt M (1999) Biogenic volatile organic compounds (VOC): an overview on emission, physiology and ecology. J Atmos Chem 33:23–88

Kovats RS, Hajat S (2008) Heat stress and public health: a critical review. Annu. Rev. Public Health 29:41–56

Liu C, Li X (2012) Carbon storage and sequestration by urban forests in Shenyang, China. Urban For Urban Green 11:121–128

Luederitz C, Brink E, Gralla F, Hermelingmeier V, Meyer M, Niven L, Panzer L, Partelow S, Rau A-L, Sasaki R, Abson DJ, Lang DJ, Wamsler C, von Wehrden H (2015) A review of urban ecosystem services: six key challenges for future research. Ecosyst Serv 14:98–112

Manning WJ (2008) Plants in urban ecosystems: essential role of urban forests in urban metabolism and succession toward sustainability. Int J Sustain Dev World Ecol 15:362–370

McDonnell MJ, MacGregor-Fors I (2016) The ecological future of cities. Science 352(80):936–938

McPherson EG, Xiao Q, Aguaron E (2013) A new approach to quantify and map carbon stored, sequestered and emissions avoided by urban forests. Landsc Urban Plan 120:70–84

MEA (Millennium Ecosystem Assessment) (2005) Ecosystems and human well-being: synthesis. Island Press, Washington, DC. ISBN 1-59726-040-1

Norton BA, Coutts AM, Livesley SJ, Harris RJ, Hunter AM, Williams NSG (2015) Planning for cooler cities: a framework to prioritise green infrastructure to mitigate high temperatures in urban landscapes. Landsc Urban Plan 134:127–138

Nowak DJ, Stevens JC, Sisinni SM, Luley CJ (2002) Effects of urban tree management and species selection on atmospheric carbon dioxide. J Arboric 28:113–122

Nowak DJ, Crane DE, Stevens JC (2006) Air pollution removal by urban trees and shrubs in the United States. Urban For Urban Green 4:115–123

Nowak DJ, Crane DE, Stevens JC, Hoehn RE, Walton JT (2008) A ground-based method of assessing urban forest structure and ecosystem services. Arboric Urban For 34:347–358

Nowak DJ, Hirabayashi S, Bodine A, Hoehn R (2013a) Modeled PM2.5 removal by trees in ten U.S. cities and associated health effects. Environ Pollut 178:395–402

Nowak DJ, Greenfield EJ, Hoehn RE, Lapoint E (2013b) Carbon storage and sequestration by trees in urban and community areas of the United States. Environ Pollut 178:229–236

Pataki DE, Emmi PC, Forster CB, Mills JI, Pardyjak ER, Peterson TR, Thompson JD, Dudley-Murphy E (2009) An integrated approach to improving fossil fuel emissions scenarios with urban ecosystem studies. Ecol Complex 6:1–14

Pataki DE, Carreiro MM, Cherrier J, Grulke NE, Jennings V, Pincetl S, Pouyat RV, Whitlow TH, Zipperer WC (2011) Coupling biogeochemical cycles in urban environments: ecosystem services, green solutions, and misconceptions. Front Ecol Environ 9:27–36

PECQ (2011) The energy, climate change and air quality plan of Barcelona (PECQ) 2011–2020. Barcelona City Council. Retrieved from: http://www.covenantofmayors.eu/about/signatories_en.html?city_id=381&seap

Pino J (2007) Primera proposta de bases cartogràfiques, criteris i mètodes per a l'avaluació de l'estat ecològic del bosc a l'àmbit SITxell i de les seves tendències a curt termini. Unpublished report for the Barcelona Regional Council

Pouyat RV, Yesilonis ID, Nowak DJ (2006) Carbon storage by urban soils in the United States. J Environ Qual 35:1566–1575

Pugh TAM, Mackenzie AR, Whyatt JD, Hewitt CN (2012) Effectiveness of green infrastructure for improvement of air quality in urban street canyons. Environ Sci Technol 46:7692–7699

Revi A, Satterthwaite DE, Aragón-Durand F, Corfee-Morlot J, Kiunsi RBR, Pelling M, Roberts DC, Solecki W (2014) Urban areas. In: Field CB, Barros VR, Dokken DJ, Mach KJ, Mastrandrea MD, Bilir TE, Chatterjee M, Ebi KL, Estrada YO, Genova RC, Girma B, Kissel ES, Levy AN, MacCracken S, Mastrandrea PR, White LL (eds) Climate change 2014: impacts, adaptation, and vulnerability. Part A: Global and sectoral aspects. Contribution of working group II to the fifth assessment report of the intergovernmental panel on climate change. Cambridge University Press, Cambridge/New York, pp 535–612

Robine J-M, Cheung SLK, Le Roy S, Van Oyen H, Griffiths C, Michel J-P, Herrmann FR (2008) Death toll exceeded 70,000 in Europe during the summer of 2003. C R Biol 331:171–178

Salmond JA, Williams DE, Laing G, Kingham S, Dirks K, Longley I, Henshaw GS (2013) The influence of vegetation on the horizontal and vertical distribution of pollutants in a street canyon. Sci Total Environ 443:287–298

Satterthwaite D (2008) Cities' contribution to global warming: notes on the allocation of greenhouse gas emissions. Environ Urban 20:539–549

Scholes R, Reyers B, Biggs R, Spierenburg M, Duriappah A (2013) Multi-scale and cross-scale assessments of social–ecological systems and their ecosystem services. Curr Opin Environ Sustain 5:16–25

Seitzinger S, Svedin U, Crumley C, Steffen W, Abdullah S, Alfsen C, Broadgate W, Biermann F, Bondre N, Dearing J, Deutsch L, Dhakal S, Elmqvist T, Farahbakhshazad N, Gaffney O, Haberl H, Lavorel S, Mbow C, McMichael A, deMorais JF, Olsson P, Pinho P, Seto K, Sinclair P, Stafford Smith M, Sugar L (2012) Planetary stewardship in an urbanizing world: beyond city limits. Ambio 41:787–794

Selmi W, Weber C, Rivière E, Blond N, Mehdi L, Nowak D (2016) Air pollution removal by trees in public green spaces in Strasbourg city, France. Urban For Urban Green 17:192–201

Setälä H, Viippola V, Rantalainen A-L, Pennanen A, Yli-Pelkonen V (2013) Does urban vegetation mitigate air pollution in northern conditions? Environ Pollut 183:104–112

Statistical Institute of Catalonia (2015) Data on population by territory. Retrieved from: http://www.idescat.cat/en/

Strohbach MW, Arnold E, Haase D (2012) The carbon footprint of urban green space—a life cycle approach. Landsc Urban Plan 104:220–229

Tang YJ, Chen AP, Zhao SQ (2016) Carbon storage and sequestration of urban street trees in Beijing, China. Front Ecol Evol 4:53

TEEB (The Economics of Ecosystems and Biodiversity) (2010) The economics of ecosystems and biodiversity. In: Kumar P (ed) Ecological and economic foundations. Earthscan, London/Washington, DC

UN (United Nations) (2015) World urbanization prospects: the 2014 revision, ST/ESA/SER.A/366. United Nations, Department of Economic and Social Affairs, Population Division, New York

Vaccari FP, Gioli B, Toscano P, Perrone C (2013) Carbon dioxide balance assessment of the city of Florence (Italy), and implications for urban planning. Landsc Urban Plan 120:138–146

Velasco E, Roth M, Norford L, Molina LT (2016) Does urban vegetation enhance carbon sequestration? Landsc Urban Plan 148:99–107

Villamagna AM, Angermeier PL, Bennett EM (2013) Capacity, pressure, demand, and flow: a conceptual framework for analyzing ecosystem service provision and delivery. Ecol Complex 15:114–121

Vos PEJ, Maiheu B, Vankerkom J, Janssen S (2013) Improving local air quality in cities: to tree or not to tree? Environ Pollut 183:113–122

Wania A, Bruse M, Blond N, Weber C (2012) Analysing the influence of different street vegetation on traffic-induced particle dispersion using microscale simulations. J Environ Manag 94:91–101

WHO (World Health Organization) (2005) Air quality guidelines. Global update 2005. World Health Organization, Regional Office for Europe, Copenhagen, p 485

WHO (World Health Organization) (2014) Burden of disease from Ambient Air Pollution for 2012 — summary of results, World Health Organization. Retrieved from: http://www.who.int/phe/health_topics/outdoorair/databases/AAP_BoD_results_March2014.pdf

Wolff S, Schulp CJE, Verburg PH (2015) Mapping ecosystem services demand: a review of current research and future perspectives. Ecol Indic 55:159–171

Yang J, McBride J, Zhou J, Sun Z (2005) The urban forest in Beijing and its role in air pollution reduction. Urban For Urban Green 3:65–78. doi:10.1016/j.ufug.2004.09.001

Yin S, Shen Z, Zhou P, Zou X, Che S, Wang W (2011) Quantifying air pollution attenuation within urban parks: an experimental approach in Shanghai, China. Environ Pollut 159:2155–2163

Zhao C, Sander HA (2015) Quantifying and mapping the supply of and demand for carbon storage and sequestration service from urban trees. PLoS One 10:e0136392

Zulian G, Polce C, Maes J (2014) ESTIMAP: a GIS-based model to map ecosystem services in the European union. Ann Bot 4:1–7

Chapter 10
Nature-Based Solutions and Buildings – The Power of Surfaces to Help Cities Adapt to Climate Change and to Deliver Biodiversity

Vera Enzi, Blanche Cameron, Péter Dezsényi, Dusty Gedge, Gunter Mann, and Ulrike Pitha

Abstract By 2020, according to United Nations and European Union reports, 75% of Europe's population will be living in cities – that's around 365 million citizens. The majority of our cities are hot, dry, polluted and impermeable and increasingly densely populated. The pressure for new development means hard, impermeable surfaces are replacing urban green space and natural habitats. At the same time, climate change is bringing more frequent and extreme weather events such as summer storms, flash flooding and heatwaves.

New developments must be resilient. But we also need to retrofit our existing building stock – to adapt to the impacts of climate change. This challenge is also a

V. Enzi (✉)
Austrian Greenroof and Livingwall Association VfB- GRÜNSTATTGRAU, Vienna, Austria

European Federation of Green Roof and Wall Associations EFB, Vienna, Austria

The Urban Green Infrastructure Competence Centre, Green4Cities GmbH, Vienna, Austria
e-mail: vera.enzi@gruenstattgrau.at

B. Cameron
Bartlett School of Architecture, University College London, London, UK
e-mail: becameron@hotmail.co.uk

P. Dezsényi
Hungarian Greenroof and Livingwall Association ZEOSZ, Budapest, Hungary

Deep Forest Kft., Budapest, Hungary
e-mail: pdezsenyi@deepforest.hu

D. Gedge
European Federation of Green Roof and Wall Associations EFB, Vienna, Austria

Livingroofs Enterprises Ltd, London, UK

The Green Infrastructure Consultancy Ltd, London, UK
e-mail: dusty@dustygedge.co.uk

© The Author(s) 2017   159
N. Kabisch et al. (eds.), *Nature-based Solutions to Climate Change Adaptation in Urban Areas*, Theory and Practice of Urban Sustainability Transitions,
DOI 10.1007/978-3-319-56091-5_10

chance – to green cities and to create habitats for species which in turn provide us with the ecosystem services and benefits cities will rely on for health, well-being and prosperity through the twenty first century. When designed in an integrative and inclusive way, nature-based solutions such as green roofs, green walls, rain gardens, street trees and other urban green infrastructure generate a wide range of benefits.

As well as providing habitats for species, urban greening helps to keep cities cool during summer heat waves, reducing the Urban Heat Island Effect, to manage surface water flooding due to heavy rains and to improve air quality. Green infrastructure also offers an attractive economic Return On Investment (ROI) and a range of other benefits to society, such as connection with nature, and mental and physical health. High quality green infrastructure can also reduce noise pollution, a major cause of stress for city dwellers. Greening a building can help cut heating and cooling costs too, saving energy and other resources. Green cities give better quality of life, meaning healthier, happier citizens, higher productivity at work and a reduction in absence from work due to illness.

This paper focuses on the microclimate benefits of integrating high quality green infrastructure as part of adapting cities to climate change. It estimates market potential and related factors such as energy use, evapotranspiration and water management. It explains through best practise examples how green roofs and green walls designed for nature can contribute to urban biodiversity networks. And it shows how twenty first century nature-based cities can be natural, healthy and resilient.

Keywords Green infrastructure • Nature based solutions • Living walls • Green walls • Green roofs • Raingardens • Permeable surfaces • Urban heat island mitigation • Flood risk reduction • Quality of life • Health and social benefits • Re-naturing cities • Urban retrofit • Biodiversity • Energy • Rainwater management • Climate change resilience • Return of invest • Ecosystem services • Ecosystem disservices • Air quality • Noise reduction

G. Mann
Optigrün International AG, Krauchenwies, Germany

Fachvereinigung Bauwerks Begrünung – FBB, Saarbrücken, Germany
e-mail: gunter.mann@t-online.de

U. Pitha
University of Natural Rescources and Life Sciences Vienna BOKU, Institute of Soil-Bioengineering and Landscape Construction, Vegetation Engineering Group, Vienna, Austria

The Urban Green Infrastructure Competence Centre, Green4Cities GmbH, Vienna, Austria
e-mail: ulrike.pitha@boku.ac.at

10.1   Greening the Urban Market: Now We're Growing!

By 2020, 75% of Europe's population will be living in cities – a total of about 365 million citizens (United Nations 2014). Urban environments are becoming increasingly dense with ever more demand on space for development. The majority of our cities are hot, dry, polluted and impermeable. Pressure for new development means hard, impermeable surfaces are replacing urban green space and natural habitats. At the same time, climate change is bringing more frequent and extreme weather events such as summer storms, flash flooding and heatwaves (EEA 2012).

New developments must be resilient. But we also need to retrofit our existing building stock – to adapt to the impacts of climate change. This challenge is also a chance – to green cities and to create habitats for species which in turn provide us with the ecosystem services and benefits cities will rely on for health, well-being and prosperity through the twenty first century. Cities are growing, but it is in our hands to grow them in a green, sustainable and resilient way.

Most current business forecasts predict that Europe will continue to grow physically and in market terms, and will remain an attractive global trading partner. Therefore, expanding sectors such as innovation and employment is an essential part of European politics (European Commission 2012).

The European Commission also recognizes the value of ecosystem services, the benefits provided by green infrastructure such as green roofs, green walls, rain gardens, street trees parks, gardens and more. In 2013, the Commission published its Green Infrastructure Strategy, Europe's Natural Capital (European Union 2013), followed by a research and innovation policy agenda for nature-based solutions & re-naturing cities (European Commission 2015) and the final Report on Supporting the Implementation of Green Infrastructure (European Commission 2016a) (Fig. 10.1).

How are businesses and commercial success linked to Europe's urban green infrastructure agenda? Even at a conservative estimate, the green roof industry produces promising figures. The most detailed market report comes from Germany. The German market, along with Switzerland and Austria, is the most mature and therefore has the most accurate data. Up to 2015, 86 million m^2 of green roofs (see Table 10.1) had been installed in Germany and many flat roofs are already greened (EFB 2015).

Since 2008, The German Green Roof and Wall Association (Fachvereinigung Bauwerksbegrünung FBB) has been constantly monitoring trends that show a market increasing by an average of 5% per year. Across Austria, Switzerland and Germany, a minimum of 10.3 million m^2 of green roofs are installed each year, driven by regulations and policies and the efforts of around 200 small to medium sized enterprises (EFB 2015).

Outside these three main European markets, several other cities, such as London, Rotterdam and Paris, are showing significant increases in the installation of green roofs particularly driven by policy (e.g., Greater London Authority 2008).

The majority of companies involved in the green roof industry also have the knowledge and skills to contribute to the internal and external vertical greening of our building stock (EFB 2015).

Fig 10.1 Green design for Londons Roofs (Source: Arup, on behalf of the London Sustainable Development Commission)

Independent market research estimates 2017s vertical greening market at 680 million Euros, a figure equating to the installation of around 1 million m^2 of green walls (Caroles 2015). Further, aside from the capital market, the revenue market associated with building vegetation maintenance is also set to increase, providing long-term, secure and sustainable new jobs.

Currently, concern over green infrastructure maintenance costs such as for green roofs is a perceived barrier to faster uptake. Comparisons however show that maintenance costs for vegetated envelopes on buildings are not actually significantly higher than those of comparable conventional building envelopes, such as a glass

Table 10.1 Trends in European green roof market

Target country	Green roof stock total m² (2014)	Green roofs new/year m²	Ratio extensive %	Ratio intensive %	Yearly sales figures €
Austria	4.500.000	500.000	73	27	27.350.000
Germany	86.000.000	8.000.000	85	15	254.000.000
Hungary	1.250.000	100.000	35	65	5.662.500
Scandinavia (S, N, DK)		600.000	85	15	16.050.000
Switzerland		1.800.000	95	5	51.300.000
UK	3.700.000	250.000	80	20	28.000.000
	95.450.000	11.250.000			382.362.500

Trend: growing (FBB DE)
Source: European Federation of Green Roofs and Walls – EFB 2015

facade compared to a green wall (Pfoser 2013). With green roofs, once reduced energy demand and a longer life expectancy of the envelope are taken into account, the overall cost benefit calculation becomes positive (Hämmerle and EFB 2007).

As green roofs and walls are intrinsic nature-based solutions they also have the potential to quantitatively and qualitatively improve biodiversity at a local and regional level. This improvement will be dependent on the design and systems used but has already been realised in several cities in Europe at the building level (EFB 2015).

10.2   "Green" Versus "Grey" Solutions for Climate Change Adaptation and Mitigation

10.2.1   Extreme Weather – Excess Heat Events and Energy

Today, cities provide homes to 50% of the world's population on just 2% of the Earth's surface (United Nations 2014). At the same time, cities are responsible for 80% of global CO_2 emissions and two thirds of world energy consumption (UNEP 2016). Today, buildings are responsible for 40% of European energy consumption and 36% of CO_2 emissions (European Commission 2016b). Energy efficiency could clearly have a major positive impact and is an integral part of European climate change mitigation policy.

Energy demand for heating and especially cooling is still on the increase worldwide, due to increased development and more extreme climate conditions (Pfoser 2013). In fact, during the European heat wave of summer 2003, nearly 70,000 European citizens died from heat-related stress (Robine et al. 2007).

This clearly shows the danger of sealed urban surfaces and "grey" densification, resulting in increased Urban Heat Island Effect (UHIE). With an increase in the UHIE, there is a general increase in energy consumption because of the increased need for cooling. We can control the internal temperature of some of our buildings

Fig. 10.2 "Cool" facades
in Europe's capitals
(Source: Vera Enzi)

through cooling units yet this can be energy inefficient, leading to the release of more greenhouse gases into the atmosphere which in turn intensifies climate change. Furthermore, this does not protect the most vulnerable in our society: the very young or old, the sick and the financially vulnerable who cannot afford air conditioning units for their home. Increasing the supply of air conditioning units is unlikely to lead to sustainable or long-term energy efficiency. Furthermore, with increasing extreme heat events, the current reflective surfaces of cities only add intensity to these events.

There is a direct link between energy efficiency and reduction in the Urban Heat Island. Nature-based solutions such as green roofs and walls (see Fig. 10.2) can have a positive impact on ameliorating the Urban Heat Island and therefore help to increase energy efficiency, explained in detail in Chap. 3, below.

10.2.2   *Urban Flooding*

Most European cities also face another extreme weather threat: heavy rain. Stormwater incidents are leading to severe infrastructure-related financial losses and property damage (European Climate Adaptation Platform, Case Study Copenhagen

2016). Pressure to find solutions is placed mostly on the public sector. Cities often do not fully recognize or exploit the urban environment's potential to help manage rainwater. "Grey" drainage solutions are often optimised to drain water away from urban areas as quickly as possible (see also Davis and Naumann, this volume).

Climate change not always leads to changes in the overall amount of precipitation but often to changes in rainfall patterns. Rainwater falls more heavily and in more concentrated time periods, with months of drought between. So even though our urban environments receive a high level of heavy precipitation, it is drained away very effectively, requiring compensatory activity such as irrigation technology for urban vegetation during times of drought.

It is not just vegetation that needs water. The city itself also needs to keep water in the urban climate cycle to help reduce the UHIE (Magistrat der Stadt Wien 2015) and maintain healthy levels of humidity. This linkage leads to the conclusion that goals to reduce the Urban Heat Island Effect AND stormwater resilience goals overlap.

10.2.3   Linkages Between Water and Energy

Energy never disappears. Heat does not stop at the building envelope. Solar radiation, wind intensity and direction, building materials, trees, plants and soil help determine urban microclimate conditions. Solar energy absorbed and reflected by mineral and insulated, single-beneficiary urban surfaces, along with the heat/energy emitted by air conditioning units, contribute to ever hotter city environments. We have to consider a city's energy balance and energy efficiency from an integrated perspective.

The basic precept is that nature-based solutions and permeable surfaces (essentially soil, water and plants) transform heat/energy. The cooling evapotranspiration by vegetation and soil help regulate surrounding microclimates – this is why water is required. Thus, blue and green infrastructure form a unit to provide ecosystem services to the urban microclimate. By implication, if we can prevent a building envelope heating up, less energy will be needed to cool the inside (Verband für Bauwerksbegrünung 2013; Pitha 2015).

Nature-based solutions offer the opportunity to rewire the city to help overcome many of the issues faced due to climate change ensuring cities become more resilient. Conventional surfaces are protective – yet they lack the multitude of other benefits and services green infrastructure can supply. Green surfaces, as shown in an implemented Pilot Project of Green Roofs and Walls for Zero-Emission (see Fig. 10.3) are the alternative as will be further elaborated on in the following sub-section.

10.3   The Power of Surfaces – Changing the Urban Skin to Green

In comparison with conventional building surfaces, vegetated surfaces have a living relationship to the weather. When solar radiation hits plants, they start to photosynthesise. Plants absorb CO_2 and oxygen is produced. But that's not all – plants

Fig. 10.3 Zero-emission Boutique Hotel Stadthalle, Vienna (Source: Michaela Reitterer)

also evapo-transpirate, they "sweat". Accumulated moisture is evaporated into the environment, helping to regulate humidity and temperature.

As with the evaporation effects of soils and substrates, these plant processes need energy to transform water from a liquid into a gas. This energy is extracted from the environment, cooling the surroundings. But the plant does not just cool the environment, it also cools itself. The surface temperature of a leaf for example will never exceed the surrounding air temperature, thus causing very little sensible heat radiation. By comparison, sheet metal and black roofs can reach over 80 °C on a hot summer day (Verband für Bauwerksbegrünung 2013). A third positive impact is the increase in air humidity. Thus, evapo-transpirating plants and substrates can contribute considerably to human comfort in urban environments during periods of heat excess.

Multiple research results show that plants increase their cooling effects as air temperatures increase. A living wall of 850m^2 on a public building in Vienna (see Fig. 10.4) on a hot summer day shows cooling equal to more than 80 air conditioning units of 3000 Watts each over an 8 hour operating period – a total of 712kWh (Scharf et al. 2012). This living wall also produces enough oxygen for 40 people per day (Magistrat der Stadt Wien 2017) – comparable to four 100-year old Fagus trees. It shows the potential of living walls in places where space constraints do not allow conventional approaches to greening – for example planting trees.

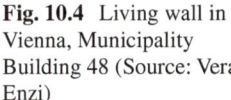
Fig. 10.4 Living wall in Vienna, Municipality Building 48 (Source: Vera Enzi)

10.3.1   Multiple Benefits of Green Walls and Roofs

There are many reasons to invest in green infrastructure technologies such as green roofs and walls, although public and private sector motivations can differ. The public sector tends to favour investment that benefits public sector challenges, such as the Urban Heat Island Effect, urban microclimate issues, stormwater and rainwater management, air quality, ground and air transport noise, fine particulate air pollution, recreation and community activities, social cohesion, health and biodiversity issues. Figure 10.5 by Pfoser and Jakobs AG (2013) clearly shows different applications of greenery on building and the linked benefits for public and private sectors.

Studies show that intensive green roofs and especially living walls can provide a valuable service, dissolving Urban Heat Island hotspots in dense urban areas because they can change the energy regime at street level. Simulations show a reduction of PET (Physiological Equivalent Temperature, a measure of human comfort) of up to −13 °C (Verband für Bauwerksbegrünung 2013).

Living walls can also reduce noise pollution by between 1 and 10 dB and green roofs especially can buffer noise pollution from air traffic sources (Pfoser 2013). Climbing plants such as Ivy *(Hedera helix)* and Veitchii *(Parthenocissus tricuspidata)*

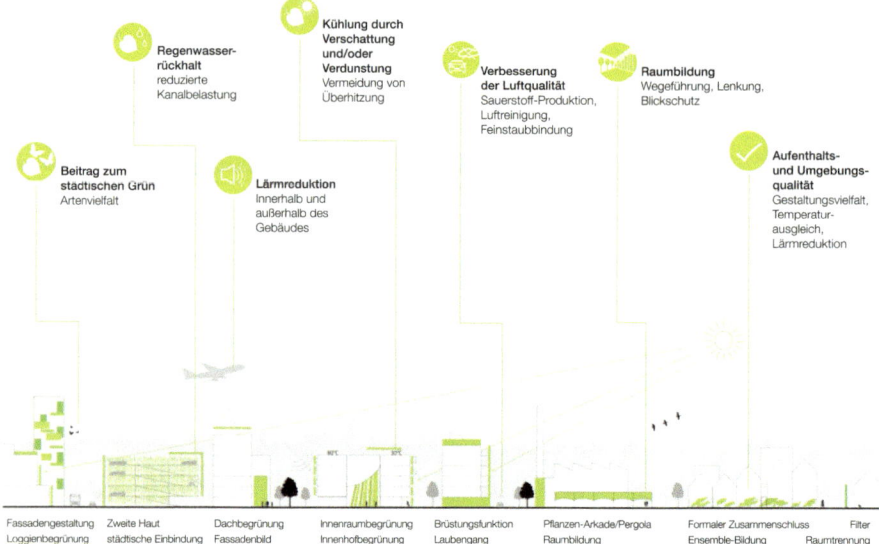

Fig. 10.5 Reasons to green buildings/Motivation Gebäudebegrünung (Source: Pfoser and Jakobs AG 2013)

can bind 1.7 kg/m²/a of urban fine particulate pollutants on their leaf surfaces (Thönessen 2002, 2006; Ottelé 2011).

Numerous different citizen engagement projects all over Europe show that urban green spaces play a vital role in the sustainable development and cohesion of our society. Green roofs at ground and other levels can serve as versatile urban gardening and recreation landscapes. An excellent example is an 800 m² large green roof in Paris, Gymnasium Deshaye that since its creation has become the focal point for the community (see Fig. 10.6) and was implemented in line with the Greening Programme of the Paris Mayors Office (Direction des espaces verts 2014).

Many different studies around the globe at city and national scales indicate that professional urban farming and agriculture will soon be moving into our cities and onto our buildings (e.g., Mann 2016; Orsini et al. 2014) (see Figs 10.7 and 10.8).

The next generations are learning about urban nature and biodiversity. Green Roofs and Living Walls providing public access serve as good examples.

10.3.2   Green Building Technology as an Attractive Investment

As the vast majority of urban buildings are in private ownership, the building-related benefits of green infrastructure investment are crucial. Private investment is usually based on financial benefits, for example cost savings in utilities such as heating and cooling, increased property values (Government of the Netherlands 2013) and the extended lifespan of building materials (Pfoser 2013).

Fig. 10.6 Gymnasium green roof in Paris (Source: Dusty Gedge)

Fig. 10.7 Urban gardening roof "Oase22", Vienna (Source: VfB)

Fig. 10.8 Urban farming roof in the US (Source: Gunter Mann)

Tax incentives especially can have a major impact on green investment on roof level, as the intelligent model of split waste water tax (GAG) in Germany has demonstrated (FBB 2016). The strategy encouraged private property owners to manage their rainwater in a decentralized manner – on their own property. It effectively shared the responsibility of a public sector challenge with private property owners (FBB 2016). An improved extensive to semi-intensive green roof can hold up to 137 l/m², a value comparable to one full standard bathtub. In heavy rain, the public drainage system is discharged. The retained water re-enters the urban climate cycle via evapotranspiration, and the substrate and vegetation turn the water into biomass and clean, cool air (Verband für Bauwerksbegrünung 2013).

Compared to installing a gravel roof, this is an easy business case for property owners (Pfoser 2013). Their Return On Investment (ROI) is increased, payback time is reduced significantly and they can profit from additional, building-related benefits like cost savings in heating and cooling energy and the extended lifespan of their property. As an example, the building's envelope especially is exposed to extreme temperatures, causing material damage, leading to recurring renovation costs. Green roofs and walls act as buffers for extreme temperatures. The maximum daily material temperature variation of a bitumen roof is 63 °C, compared to a simple extensive green roof with 19 °C variation. Heat transfer into the building is slowed down or blocked out significantly, and the internal temperature of rooms under a green roof can be 3–4 °C less than the reference (Köhler 2012). The thicker the vegetation and substrate layer the greater the impact (Verband für Bauwerksbegrünung 2013).

Fig. 10.9 Green wall takes on the function of external shading. Retrofitted building from the 60ies. MA 31 Vienna (Source: Vera Enzi)

The maintenance costs of extensive and semi-intensive green roofs are comparatively low, less than €17/m^2 for green roofs over 1000 m^2 for 40 years of maintenance (Hämmerle and EFB 2007).

Living walls and climbing plants can even replace other costs by taking on the role of technical wall system parts such as external shading elements (Pfoser 2013). The Faculty of Physics at Technical University of Berlin Adlershof (installed in 2008) and the Municipality Building 31, Vienna Water (installed in 2015, see Fig. 10.9) provide all their summer external shading needs with vegetation. In winter, the leaves fall and the buildings profits from warming sun energy.

Some experts think, that the shading effects of living walls result in higher heating costs during winter. A detailed analysis by the Technical University of Vienna in 2015 showed the opposite: Living wall systems and climbers can reduce energy transmissions in winter by a minimum of 20% (Korjenic and Tudiwer 2016). Energy transmissions are reduced by up to 0.19 W/m^2 (Scharf et al. 2012).

Urban space is a scarce resource. Especially at roof level, green space is still less valued than other technologies competing for space, encouraged by financial incentives and policies, such as solar electric and solar hot water systems. However, green roofs and energy generation work very well together (see Figs 10.10 and 10.11). Some European system suppliers already offer them as a single unit. The green roof

Fig. 10.10 Energy generation and green roofs in Switzerland (Source: Dusty Gedge)

Fig. 10.11 Energy-Greenroof in Germany (Source Gunter Mann, Optigrün AG)

substrate provides the ballast required to keep photovoltaic (PV) panels weighted on the roof. The evapotranspiration of plants and substrate keeps solar panels cool and can increase their productivity rate by up to 20% (Mann, 2013). This result refers to the decrease in solar panels' productivity when operating at air temperatures over 25 °C. Compared with blackroofs, vegetated roofs reduce reflected solar radiation and therefore heating effects for the solar panels by up to 40 °C (BUND 2008).

Demand for dense urban development and the opportunity to sustainably retrofit our existing building stock clearly points towards green roof and wall technologies becoming an automatic part of this process. Designed in an integrative and inclusive way, these multi-beneficial nature-based solutions offer an attractive Return On Investment (ROI) as well as many social and economic benefits (Government of the Netherlands 2013).

10.3.3 Disservices of Green Building Technology

Green buildings offer multiple benefits for investors, communities, environment and nature. However, it is important to take into consideration the potential disservices of this technology, even if they seem to be marginal or they are just disadvantageous for specific stakeholders (Baggethun and Barton 2013).

Green roofs and especially living walls may have a higher investment price compared to the majority of traditional building envelope technologies. Nevertheless, they provide a significantly higher benefit value, a study undertaken in Hongkong (Peng and Jim 2015) showed a Return of Invest time of 6.8 years for extensive and 19.5 years for intensive green roofs taken a 40-years lifetime into consideration. Policy schemes, incentives by the state and a broader transfer of benefits into financial terms can significantly change the cost-benefit calculation (Bianchini and Hewage 2012).

Green roof and wall technologies can also require more frequent maintenance than traditional facades and roofs. Especially in the first one or two years after installation – during the establishment period of the ecosystem – the lack of proper care can lead to poor results and unhappy customers (Magistrat der Stadt Wien 2015; Mann 2015).

In contrast to living walls, providing cooling effects for public due to their close distance to street spaces and ground level, green roofs cool on roof level only and extensive roofs don't cool effectively during heatwaves (Rittel et al. 2011). Some authors state a missing acceptance of green walls near windows and consider irrigation needs of living walls not as a contribution to the local climate but as a lack of resource efficiency (Rittel et al. 2011; Mann 2015).

And finally, the fear of insects, rodents, etc.: human biophobia in general is something to take into consideration when applying these technologies. Studies show, that co-creation and co- implementation can help to create high levels of citizen acceptance and identification to overcome potential fears (Davies 2015).

Unfortunately, there is still a lot of incorrect information about other disservices. One of the typical examples is that green roofs are a hazard for waterproofing. It is the opposite, because green roofs protect, thus considerably prolong the lifetime of waterproofing (Pfoser 2013).

10.4   Technology Versus Biodiversity? Or Technology Delivering Biodiversity?

There is often a perception that technology within the built environment acts against biodiversity. However, the development of green roof and wall technologies has always been firmly based in an ecological approach (Mann 1996). Since the birth of the green roof and wall movement in Germany, Switzerland and Austria, delivering nature has been central to the development of these industries. Fortunately, these industries have also developed guidelines and standards over the last 30 years to ensure the delivery of ecological and biodiversity benefits:

Technical standards for green roofs have been published regularly since 1990 such as the German FLL Guidelines (FLL 2000; 2008; 2011), Austrian ÖNORM L1131 (ON 2010) and the Swiss Norm SiA 312 (SIA 2013).

Traditionally built green walls, focusing on climbing plants and their use, have had their official FLL Guideline since 2000 (FLL 2008). In 2013, Vienna published their first living wall guideline followed by a second edition in spring 2017 (Magistrat der Stadt Wien 2017). At green roof level, there are point system models to evaluate the quality of the installed roof and the quality of products, including biodiversity aspects in the annex of ÖNORM L1131 (ON 2010).

Common building certification standards such as LEED and BREEAM have begun to include nature-based solutions and rainwater management in their scoring systems (BREEAM 2016), but the level of detail and possibilities for vegetation technologies is still too limited to have a significant high quality impact, experts involved in this Article say. Nevertheless, there are some certified pilot buildings in Europe dedicated to nature and supporting specific species, for example the 3 level Green House Project in Budapest, see Fig. 10.12. (Skanska 2012).

Green roof and wall experts know that diversity in structures and species on buildings generates long-term ecological stability and therefore can also reduce maintenance requirements once the roof or wall is established. All green roof and wall solutions implicitly provide different habitat functions for their bird and insect users (Mann 1994), and contribute in some way to the urban ecological habitat network, serving as stepping stones for species such as insects and birds as do parks at ground level (Mann 1998) (Fig. 10.13).

Moreover, there is the potential to deliver targeted specific biodiversity measures. The more detailed the local urban nature development strategies and programmes are the more customised service implementation projects can deliver. Nature conservation or ecological compensation projects can be located at roof level too, for example the orchid habitat conservation roof in Switzerland (Brenneisen 2002) or the 4 hectare biodiverse green roof on a shopping mall in Basel (Brenneisen et al. 2010, see Fig. 10.14).

Extensive green roofs by their technological constraints can provide opportunities to create specific habitats, especially those associated with dry grassland communities. Whilst there has been a focus on "productising" green roof technologies to meet the construction industry's need for homogeneity, over the last 20 years,

Fig. 10.12 LEED
Platinum certified "Green
House" Budapest,
biodiverse hybrid green
roof 7th floor (Source:
Peter Dezsényi)

Fig. 10.13 Wildflower roof with insect hotel in London (Source: Dusty Gedge)

approaches have been developed to target the replication of ecological circumstances at ground level. This approach was initially started in Switzerland, where
policies at local level were developed (Brenneisen et al. 2010).

Fig. 10.14 A 4 hectare biodiverse green roof on a shopping mall in Basel (Source: Péter Dezsényi)

These policies targeted the need to create dry grassland communities at roof level. This approach was embraced in London (see Fig. 10.14) and now provides the basis of the planning approach to extensive green roofs in the UK capital (Greater London Authority 2008).

The good and bad implementation experiences with green roofing policies in Switzerland, the UK, and also in Austria have shown that extensive green roof technologies should be implemented in combination with ecological performance criteria and their continuous assessment (comment by the authors of this article).

Scientists from all over the world have been monitoring various types of green roofs and walls for several decades, surveying their ecological development and performance. Their knowledge has resulted in ecological principles for designing biodiverse green roofs, implemented in the Swiss green roof standard SiA 312 (SIA 2013) and a free online guideline, providing best practice examples to support invertebrates at roof level, published by the Invertebrate Conservation Trust (Buglife 2009) in the UK. This guideline and the launch of the National Pollinators Support Strategy UK, including the urban context (Department for environmental and rural affairs UK 2014) have led to a certain number of projects, in particular to support urban pollinators such as wild bees (GEDGE, D.; GRANT, G. GREENINFRASTRUCTURECONSULTANCY).

There are numerous projects across Europe and elsewhere in the world where biodiversity has been delivered at roof and wall level (URBANHABITATS 2006). There is, however, a general perception within the nature conservation community that these technologies are ecologically limited. This perception needs to be challenged and transformed so that in co-creation and co-operation with citizens, municipalities and planners across Europe approaches can be developed to ensure that nature-based solutions on the building envelopes do deliver biodiversity at the local and regional level.

10.5   Nature Provides the Power to Re-wire the City

Leading European cities such as London, Vienna, Budapest, Copenhagen, Malmö and Paris are setting strategies and implementing policies in line with green infrastructure and biodiversity, encouraging nature-based solution investments in the urban realm (e.g., Greater London Authority 2016; Magistrat der Stadt Wien 2015). Nevertheless, it is a long-term, complex process, as participants at the 1st European Urban Green Infrastructure Conference (EUGIC), held in Vienna, in 2015 stated:

> "Green infrastructure in the urban agenda is currently about plumbing, it is necessary to share knowledge (…)",

commented Juliet Lindgren of Malmö City Architecture Department.

> "But the positive feedback of people makes you believe that you are doing the right thing",

her colleagues Jürgen Preiss from Vienna City and Peter Massini from the Greater London Authority added (EUGIC 2015).

Certain implementation barriers such as technical knowledge gaps, missing internal collaboration links between different municipality departments (e.g., urban greening and water) and a current absence of strong communication strategies towards citizens were also identified at the EUGIC conference.

A growing number of small to mega scale cities in Europe and beyond have been already setting out their green infrastructure strategies, followed by legislation processes and funding in regards to green roof and walls, e.g., the city of Hamburg (Behörde für Umwelt und Energie 2015). Some cities are already by far advanced and could be recognized as frontrunners, e.g., Green Capital award winning Victoria-Gasteiz in Spain.

A clear knowledge gap and barrier to successfully mainstreaming of green infrastructure was identified in quantitative and qualitative, integrated short- and long term monitoring data of all considerable economic, ecologic and societal benefits and disservices of strategic larger scale nature-based solution implementation in Europe. The European Commission has therefore launched a rich bundle of Horizon 2020 calls, especially call SCC-02-2016-2017 demonstrating innovative nature-based solutions in cities, fostering demonstration and implementation actions, is expected to create a significant impact on implementation, research and communication.

On the other hand, the way cities will approach the challenge are context-specific: starting by recognising international frameworks and targets (IEEP 2011), followed by national specific strategies/governance plans (Buijs et al. 2016) complemented by an analysis of local target challenge areas resulting in specific implementation plans and monitoring systems (Madueira et al. 2011) and creation of a common knowledge base on green infrastructure existing stock and potentials of implementation in certain built structures (e.g., Urban Green Stock and potential Cadastre of the City of Vienna[1]).

---

[1] https://www.wien.gv.at/umweltschutz/umweltgut/index.html

Fig. 10.15 Birds nesting
on an extensive green roof
in Germany (Source:
Gunter Mann)

An understanding of general and local nature-based solutions state of the art and technology readiness is crucial (Derzken et al. 2015) resulting in targeted planning and implementation methods (Davies 2015), respecting the overall goal to create a functional, interlinked green infrastructure network (Hansen et al. 2016) in respect of ecological (Elmquist et al. 2015), economical (Peng and Jim 2015) and societal (Baggethun and Barton 2013) maximised impact (Fig. 10.15).

Nature has the power to re-wire the city, delivering multiple benefits across the sustainability, ecological and well-being agendas. This paper has shown the innovative capacity and impact with nature-based solutions on cities buildings:

- Green roofs and green walls are technologies classified as nature-based solutions in the context of urban green infrastructure. There is an active European market in the available technology
- The vision of cities resilient to climate change can only be accomplished by choosing "green over grey"
- Many publications on the measurable public and private benefits of investing in urban green are available

- Certain clever incentives of governments, such as split waste water taxation in Germany, could speed up implementation by generating attractive and simple business cases and Return On Investment (ROI)
- Policy and legislation should be closely tied to ecological performance and quality benchmarks; existing evaluation models could be used
- Financial barriers such as higher installation and maintenance costs, technical barriers like retrofitting, and knowledge barriers in planning and legislation currently exist
- Trendsetters have recognized the potential of ecologically improved technologies for green roof and living walls that deliver biodiversity

We have shown how urban greening helps to keep cities cool in heat waves, to manage surface water flooding, to improve air quality as well as to provide habitats for species. Green infrastructure offers an attractive economic Return On Investment (ROI) and a range of other benefits to society, such as connection with nature, and mental and physical health.

High quality green infrastructure can also reduce noise pollution, a major cause of stress for city dwellers. Greening a building can help cut heating and cooling costs too, saving energy and other resources.

Green cities give better quality of life, meaning healthier, happier citizens, higher productivity at work and a reduction in absence from work due to illness.

This paper has focused on the microclimate benefits of integrating high quality green infrastructure as part of adapting cities to climate change. It has explained

Fig. 10.16 A new and old concept for Biodiversity: LEED Platinum certified "Green House" Budapest, biodiverse hybrid Green Roof 7th floor (Source: Peter Dezsényi).

through best practise examples how green roofs and green walls designed for nature can contribute to urban biodiversity networks. And it has shown how cities designed with nature-based solutions can provide the ecosystem services needed for natural, healthy and resilient cities in the twenty first century (Fig. 10.16).

References

Austria Standards Institute ON (2010) Begrünung von Dächern und Decken auf Bauwerken – Anforderungen an Planung, Ausführung und Erhaltung. https://shop.austrian-standards.at/action/de/public/details/362996/OENORM_L_1131_2010_06_01 Accessed 24 Sept 2016

Baggethun E, Barton D (2013) Classifying and valuing ecosystem services for urban planning. Ecol Econ 86(2013):235–245

Behörde für Umwelt und Energie (2015) Auf die Dächer, – fertig – Grün! Die Hamburger Gründach Förderung http://www.hamburg.de/gruendach/4364756/gruendachfoerderung/. Accessed 30 Sept 2016

Bianchini F, Hewage K (2012) Probabilistic social cost-benefit analysis for green roofs: a lifecycle approach. Build Environ 58(2012):152e162

Brenneisen S (2002) Vögel, Käfer und Spinnen auf Dachbegrünungen-Nutzungsmöglichkeiten und Einrichtungsoptimierungen. Geografisches Institut Basel/Baudepartment des Kantons Basel Stadt

Brenneisen S et al (2010) Ökologischer Ausgleich auf dem Dach: Vegetation und bodenbrütende Vögel. Zürcher Hochschule für angewandte Wissenschaften, Zürich

Buglife – Invertebrates Conservation Trust (2009) Creating green roofs for invertebrates. https://www.buglife.org.uk/sites/default/files/Creating%20Green%20Roofs%20for%20Invertebrates_Best%20practice%20guidance.pdf. Accessed 24 Sept 2016

Buijs A et al (2016) Innovative Governance of Urban Green Spaces-Learning from 18 Innovative examples across Europe, GREEN SURGE PROJECT, Deliverable 6.2

BUND; Bund für Umwelt und Naturschutz Deutschland (2008) Auf einem Dach: Begrünung und Photovoltaik. http://www.bund.net/nc/service/oekotipps/detail/artikel/auf-einem-dach-begruenung-und-photovoltaik/ Accessed 24 Sept 2016

Caroles J for Lux Research (2015) http://www.luxresearchinc.com/news-and-events/press-releases/read/innovation-will-drive-costs-green-roofs-and-walls-28-2017. Accessed 24 Sept 2016. http://www.greenroofs.com/blog/tag/building-integrated-vegetation-redefining-the--landscape-or-chasing-a-mirage/. Accessed 24 Sept 2016. https://portal.luxresearchinc.com/research/report_excerpt/17494. Accessed 24 Sept 2016

Davies C (2015) Green Infrastructure Planning and Implementation: the Status of European green space planning and implementation based on an analysis of selected European city-regions, GREEN SURGE PROJECT, Deliverable 5.1

Department for Environmental Food & Rural Affairs UK 2014 The National Pollinator Strategy: for bees and other pollinators in England https://www.gov.uk/government/uploads/system/uploads/attachment_data/file/409431/pb14221-national-pollinators-strategy.pdf. Accessed 25 Sept 2016

Derzken M et al (2015) Green infrastructure for urban climate adaptation: how do resident's view on climate impacts and green infrastructure shape adaptation preferences? Landsc Urban Plan 157(2017):106–130

Direction Des Espaces Verts Et De L'environement Paris (2014) Greening Programme of the Mayor of Paris 2014–2020 http://www.paris.fr/duvertpresdechezmoi. Accessed 24 Sept 2016

Elmqvist T et al (2015) Benefits of restoring ecosystem services in urban areas. Curr Opin Environ Sustain 2015(14):101–108

EUGIC, 1st European Urban Green Infrastructure Conference, Vienna (2015) Conference Proceedings http://urbangreeninfrastructure.org/conference-brochure/#fb0=1. Accessed 24 Sept 2016

European Climate Adaptation platform (2016) The economics of managing heavy rains and stormwater in Copenhagen – The Cloudburst Management Plan (2016). http://climate-adapt.eea. europa.eu/metadata/case-studies/the-economics-of-managing-heavy-rains-and-stormwater-in-copenhagen-2013-the-cloudburst-management-plan. Accessed 24 Sept 2016

European Commission (2012) Europe 2020: Europe's growth strategy ISBN 978-92-79-23972-4

European Commission (2015) Towards an EU research and innovation policy agenda for nature-based solutions & re-naturing cities. http://www.vhg.org/media/rtf/Kennisbank/2015_0739_DG_RTD_WEB-Publication_A4_NBS_long_version_20150310.pdf. Accessed 24 Sept 2016

European Commission (2016a) Supporting the Implementation of Green Infrastructure-Final Report http://ec.europa.eu/environment/nature/ecosystems/docs/green_infrastructures/GI%20Final%20 Report.pdf?utm_content=buffer6f1bb&utm_medium=social&utm_source=twitter.com&utm_campaign=buffer&utm_source=&utm_medium=&utm_campaign=. Accessed 24 Sept 2016

European Commission (2016b) Energy Efficiency-BUILDINGS https://ec.europa.eu/energy/en/topics/energy-efficiency/buildings Accessed 24 Sept 2016

European Environmental Agency-EEA (2012) Urban adaptation to climate change in Europe Challenges and opportunities for cities ISBN: 978-92-9213-308-5

European Federation of Green Roof and Living Wall Associations-EFB (2015) White Paper 2015. http://www.efb-greenroof.eu/EFB_WhitePaper_2015.pdf. Accessed 24 Sept 2016

European Union (2013) Building a Green Infrastructure for Europe. http://ec.europa.eu/environment/nature/ecosystems/docs/green_infrastructure_broc.pdf. Accessed 24 Sept 2016

Fachvereinigung Bauwerksbegrünung e.V. FBB (2016) Ergebnisse der bundesweiten Umfrage zur Förderung von Gebäudebegrünung der FBB 2016. http://www.gebaeudegruen.info/gruen/dach-begruenung/wirkungen-vorteile-fakten/foerderung-2016/?key=1-1. Accessed 27 Sept 2916

Forschungsgesellschaft Landschaftsentwicklung Landschaftsbau e.V. FLL (2000) Fassadenbegrünungsrichtlinie – Richtlinie für die Planung, Ausführung und Pflege von Fassadenbegrünungen mit Kletterpflanzen Artikelnummer: 12020001

Forschungsgesellschaft Landschaftsentwicklung Landschaftsbau e.V. FLL (2008) Richtlinie für die Planung, Ausführung und Pflege von Dachbegrünungen ISBN: 978-3-940122-08-7

Forschungsgesellschaft Landschaftsentwicklung Landschaftsbau e.V. FLL (2011) Richtlinien für die Planung, Ausführung und Pflege von Innenraumbegrünungen, 2011 ISBN: 978-3-940122-25-4

Green Surge (2016) BREEAM Certification UK, Strategic Ecology Framework. http://www.breeam.com/filelibrary/Response-Document/SEF-External-Consultation-Response-Document--April-2016-.pdf. Accessed 25 Sept 2016

Government of the Netherlands (2013) Biodiversity – what's it worth to you? 10 good reasons for Biodiversity Action Plans. https://www.youtube.com/watch?v=pYFQt3k_Cks. Accessed 24 Sept 2016

Greater London Authority (2008) Living roofs and walls; technical report: supporting London plan policy. https://www.london.gov.uk/sites/default/files/living-roofs.pdf. Accessed 24 Sept 2016)

Greater London Authority (2016) The London Plan; Spatial development Strategy for London consolidated with alterations since 2011 https://www.london.gov.uk/sites/default/files/the_london_plan_malp_final_for_web_0606_0.pdf. Accessed 24 Sept 2016

Hansen R et al (2016) Advanced urban green infrastructure planning and implementation: innovative approaches and Strategies from European cities, GREEN SURGE PROJECT, Deliverable 5.2

Hämmerle F, EFB (2007) Zur Wirtschaftlichkeit von Gründächern. http://www.efb-greenroof.eu/verband/fachbei/Die%20Wirtschaftlichkeit%20von%20Gruendaechern.pdf. Accessed 24 Sept 2016

Institute For Environmental Policy-Ieep (2011) Green infrastructure implementation and efficiency final report: ENV.B.2/SER/2010/0059

Köhler M (2012) Handbuch Bauwerksbegrünung ISBN 978-3-481-02968-5

Korjenic A, Tudiwer D (2016) The effect of living wall systems on the thermal resistance of the façade. Energy and Buildings, Accepted Manuscript, 2016 (Forschungsbericht im Auftrag der Magistratsabteilung 22 der Stadt Wien – 154330/2015: Erforschung von Grünfassaden hinsichtlich deren wärmedämmenden Wirkung mittels flächigen Wärmeflussmessungen)

Madueira H et al (2011) Green structure and planning evolution in Porto. Urban For Urban Green 10(2011):141–149

Magistrat der Stadt Wien, Preiss J et al (2015) Urban heat Islands Strategieplan Wien https://www. wien.gv.at/umweltschutz/raum/pdf/uhi-strategieplan.pdf. Accessed 24 Sept 2016

Magistrat der Stadt Wien, Preiss J. et al (2017) Fassadenbegrünungsleitfaden der Stadt Wien, 2. Edition Magistratsabteilung 22 für Umweltschutz der Stadt Wien Status: Not yet published, official date May 2017

Mann G (1994) Ökologisch-faunistische Aspekte begrünter Dächer in Abhängigkeit vom Schichtaufbau. Diplomarbeit Universität Tübingen

Mann G (1996) Die Rolle begrünter Dächer in der Stadtökologie. Biologie in unserer Zeit 5:292–299

Mann G (1998) Vorkommen und Bedeutung von Bodentieren (Makrofauna) auf begrünten Dächern in Abhängigkeit von der Vegetationsform. Dissertation Universität Tübingen

Mann G (2013) SolarGrünDächer. Das Dach zweifach nutzen. Greenbuilding 6

Mann G (2015) Begrünte Dächer als Ausgleichsflächen Stadt und Grün Heft 1/2015

Mann G (2016) Urban farming – natürlich auf Dächern. – Transforming Cities 3. Trialog Publishers Verlagsgesellschaft München

Orsini F et al (2014) Exploring the production capacity of rooftop gardens (RTGs) in urban agriculture: the potential impact on food and nutrition, security, biodiversity and other ecosystem services in the city of Bologna. Springer Science+Business Media Dordrecht and International Society for Plant Pathology 2014

Ottelé M (2011) The Green Building Envelope – Vertical Greening Dissertation, Technical University of Delft, Netherlands ISBN: 978-90-9026217-8

Peng L, Jim C (2015) Economic evaluation of green-roof environmental benefits in the context of climate change: the case of Hong Kong. Urban For Urban Green 14(2015):554–561

Pfoser N (2013) Gebäude Begrünung Energie. Potenziale und Wechselwirkungen. Abschlussbericht. https://www.baufachinformation.de/literatur/Geb%C3%A4ude-Begr%C3%BCnung-Energie/2013109006683. Accessed 24 Sept 2016

Pitha U (2015) Blooming cityscapes – vegetated technical solutions for liveable and sustainable urban areas focusing on permeable paving systems, green roofs and green walls Habilitationsschrift, University of Life Sciences Vienna (BOKU)

Rittel, Wilke, Heiland (2011) Anpassung an den Klimawandel in städtischen Siedlungsräumen – Wirksamkeit und Potenzial kleinräumiger Maßnahmen in verschiedenen Stadtstrukturtypen. Die Natur der Stadt im Wandel des Klimas. CONTUREC (Hrsg)

Robine J et al (2007) Report on excess mortality in Europe during summer 2003. http://ec.europa. eu/health/ph_projects/2005/action1/docs/action1_2005_a2_15_en.pdf. Accessed 24 Sept 2016

Scharf B, Pitha U, Oberarzbacher S (2012) Living walls: more than scenic beauties, IFLA World Congress Cape Town. http://www.academia.edu/6649534/Living_Walls_more_than_scenic_beauties. Accessed 24 Sept 2016

Schweizerischer Ingenieur- und Architektenverein sia (2013) Begrünung von Dächern, SIA 312, SN 564312. http://shop.sia.ch/normenwerk/architekt/sia%20312/d/D/Product. Accessed 25 Sept 2016

Skanska (2012) Green House, Budapest. http://group.skanska.com/projects/57306/Green-House-Budapest. Accessed 25 Sept 2016

Thönessen M (2002) Elementdynamik in Fassaden begrünendem Wilden Wein Köllner Geografische Arbeiten, Heft 78

Thönessen M (2006) Staubfilterung und immisionshistorische Aspekte am Beispiel fassadenbegrünenden Wilden Weins (Parthenocissus tricuspidata). Originalarbeit, ecomed Verlag, Landsberg-Tokyo-Mumbai-Seoul-Melbourne-Paris

Unep-Dtie Initiative (2016) Cities and Buildings Projects Online here. http://www.unep.org/SBCI/pdfs/Cities_and_Buildings-UNEP_DTIE_Initiatives_and_projects_hd.pdf. Accessed 24 Sept 2016

United Nations (2014) Highlights of world urbanisation prospects. https://esa.un.org/unpd/wup/Publications/Files/WUP2014-Highlights.pdf. Accessed 24 Sept 2016

URBANHABITATS.org (2006) Green Roofs and Biodiversity, Urban Habitats, Volume 4, Number 1 ISSN 1541-7115. http://www.urbanhabitats.org/v04n01/urbanhabitats_v04n01_pdf.pdf. Accessed 24 Sept 2016

Verband Für Bauwerksbegrünung Österreich-VfB (2013) Forschungsbericht Projekt GrünStadtKlima, Grüne Bauweisen für die Städte der Zukunft. http://www.gruenstadtklima.at/download/leitfaden_GSK.pdf. Accessed 24 Sept 2016

Project Links

Definition of Nature Based Solutions European Commission 2016. https://ec.europa.eu/research/environment/index.cfm?pg=nbs. The economics of managing heavy rains and stormwater in Copenhagen: http://www.e-pages.dk/tmf/132/http://www.klimatilpasning.dk/media/665626/cph_-_cloudburst_management_plan.pdfhttp://klimakvarter.dkhttp://base-adaptation.eu/implementation-copenhagen-cloudburst-strategy-copenhagen-denmark. Victoria-Gasteiz: Winner of the European Green Capital Award: http://ec.europa.eu/environment/europeangreencapital/wp-content/uploads/2011/04/European-Green-Capital-Award-2012-13-nuevo-estandar.pdf

SCC-02-2016-2017 Demonstrating innovative Nature Based Solutions in Cities. http://ec.europa.eu/research/participants/portal/desktop/en/opportunities/h2020/topics/scc-02-2016-2017.html. GREENINFRASTRUCTURECONSULTANCY GEDGE, D., GRANT, G.: Projects for Pollinators. www.greeninfrastructureconsultancy.com

Part III
Health and Social Benefits of Nature-Based Solutions in Cities

Chapter 11
Effects of Urban Green Space on Environmental Health, Equity and Resilience

Matthias Braubach, Andrey Egorov, Pierpaolo Mudu, Tanja Wolf, Catharine Ward Thompson, and Marco Martuzzi

Abstract Modern urban life style is associated with chronic stress, insufficient physical activity and exposure to anthropogenic environmental hazards. Urban green space, such as parks, playgrounds, and residential greenery, can promote mental and physical health and reduce morbidity and mortality in urban residents by providing psychological relaxation and stress alleviation, stimulating social cohesion, supporting physical activity, and reducing exposure to air pollutants, noise and excessive heat.

This chapter summarizes the pathways that link green spaces to health and well-being, and discusses available evidence of specific beneficial effects such as improved mental health, reduced risks of cardiovascular disease, obesity, diabetes and death, and improved pregnancy outcomes

Specific attention is given to benefits of urban green space for disadvantaged groups and their impacts on health equity. Potential health risks associated with urban green spaces are also discussed along with approaches to reducing or eliminating these risks through proper design and maintenance of green spaces.

Keywords Urban health • Environment • Environmental health • Resilience • Health • Public health • Health impacts • Health impact mechanisms • Relaxation • Restoration • Immune system • Physical activity • Obesity • Diabetes • Mental health • Morbidity • Mortality • Side effects • Prevention • Equity • Social cohesion • Social capital • Vulnerable groups • Urban planning • Public spaces • Green spaces • Evidence review • Epidemiological studies

M. Braubach (✉) • P. Mudu • T. Wolf • M. Martuzzi
European Centre for Environment and Health, WHO Regional Office for Europe,
Bonn, Germany
e-mail: braubachm@who.int; mudup@who.int; wolft@who.int; martuzzim@who.int

A. Egorov
Office of Research and Development, U.S. Environmental Protection Agency,
Chapel Hill, North Carolina, USA
e-mail: egorov.andrey@epa.gov

C. Ward Thompson
OPENspace Research Centre, University of Edinburgh, Edinburgh, UK
e-mail: c.ward-thompson@ed.ac.uk

© The Author(s) 2017   187
N. Kabisch et al. (eds.), *Nature-based Solutions to Climate Change Adaptation in Urban Areas*, Theory and Practice of Urban Sustainability Transitions,
DOI 10.1007/978-3-319-56091-5_11

11.1 Introduction

The previous chapters have summarized evidence that nature-based solutions to the common problems of urbanization can provide ecosystem services and enhance climate change adaptation in urban settings. Nature-based solutions can also improve the health and well-being of urban residents through salutogenic elements in the urban environment facilitating psychological relaxation and stress relief, providing enhanced opportunities for physical activity and reducing exposure to noise, air pollution and excessive heat. Many epidemiological studies have demonstrated various positive health effects of urban green spaces, including reduced depression and improved mental health, reduced cardiovascular morbidity and mortality, improved pregnancy outcomes and reduced rates of obesity and diabetes (reviewed by WHO Regional Office for Europe 2016). Thus, providing urban green space is a nature-based solution with a variety of known health and well-being benefits. While urban green space can also be associated with health hazards, such as increased exposure to allergenic pollen, infections transmitted by arthropod vectors such as ticks or mosquitoes, and risk of injuries, potential detrimental effects can be eliminated or minimized through proper design, maintenance and operation of green space (Lõhmus and Balbus 2015).

It is important to note that disadvantaged population groups often live in neighbourhoods with reduced availability of green space. Studies have shown that socioeconomically disadvantaged individuals tend to benefit the most from improved access to urban greenery. Thus, reducing socioeconomic disparities in the availability of urban green space may help to reduce inequalities in health related to income, minority status, disability and other socioeconomic and demographic factors (Allen and Balfour 2014).

Providing equitable access to green space is an important goal of health-oriented urban policies. Targets related to improving access to green space have been included in international agreements and declarations. The Health2020 strategy calls for the development of resilient and supportive local environments in the WHO European Region (WHO Regional Office for Europe 2013). The Parma Declaration on Environment and Health adopted by the Member States of the WHO European Region includes the commitment "…to provide each child by 2020 with access to health and safe environments and settings of daily life in which they can walk and cycle to kindergartens and schools, and to green spaces in which to play and undertake physical activity" (WHO Regional Office for Europe 2010). An urban environment assessment performed by the European Environment Agency concluded that urban green space can buffer environmental hazards and contribute to health, stimulating EU Member States to develop national policy targets (European Environment Agency 2015). The Sustainable Development Goals (United Nations 2016) adopted by the United Nations in 2015 include the following goal: "By 2030, provide universal access to safe, inclusive and accessible, green and public spaces, particularly for women and children, older persons and persons with disabilities" (Target 11.7).

This section discusses pathways leading to health benefits, and summarizes existing epidemiological evidence of public health benefits of urban green space, with a particular focus on equity issues. In this context, it is important to note that different studies have used varying definitions of urban green space. Therefore, for the purpose of this chapter, the term "urban green space" is used inclusively to mean any type of greenery in urban settings, without distinctions regarding size, quality, and public or private ownership.

11.2 Pathways Linking Urban Green Space with Health and Well-Being

The pathways leading to beneficial health effects of green space are diverse and complex. Various models have been formulated to explain the relationship between green space and health. Hartig et al. (2014) suggested four interacting pathways through which green space can affect health and well-being: (1) improved air quality, (2) enhanced physical activity, (3) stress compensation and (4) greater social cohesion. Lachowycz and Jones (2013) proposed physical activity, engagement with nature, relaxation, and social interactions as major pathways to health. Villanueva et al. (2015) argued that urban green spaces mitigate the urban heat island effect providing protection from heat-related health hazards, improve social capital and cohesion, and enhance physical activity. In addition to the pathways outlined above, Kuo (2015) suggested exposure to natural microbes and enhanced immune system functioning as a major pathway linking nature and health. Relative contributions of different pathways as well as their potential synergistic effects remain to be elucidated. Meanwhile, insufficient knowledge of underlying causal pathways to health outcomes and the complex modifying social and environmental factors hinders the implementation of focused policy interventions. An in-depth interdisciplinary research is needed to close these knowledge gaps in the area of urban environment and health (Shanahan et al. 2015).

This chapter focuses on the discussion of, and the health impacts related to (1) improved relaxation and restoration, (2) improved functioning of the immune system, (3) enhanced physical activity and (4) improved social capital. Health and societal benefits related to ecosystem services (e.g. reduction of air pollution, noise, exposure to excessive heat) will be discussed in the next chapter.

11.2.1 Improved Relaxation and Restoration

The evidence of health benefits due to mental restoration and relaxation from having contact with nature and green space is well documented (Hartig et al. 2014; Hartig 2007). It has been suggested that contacts with nature (e.g. views of green space) can trigger positive effects for persons with high stress levels by shifting them to a more positive emotional state (Ulrich 1983; Ulrich et al. 1991) and that stimuli in natural settings help to restore a sense of well-being in persons suffering mental fatigue (Kaplan 1995; Kaplan 2001).

Hartig et al. (2014) noted that "substantial evidence speaks to the potential benefits of contact with nature for avoiding health problems traceable to chronic stress and attentional fatigue", but also pointed out that most previously conducted studies demonstrated only short-term restorative benefits of one-off nature experiences. For example, a study in the United Kingdom used wearable sensors to demonstrate the effects of a short walk in a green space on brain activity that might be associated with enhanced relaxation and restoration (Aspinall et al. 2015). It was also shown that walking in natural environments produces stronger short-term cognitive benefits than walking in the residential urban environment (Gidlow et al. 2016a). Cortisol measures have demonstrated that gardening alleviated acute stress faster than reading (van den Berg and Custers 2011). It has also been demonstrated that exposure to green space reduces neural activity in the subgenual prefrontal cortex and alleviates depression symptoms (Bratman et al. 2015).

Recent studies have also provided evidence of chronic stress alleviation by green space. Using diurnal cortisol patterns was an innovative approach applied in the United Kingdom to demonstrate that exposure to green space reduced chronic stress in adults living in deprived urban neighbourhoods (Roe et al. 2013; Ward Thompson et al. 2012; Beil and Hanes 2013). Similar relationships between green space and stress reduction have been shown using hair cortisol as a biomarker of chronic stress (Honold et al. 2015; Gidlow et al. 2016b).

11.2.2 Improved Functioning of the Immune System

Kuo (2015) suggested a central role for enhanced immune functioning in the pathways between nature and health. Associations between visiting forests and beneficial immune responses, including expression of anti-cancer proteins, have been demonstrated in Japan (Li et al. 2008). This suggests that immune systems benefit from direct exposure to natural environments or through contacts with certain factors in the green space. It has also been shown that children with the highest exposure to specific allergens or bacteria during their first year of life were least likely to have recurrent wheeze and allergic sensitization (Lynch et al. 2014). Living in residential areas with more street trees was shown to be associated with lower asthma prevalence (Lovasi et al. 2008). One hypothesized immunological pathway is exposure to commensal microorganisms in biodiverse natural environments (Rook 2013), which can play an immunoregulatory role. Studies have demonstrated that increased biodiversity in the environment around homes is linked with reduced risk of allergy (Ruokolainen et al. 2015; Hanski et al. 2012). Greater exposure to commensal microorganisms, especially in the early life, may lead to more diverse skin and gut microbiomes, and provide protection against allergy and autoimmunity

(Kondrashova et al. 2013). It has also been suggested that the human microbiome associated with natural environmental may improve mental health (Logan 2015).

11.2.3   Enhanced Physical Activity and Improved Fitness

Physical activity has been shown to improve cardiovascular health, mental health, neurocognitive development, and general well-being and to prevent obesity, cancer, and osteoporosis (Owen et al. 2010). Providing attractive and accessible urban environments may encourage people to spend more time outdoors and facilitate physical activity (Bedimo-Rung et al. 2005). The quality of the urban green space and presence of specific amenities are important factors facilitating physical activity in older adults (Aspinall et al. 2010; Sugiyama and Ward Thompson 2008). For urban residents with mental health problems, physical activity in green space may be particularly therapeutic (Roe and Aspinall 2011). Other populations or subgroups may benefit, in a similar way, from green space that makes outdoor activity enjoyable and easy, and encourages less sedentary lifestyles.

Hartig et al. (2014) summarized available evidence for an association between green space and physical activity levels in three domains: work, active transport and leisure. While access to green spaces has been linked to active leisure, associations between greenness and active commuting (such as walking and cycling) are inconsistent because very green living environments can be highly car-dependent for transport (Bancroft et al. 2015).

Numerous studies in multiple countries have demonstrated that recreational walking, increased physical activity and reduced sedentary time were associated with access to, and use of green space in working age adults, children and senior citizens (Epstein et al. 2006; Kaczynski and Henderson 2007; Kaczynski et al. 2008; Sugiyama and Ward Thompson 2008; Cochrane et al. 2009; Almanza et al. 2012; Lachowycz et al. 2012; Astell-Burt et al. 2013; Schipperijn et al. 2013; Lachowycz and Jones 2014; Sugiyama et al. 2014; Gardsjord et al. 2014; James et al. 2015; Sallis et al. 2016).

Almanza et al. (2012) used satellite images and GPS and accelerometer data from children in several communities in California, the United States to demonstrate that increased residential greenness was positively associated with moderate to vigorous physical activity. Bjork et al. (2008) and De Jong et al. (2012), working in Sweden, found a positive association between access to high quality green space and higher levels of physical activity.

There is also accumulating evidence that physical activity in green space ("green exercise") is more restorative and beneficial for health than physical activity in non-natural environments (Barton and Pretty 2010; Bodin and Hartig 2003). Mitchell's (2013) study of the Scottish population showed an association between physical activity in natural environments and reduced risk of poor mental health, while activity in other types of environment was not linked to the same health benefit.

11.2.4   Improved Social Capital and Cohesion

Social relationships have a well-known protective effect on health and well-being, while social isolation is predictor of morbidity and mortality (Nieminen et al. 2010; Pantell et al. 2013; Yang et al. 2016). Green space can play an important role in fostering social interactions and promote a sense of community that is essential for social cohesion (Kim and Kaplan 2004) as well as for human health (Lengen and Kistemann 2012). Public urban green space has been shown to facilitate social networking and promote social inclusion in children and adolescents (Seeland et al. 2009; Ward Thompson et al. 2016). The quantity and the quality of greenery have been linked with improved social cohesion at the neighbourhood scale (de Vries et al. 2013) while shortage of green space has been associated with perception of loneliness and lack of social support (Maas et al. 2009a). However, the relationships between green space and social well-being are complex. Although observational studies have demonstrated positive effects on well-being, characterizing the underlying mechanisms remains a challenge (Hartig et al. 2014).

11.3   Health Benefits of Green Space and Potential Health Risks

This section summarizes the available epidemiological evidence of health benefits linked to green space through pathways which were discussed in the previous section. It is important to note that each pathway can lead to more than one health benefit, and that different pathways can also contribute to the same benefit. For example, cardiovascular benefits can be caused by enhanced physical activity, improved mental restoration leading to reduced chronic stress, and by reduced exposure to air pollution and noise.

11.3.1   Improved Mental Health and Cognitive Function

There is strong evidence of mental health benefits of urban green space (reviewed by de Vries 2010). Greater perceived neighbourhood greenness has been strongly associated with improved mental health (Sugiyama et al. 2008). Greater surrounding greenness has been linked to improved physical and mental health in all socio-economic strata and in both sexes in Spain (Triguero-Mas et al. 2015). The associations were stronger for the surrounding greenness measured by the Normalized Difference Vegetation Index (NDVI) than for access to geographically distinct green spaces. Further analysis also demonstrated that this association was not mediated by physical activity. This suggests that psychological relaxation was an important contributing pathway to health.

It was shown that individuals living in urban areas with more green space tend to have reduced level of stress and better well-being compared to those with poorer availability of green space (White et al. 2013). More greenery in the neighbourhood was linked to lower levels of depression, anxiety and stress (Beyer et al. 2014; Bratman et al. 2015; Reklaitiene et al. 2014; Pope et al. 2015); improved access to water bodies (blue spaces) has also been linked to enhanced mental well-being in city dwellers (Völker and Kistemann 2015). In a prospective study in the United Kingdom, moving to greener residential areas has been linked with persistent mental health improvements; the mechanisms leading to these improvements have not been elucidated (Alcock et al. 2014). While most studies relied on measures of green space availability as a proxy measure of exposure, a multicity study in Europe linked greater time spent in green spaces with improved self-reported health and vitality; the effects were consistent in all four study sites in Spain, Lithuania, The Netherlands and the United Kingdom suggesting that health benefits are independent of cultural and climatic contexts (Van den Berg et al. 2016).

A number of studies provided evidence that green spaces are especially beneficial for certain subpopulations or for disadvantaged groups. Moving to an area with better access to green space characterized as "serene" has been linked to improved mental health in a representative sample of adult Swedish women (van den Bosch et al. 2015). There is also some evidence for the beneficial effects of green space on mental health and cognitive development in children. A study in Lithuania found mental health benefits in children of mothers with low education level while the results for children of highly educated mothers were inconsistent (Balseviciene et al. 2014).

Increased usage of green and blue spaces, and greater residential greenness have been associated with improved behavioural development and reduced rate of Attention Deficit Hyperactivity Disorder (ADHD) in children (Amoly et al. 2014). Higher levels of greenness at home and school were associated with improved cognitive development (i.e., better progress in working memory and reduced inattentiveness) in schoolchildren in an observational study in Spain (Dadvand et al. 2015). A number of other studies have similarly demonstrated the positive impact of green space exposure on ADHD and related cognitive symptoms (Faber Taylor and Kuo 2011; van den Berg and van den Berg 2011; Markevych et al. 2014). There is also evidence of therapeutic benefits of engaging people with autism with nature (reviewed by Faber Taylor and Kuo 2006).

11.3.2 Reduced Cardiovascular Morbidity and Mortality

Improved access to green space was linked to a reduced detrimental impact of income deprivation on cardiovascular mortality (Mitchell and Popham 2008) while a prospective study in Lithuania demonstrated that greater distance to green space was associated with increased risk of cardiovascular disease (Tamosiunas et al. 2014). Many studies have also provided evidence that the risk of cardiovascular mortality is lower in areas with higher residential greenness (reviewed by Gascon et al. 2016).

A randomized intervention study in Lithuania by Grazuleviciene et al. (2015b) demonstrated that walking in a park had a stronger effect on reducing diastolic blood pressure than similar amount of walking along a busy urban street, suggesting a potential biological mechanism of long-term clinical benefits. An observational study in Australia demonstrated that a greater variability in landscape combining trees and open green space was associated with reduced risk of cardiovascular disease and stroke (Pereira et al. 2012). The authors hypothesized that variability in neighbourhood greenness reflects two factors promoting physical activity - an aesthetically pleasing natural environment and access to urban destinations.

11.3.3 Reduced Prevalence of Obesity and Type 2 Diabetes

A systematic review of 60 studies from the United States, Canada, Australia, New Zealand and Europe on the relationships between green space and obesity indicators found that the majority (68%) of papers showed that green space is associated with reduced obesity; the relationships could be modified by age and socioeconomic status (Lachowycz and Jones 2011). A more recent study in Spain confirmed these conclusions and showed that living in greener residential areas and living in closer proximity to forests was linked with less sedentary time and reduced risks of overweight and obesity in children (Dadvand et al. 2014a).

Furthermore, there is some evidence that using green space for gardening may influence physical activity, improve social well-being and encourage eating healthy food, thereby reducing obesity. A pilot intervention study using community gardening and education in nutrition in a southern state in the United States found that 17% of obese and overweight children had improved their BMI classification by the end of the seven-week-long programme (Castro et al. 2013).

It is well-known that type 2 diabetes mellitus can be prevented by reducing obesity and improving physical activity. Exposure to air pollution is another risk factor for systemic diseases including cardiovascular disease and type 2 diabetes acting through oxidative stress and systemic inflammation mechanisms. Cross-sectional observational studies in The Netherlands, Australia and the United Kingdom demonstrated statistically significant associations between neighbourhood greenness and reduced odds of having type 2 diabetes mellitus (Astell-Burt et al. 2014; Maas et al. 2009b; Bodicoat et al. 2014). A study in Germany demonstrated an inverse association between neighbourhood greenness (measured by NDVI) and insulin resistance in adolescents (Thiering et al. 2016). The observed protective effect of green space was partially explained by a reduced exposure to traffic-related nitrogen dioxide.

11.3.4 Improved Pregnancy Outcomes

In a systematic review, Dzhambov et al. (2014) summarized evidence showing that residential access to green space is associated with reduced risk of low birth weight. Low birth weight is one of main predictors of neonatal and infant mortality, as well as long-term adverse effects in childhood and beyond. More recent studies (Agay-Shay et al. 2014; Markevych et al. 2014; Dadvand et al. 2014b) provided further evidence of beneficial effects of green space on birth weight. A larger distance to a city park was also linked with increased risk of preterm birth and with reduced gestational age at birth (Grazuleviciene et al. 2015a), while improved availability of green space was linked with reduced risk of preterm births (Laurent et al. 2013).

11.3.5 Reduced Mortality and Increased Life Span

A recent meta-analysis of previously conducted studies demonstrated that increased availability of green space is linked with a reduction of mortality (reviewed by Gascon et al. 2016).

For example, a study in Japan has shown that the five-year survival rate in elderly individuals was positively associated with having access to green space suitable for taking a stroll and with parks and tree-lined streets near the residence (Takano et al. 2002). Another study of pre-retirement age population in England showed that a greater amount of green space in the neighbourhood was associated with reduced all-cause mortality (Mitchell and Popham 2008). The study reinforced earlier findings based on the 2001 census population of England, which found that a higher proportion of green space in an area was associated with better self-reported health (Mitchell and Popham 2007).

A longitudinal study in Canada found that increased residential green space was associated with a reduction in mortality (Villeneuve et al. 2012); the strongest effect was on mortality from respiratory diseases. In Spain, Xu et al. (2013) showed that perceived greater neighbourhood greenness was associated with reduced mortality risk during heat waves due to an urban heat island mitigating effect of green space.

In the United States, residential proximity to green space has been associated with a reduced risk of stroke mortality (Hu et al. 2008) and with higher survival rates after ischemic stroke (Wilker et al. 2014). In contrast to the above findings, Richardson et al. (2012) did not find an association between availability of green space and overall mortality in the 49 largest cities in the United States. The authors suggested this might be due to the sprawling nature of these cities and high levels of car dependency.

11.3.6 Potential Adverse Health Effects

Greater availability and enhanced use of green space may also be associated with exposure to health hazards (reviewed by Lõhmus and Balbus 2015), although the evidence on adverse health impacts of urban green space is based on fewer studies and it is less consistent than the evidence of health benefits.

Potential adverse health effects are associated with increased exposure to pesticides, allergenic pollen, arthropod vectors of infectious diseases, infectious agents in soils contaminated with animal faeces, and increased risk of injuries (reviewed by WHO Regional Office for Europe 2016). It should be noted, however, that associations between green space and allergy are inconsistent, with research in some areas showing that green space is linked to an increased risk of allergy, while similar investigations in other geographic areas show strong protective effects (Fuertes et al. 2016). The latter are corroborated by studies that found associations between increased biodiversity around homes and reduced atopic sensitization (Ruokolainen et al. 2015; Hanski et al. 2012). This demonstrates the need for more in-depth studies quantifying exposure to pollen, addressing potential confounding and characterizing mechanisms of age-specific adverse and beneficial health effects. It should also be noted that activities involving risk-taking and exploratory behaviour are important for normal development of children and that environments supportive of risky play promote increased play time, social interactions, creativity and resilience (Brussoni et al. 2015).

Most potential detrimental effects can be eliminated or minimized through proper design, maintenance and operation of green space (Lõhmus and Balbus 2015). Thus, the design of green spaces such as of parks, green trails and playgrounds should take in account these potential side effects and take measures to minimize the risk of allergens or serious injuries.

11.4   Benefits in Disadvantaged Groups and Reduction of Health Inequality

Green space is not equally available or accessible to all population groups (WHO Regional Office for Europe 2012), with low-income communities often having less green space or being exposed to poorly maintained, vandalized or unsafe green areas. Socioeconomic inequalities in access to green space and resulting health benefits may therefore contribute to inequalities in health (The Marmot Review 2010).

It has been suggested that addressing the 'upstream' determinants of health – e.g. by making changes to the built environment – has the potential to reduce population-wide health inequalities; especially the availability of good quality green space across the social gradient is considered essential to tackle health inequalities (Allen and Balfour 2014). While the equity aspects of access to quality green space are discussed in detail in the following chapter by Kabisch et al., this section addresses potential equity issues regarding the health benefits of green space.

There is accumulating evidence showing that urban green space may be 'equigenic' (Mitchell et al. 2015), i.e. that the health benefits linked with access to green space may be strongest among the disadvantaged groups. The study found socioeconomic inequalities in mental health to be substantially lower amongst people reporting good access to recreational/green areas compared to those with poorer access. Other studies showed that populations exposed to the greenest environments had the lowest level of health inequality related to income deprivation (Mitchell and Popham 2008; Lachowycz and Jones 2014). Pope et al. (2015) identified statistically significant associations between reported access to, and better quality of, green space and reduced psychological distress in a deprived urban population in the US. In a large European epidemiological study, Mitchell et al. (2015) found that socioeconomic inequality in mental well-being was 40% narrower among respondents reporting good access to green space, compared with those with poorer access.

Improving the availability of good quality green space in disadvantaged urban neighbourhoods contributes to addressing health inequalities (Allen and Balfour 2014). For example, improvements in access to woodland green space near deprived urban communities in Scotland, United Kingdom had a positive impact on green space use and may have contributed to improvements in activity levels and perceived quality of life (Ward Thompson et al. 2013).

Sreetheran and van den Bosch (2014), in a systematic review of English language literature, found that being an ethnic minority and living in low-income neighbourhoods affects feelings of insecurity in urban green space. Research findings also suggest that, in some settings, beneficial effects of exposure to green space on birth weight may be pronounced only in the lowest socioeconomic position group (Dadvand et al. 2012) or in specific cultural or ethnic groups (Dadvand et al. 2014b). A French study indicated that both greenness level and socioeconomic deprivation partially explained the distribution of neonatal mortality (Kihal-Talantikite et al. 2013). Reducing barriers to the use of green space and taking measures to facilitate health-beneficial behaviours is likely to be an effective and far reaching strategy for public health improvement (Public Health England 2014). A WHO report on addressing inequities in overweight and obesity calls for urban planning policies to set minimum green space requirements in residential developments (WHO Regional Office for Europe 2014).

11.5   Conclusion

The available evidence summarized in this chapter suggests that potential causal pathways leading to public health benefits of urban green spaces include psychological relaxation and stress reduction, improved social cohesion and psychological attachment to the home area, immune system benefits and enhanced physical activity. Green space can also provide ecosystem services associated with reduced exposures to noise, air pollution and excessive heat, which are discussed in the following chapter.

The evidence for health benefits due to relaxation, stress reduction and other psychological effects appears to be very consistent. Many studies have demonstrated associations between greenery in close proximity to residence and health benefits suggesting that being in green space can produce health benefits regardless of the level of physical activity. These health benefits depend on the overall greenness of residential areas and can be provided by adequate urban planning mechanisms. The health benefits mediated by physical activity in green spaces depend on the availability of public green spaces suitable for active leisure and physical play. Green space can also contribute to the reduction of environmental and health inequalities by providing all population groups with equal opportunities to engage in and benefit from natural environments, and with equal ecosystem services, such as buffering of air pollution and noise.

In general, the health benefits of urban green space outweigh its potential detrimental effects, such as allergies to pollen, infections and injuries. Most negative effects are typically associated with poorly designed or poorly maintained green space; they can be reduced or prevented by proper planning, design and maintenance of urban green space (e.g. planting non-allergenic species, controlling disease vectors, improving safety of playgrounds). Measuring the availability, accessibility and quality of green space, and monitoring green space usage by specific population groups are essential steps for providing information in support of evidence-based targeted interventions. Such interventions will include measures that aim to remove barriers to green space utilization, and enhance their utilization by specific population groups, such as children, elderly, working age adults, pregnant women, cultural and ethnic minorities, and individuals with mental illness, cognitive impairments or physical limitations. Examples of such measures can include the provision of greenery in deprived neighbourhoods, physical interventions to enhance access and use of urban green space (such as attractive and welcoming entrances or well-drained and paved footpaths and benches), the regular provision of attractive, nature-based play grounds, and improving safety in urban parks. Improving the availability of green spaces in under-served and socioeconomically disadvantaged communities may help to reduce health inequalities in urban populations.

In addition to the documented health benefits described in this chapter, urban green space may also provide important economic and ecological co-benefits such as reduction of fossil fuel usage through enhanced use of cycling and walking and wildlife habitat supporting biodiversity in urbanized areas. Overall, cities that build and maintain well-connected, attractive green spaces are likely to have healthier, happier and more productive citizens with fewer demands for health services.

Developing and applying harmonized approaches to measuring green space, and producing consistent and comparable data on many cities is essential to enable local planners and policy-makers to assess the need for improvement, and to identify specific areas where green space interventions are warranted (WHO Regional Office for Europe 2016). A review of urban green space interventions has been carried out by WHO to assess environmental and health outcomes of urban green space actions and to inform local practitioners about the aspects to consider when planning green space interventions (WHO Regional Office for Europe 2017).

References

Agay-Shay K, Peled A, Crespo AV, Peretz C, Amitai Y, Linn S, Friger M, Nieuwenhuijsen MJ (2014) Green spaces and adverse pregnancy outcomes. Occup Environ Med 71:562–569

Alcock I, White MP, Wheeler BW, Fleming LE, Depledge MH (2014) Longitudinal effects on mental health of moving to greener and less green urban areas. Environ Sci Technol 48:1247–1255

Allen J, Balfour R (2014) Natural solutions for tackling health inequalities, UCL Institute of Health Equity. http://www.instituteofhealthequity.org/projects/natural-solutions-to-tackling-health-inequalities. Accessed 13 May 2016

Almanza E, Jerrett M, Dunton G, Seto E, Ann Pentz M (2012) A study of community design, greenness, and physical activity in children using satellite, GPS and accelerometer data. Health Place 18:46–54

Amoly E, Dadvand P, Forns J, Lopez-Vicente M, Basagana X, Julvez J, Alvarez-Pedrerol M, Nieuwenhuijsen MJ, Sunyer J (2014) Green and blue spaces and behavioral development in Barcelona schoolchildren: the BREATHE project. Environ Health Perspect 122:1351–1358

Aspinall PA, Thompson CW, Alves S, Sugiyama T, Brice R, Vickers A (2010) Preference and relative importance for environmental attributes of neighbourhood open space in older people. Environ Plann B Plann Des 37:1022–1039

Aspinall P, Mavros P, Coyne R, Roe J (2015) The urban brain: analysing outdoor physical activity with mobile EEG. Br J Sports Med 49:272–276

Astell-Burt T, Feng X, Kolt GS (2013) Does access to neighbourhood green space promote a healthy duration of sleep? Novel findings from a cross-sectional study of 259 319 Australians. Br Med J Open 3(8):e003094

Astell-Burt T, Feng X, Kolt GS (2014) Is neighborhood green space associated with a lower risk of type 2 diabetes? Evidence from 267,072 Australians. Diabetes Care 37:197–201

Balseviciene B, Sinkariova L, Grazuleviciene R, Andrusaityte S, Uzdanaviciute I, Dedele A, Nieuwenhuijsen M (2014) Impact of residential greenness on preschool children's emotional and behavioral problems. Int J Environ Res Public Health 11:6757

Bancroft C, Joshi S, Rundle A, Hutson M, Chong C, Weiss CC et al (2015) Association of Proximity and Density of parks and objectively measured physical activity in the United States: a systematic review. Soc Sci Med 138:22–30

Barton J, Pretty J (2010) What is the best dose of nature and green exercise for improving mental health? A multi-study analysis. Environ Sci Technol 44:3947–3955

Bedimo-Rung AL, Mowen AJ, Cohen DA (2005) The significance of parks to physical activity and public health: a conceptual model. Am J Prev Med 28:159–168

Beil K, Hanes D (2013) The influence of urban natural and built environments on physiological and psychological measures of stress—a pilot study. Int J Environ Res Public Health 10:1250–1267

Beyer KM, Kaltenbach A, Szabo A, Bogar S, Nieto FJ, Malecki KM (2014) Exposure to neighborhood green space and mental health: evidence from the survey of the health of Wisconsin. Int J Environ Res Public Health 11:3453–3472

Bjork J, Albin M, Grahn P, Jacobsson H, Ardo J, Wadbro J, Ostergren P-O, Skarback E (2008) Recreational values of the natural environment in relation to neighbourhood satisfaction, physical activity, obesity and wellbeing. J Epidemiol Community Health 62(4):e2–e2

Bodicoat DH, O'Donovan G, Dalton AM, Gray LJ, Yates T, Edwardson C, Hill S, Webb DR, Khunti K, Davies MJ, Jones AP (2014) The association between neighbourhood greenspace and type 2 diabetes in a large cross-sectional study. Br Med J Open 4:e006076

Bodin M, Hartig T (2003) Does the outdoor environment matter for psychological restoration gained through running? Psychol Sport Exerc 4:141–153

Bratman GN, Hamilton JP, Hahn KS, Daily GC, Gross JJ (2015) Nature experience reduces rumination and subgenual prefrontal cortex activation. Proc Natl Acad Sci U S A 112:8567–8572

Brussoni M, Gibbons R, Gray C, Ishikawa T, Sandseter EB, Bienenstock A, Chabot G, Fuselli P, Herrington S, Janssen I, Pickett W, Power M, Stanger N, Sampson M, Tremblay MS (2015) What is the relationship between risky outdoor play and health in children? A systematic review. Int J Environ Res Public Health 12(6):6423–6454

Castro DC, Samuels M, Harman AE (2013) Growing healthy kids: a community garden-based obesity prevention program. Am J Prev Med 44:S193–S199

Cochrane T, Davey RC, Gidlow C, Smith GR, Fairburn J, Armitage CJ, Stephansen H, Speight S (2009) Small area and individual level predictors of physical activity in urban communities: a multi-level study in Stoke-on-Trent, England. Int J Environ Res Public Health 6:654–677

Dadvand P, de Nazelle A, Figueras F, Basagaña X, Jason S, Amoly E, Jerrett M, Vrijheid M, Sunyer J, Nieuwenhuijsen MJ (2012) Green space, health inequality and pregnancy. Environ Int 40(April):110–115

Dadvand P, Villanueva CM, Font-Ribera L, Martinez D, Basagana X, Belmonte J, Vrijheid M, Grazuleviciene R, Kogevinas M, Nieuwenhuijsen MJ (2014a) Risks and benefits of green spaces for children: a cross-sectional study of associations with sedentary behavior, obesity, asthma, and allergy. Environ Health Perspect 122:1329–1335

Dadvand P, Wright J, Martinez D, Basagaña X, Mceachan RRC, Cirach M, Gidlow CJ, De Hoogh K, Gražulevičienė R, Nieuwenhuijsen MJ (2014b) Inequality, green spaces, and pregnant women: roles of ethnicity and individual and neighbourhood socioeconomic status. Environ Int 71:101–108

Dadvand P, Nieuwenhuijsen MJ, Esnaola M, Forns J, Basagana X, Alvarez-Pedrerol M, Rivas I, Lopez-Vicente M, De Castro Pascual M, Su J, Jerrett M, Querol X, Sunyer J (2015) Green spaces and cognitive development in primary schoolchildren. Proc Natl Acad Sci U S A 112:7937–7942

De Jong K, Albin M, Skärbäck E, Grahn P, Björk J (2012) Perceived green qualities were associated with neighborhood satisfaction, physical activity, and general health: results from a cross-sectional study in suburban and rural Scania, southern Sweden. Health Place 18:1374–1380

De Vries S (2010) Nearby nature and human health: looking at the mechanisms and their implications. In: Ward Thompson C, Aspinall P, Bell S (eds) Innovative approaches to researching landscape and health. Routledge, Abingdon

De Vries S, Van Dillen SME, Groenewegen PP, Spreeuwenberg P (2013) Streetscape greenery and health: stress, social cohesion and physical activity as mediators. Soc Sci Med 94:26–33

Dzhambov AM, Dimitrova DD, Dimitrakova ED (2014) Association between residential greenness and birth weight: systematic review and meta-analysis. Urban For Urban Green 13:621–629

Epstein LH, Raja S, Gold SS, Paluch RA, Pak Y, Roemmich JN (2006) Reducing sedentary behavior: the relationship between park area and the physical activity of youth. Psychol Sci 17:654–659

European Environment Agency (2015) SOER 2015 – The European environment — state and outlook 2015. Urban systems briefing, Copenhagen. http://www.eea.europa.eu/soer-2015/europe/urban-systems#tab-based-on-indicators. Accessed 13 May 2016

Faber Taylor A, Kuo FE (2006) Is contact with nature important for healthy child development? State of the evidence. In: Spencer C, Blades M (eds) Children and their environments: learning, using and designing spaces. Cambridge University Press, Cambridge, pp 124–140

Faber Taylor AF, Kuo FEM (2011) Could exposure to everyday green spaces help treat ADHD? Evidence from children's play settings. Appl Psychol Health Well-Bring 3:281–303

Fuertes E, Markevych I, Bowatte G, Gruzieva O, Gehring U, Becker A, Berdel D, von Berg A, Bergström A, Brauer M, Brunekreef B, Brüske I, Carlsten C, Chan-Yeung M, Dharmage SC, Hoffmann B, Klümper C, Koppelman GH, Kozyrskyj A, Korek M, Kull I, Lodge C, Lowe A, MacIntyre E, Pershagen G, Standl M, Sugiri D, Wijga A, Heinrich J (2016) Residential greenness is differentially associated with childhood allergic rhinitis and aeroallergen sensitization in seven birth cohorts. Allergy. doi:10.1111/all.12915. [Epub ahead of print]

Gardsjord HS, Tveit MS, Nordh H (2014) Promoting youth's physical activity through park design: linking theory and practice in a public health perspective. Landsc Res 39:70–81

Gascon M, Triguero-Mas M, Martínez D, Dadvand P, Rojas-Rueda D, Plasència A, Nieuwenhuijsen MJ (2016) Residential green spaces and mortality: a systematic review. Environ Int 86:60–67

Gidlow CJ, Jones MV, Hurst G, Masterson D, Clark-Carter D, Tarvainen MP, Smith G, Nieuwenhuijsen M (2016a) Where to put your best foot forward: psycho-physiological responses to walking in natural and urban environments. J Environ Psychol 45:22–29

Gidlow CJ, Randall J, Gillman J, Smith GR, Jones MV (2016b) Natural environments and chronic stress measured by hair cortisol. Landsc Urban Plan 148:61–67

Grazuleviciene R, Danileviciute A, Dedele A, Vencloviene J, Andrusaityte S, Uzdanaviciute I, Nieuwenhuijsen MJ (2015a) Surrounding greenness, proximity to city parks and pregnancy outcomes in Kaunas cohort study. Int J Hyg Environ Health 218:358–365

Grazuleviciene R, Vencloviene J, Kubilius R, Grizas V, Dedele A, Grazulevicius T, Ceponiene I, Tamuleviciute-Prasciene E, Nieuwenhuijsen MJ, Jones M, Gidlow C (2015b) The effect of park and urban environments on coronary artery disease patients: a randomized trial. Biomed Res Int 2015:9

Hanski I, von Hertzen L, Fyhrquist N, Koskinen K, Torppa K, Laatikainen T, Karisola P, Auvinen P, Paulin L, Mäkelä MJ, Vartiainen E, Kosunen TU, Alenius H, Haahtela T (2012) Environmental biodiversity, human microbiota, and allergy are interrelated. Proc Natl Acad Sci U S A 109:8334–8339

Hartig T (2007) Three steps to understanding restorative environments as health resources. In: Thompson CW, Travlou P (eds) Open space: people space. Taylor & Francis, Abingdon

Hartig T, Mitchell R, De Vries S, Frumkin H (2014) Nature and health. Annu Rev Public Health 35:207–228

Honold J, Lakes T, Beyer R, Van Der Meer E (2015) Restoration in urban spaces: nature views from home, greenways, and public parks. Environ Behav 48(6):796–825

Hu Z, Liebens J, Rao KR (2008) Linking stroke mortality with air pollution, income, and greenness in Northwest Florida: an ecological geographical study. Int J Health Geogr 7:20

James P, Banay RF, Hart JE, Laden F (2015) A review of the health benefits of greenness. Curr Epidemiol Rep 2:131–142

Kaczynski AT, Henderson KA (2007) Environmental correlates of physical activity: a review of evidence about parks and recreation. Leis Sci 29:315–354

Kaczynski AT, Potwarka LR, Saelens BE (2008) Association of park size, distance, and features with physical activity in neighborhood parks. Am J Public Health 98:1451–1456

Kaplan S (1995) The restorative benefits of nature: toward an integrative framework. J Environ Psychol 15:169–182

Kaplan S (2001) Meditation, restoration, and the Management of Mental Fatigue. Environ Behav 33(4):480–506

Kihal-Talantikite et al (2013) Green space, social inequalities and neonatal mortality in France. BMC Pregnancy Childbirth 13:191

Kim J, Kaplan R (2004) Physical and psychological factors in sense of community: new urbanist kentlands and nearby orchard village. Environ Behav 36:313–340

Kondrashova A, Seiskari T, Ilonen J, Knip M, Hyöty H (2013) The 'hygiene hypothesis' and the sharp gradient in the incidence of autoimmune and allergic diseases between Russian Karelia and Finland. APMIS 121(6):478–493

Kuo M (2015) How might contact with nature promote human health? Promising mechanisms and a possible central pathway. Front Psychol 25:1093

Lachowycz K, Jones AP (2011) Greenspace and obesity: a systematic review of the evidence. Obes Rev 12:e183–e189

Lachowycz K, Jones AP (2013) Towards a better understanding of the relationship between greenspace and health: development of a theoretical framework. Landsc Urban Plan 118:62–69

Lachowycz K, Jones AP (2014) Does walking explain associations between access to greenspace and lower mortality? Soc Sci Med 107:9–17

Lachowycz K, Jones AP, Page AS, Wheeler BW, Cooper AR (2012) What can global positioning systems tell us about the contribution of different types of urban greenspace to children's physical activity? Health Place 18:586–594

Laurent O, Wu J, Li L, Milesi C (2013) Green spaces and pregnancy outcomes in Southern California. Health Place 24:190–195

Lengen C, Kistemann T (2012) Sense of place and place identity: review of neuroscientific evidence. Health Place 18(5):1162–1171

Li Q, Morimoto K, Kobayashi M, Inagaki H, Katsumata M, Hirata Y, Hirata K, Suzuki H, Li Y, Wakayama Y (2008) Visiting a forest, but not a city, increases human natural killer activity and expression of anti-cancer proteins. Int J Immunopathol Pharmacol 21:117–127

Logan AC (2015) Dysbiotic drift: mental health, environmental grey space, and microbiota. J Physiol Anthropol 34:23

Lõhmus M, Balbus J (2015) Making green infrastructure healthier infrastructure. Infect Ecol Epidemiol 2015:30082

Lovasi GS, Quinn JW, Neckerman KM, Perzanowski MS, Rundle A (2008) Children living in areas with more street trees have lower asthma prevalence. J Epidemiol Community Health 62:647–649

Lynch SV, Wood RA, Boushey H, Bacharier LB, Bloomberg GR, Kattan M, O'connor GT, Sandel MT, Calatroni A, Matsui E, Johnson CC, Lynn H, Visness CM, Jaffee KF, Gergen PJ, Gold DR, Wright RJ, Fujimura K, Rauch M, Busse WW, Gern JE (2014) Effects of early-life exposure to allergens and bacteria on recurrent wheeze and atopy in urban children. J Allergy Clin Immunol 134:593–601. e12

Maas J, Van Dillen SME, Verheij RA, Groenewegen PP (2009a) Social contacts as a possible mechanism behind the relation between green space and health. Health Place 15:586–595

Maas J, Verheij RA, De Vries S, Spreeuwenberg P, Schellevis FG, Groenewegen PP (2009b) Morbidity is related to a green living environment. J Epidemiol Community Health 63:967–973

Markevych I, Fuertes E, Tiesler CM, Birk M, Bauer CP, Koletzko S, Von Berg A, Berdel D, Heinrich J (2014) Surrounding greenness and birth weight: results from the GINIplus and LISAplus birth cohorts in Munich. Health Place 26:39–46

Mitchell R (2013) Is physical activity in natural environments better for mental health than physical activity in other environments? Soc Sci Med 91:130–134

Mitchell R, Popham F (2007) Greenspace, urbanity and health: relationships in England. J Epidemiol Community Health 61:681–683

Mitchell R, Popham F (2008) Effect of exposure to natural environment on health inequalities: an observational population study. Lancet 372:1655–1560

Mitchell RJ, Richardson EA, Shortt NK, Pearce JR (2015) Neighborhood environments and socioeconomic inequalities in mental well-being. Am J Prev Med 49:80–84

Nieminen T, Martelin T, Koskinen S, Aro H, Alanen E, Hyyppä M (2010) Social capital as a determinant of self-rated health and psychological well-being. Int J Public Health 55:531–542

Owen N, Healy GN, Matthews CE, Dunstan DW (2010) Too much sitting: the population-health science of sedentary behavior. Exerc Sport Sci Rev 38:105–113

Pantell M, Rehkopf D, Jutte D, Syme SL, Balmes J, Adler N (2013) Social isolation: a predictor of mortality comparable to traditional clinical risk factors. Am J Public Health 103:2056–2062

Pereira G, Foster S, Martin K, Christian H, Boruff BJ, Knuiman M, Giles-Corti B (2012) The association between neighborhood greenness and cardiovascular disease: an observational study. BMC Public Health 12:466–466

Pope D, Tisdall R, Middleton J, Verma A, Van Ameijden E, Birt C, Bruce NG (2015) Quality of and access to green space in relation to psychological distress: results from a population-based cross-sectional study as part of the EURO-URHIS 2 project. Eur J Pub Health pii:ckv094. [Epub ahead of print]

Public Health England (2014) Local action on health inequalities: improving access to green spaces. Public Health England, London

Reklaitiene R, Grazuleviciene R, Dedele A, Virviciute D, Vensloviene J, Tamosiunas A, Baceviciene M, Luksiene D, Sapranaviciute-Zabazlajeva L, Radisauskas R, Bernotiene G, Bobak M, Nieuwenhuijsen MJ (2014) The relationship of green space, depressive symptoms and perceived general health in urban population. Scand J Public Health 42:669–676

Richardson EA, Mitchell R, Hartig T, De Vries S, Astell-Burt T, Frumkin H (2012) Green cities and health: a question of scale? J Epidemiol Community Health 66:160–165

Roe JJ, Thompson CW, Aspinall PA, Brewer MJ, Duff EI, Miller D, Mitchell R, Clow A (2013) Green space and stress: evidence from cortisol measures in deprived urban communities. Int J Environ Res Public Health 10:4086–4103

Roe J, Aspinall P (2011) The restorative benefits of walking in urban and rural settings in adults with good and poor mental health. Health Place 17:103–113

Rook G (2013) Regulation of the immune system by biodiversity from the natural environment: an ecosystem service essential to health. Proc Natl Acad Sci U S A 110:18360–18367

Ruokolainen L, von Hertzen L, Fyhrquist N, Laatikainen T, Lehtomäki J, Auvinen P, Karvonen AM, Hyvärinen A, Tillmann V, Niemelä O, Knip M, Haahtela T, Pekkanen J, Hanski I (2015) Green areas around homes reduce atopic sensitization in children. Allergy 70:195–202

Sallis JF et al (2016) Physical activity in relation to urban environments in 14 cities worldwide: a cross-sectional study. Lancet 387:2207–2217

Schipperijn J, Bentsen P, Troelsen J, Toftager M, Stigsdotter UK (2013) Associations between physical activity and characteristics of urban green space. Urban For Urban Green 12:109–116

Seeland K, Dübendorfer S, Hansmann R (2009) Making friends in Zurich's urban forests and parks: the role of public green space for social inclusion of youths from different cultures. Forest Policy Econ 11:10–17

Shanahan DF, Lin BB, Bush R, Gaston KJ, Dean JH, Barber E, Fuller RA (2015) Toward improved public health outcomes from urban nature. Am J Public Health 105(3):470–477

Sreetheran M, Van Den Bosch CCK (2014) A socio-ecological exploration of fear of crime in urban green spaces – a systematic review. Urban For Urban Green 13:1–18

Sugiyama T, Ward Thompson C (2008) Associations between characteristics of neighbourhood open space and older people's walking. Urban For Urban Green 7:41–51

Sugiyama T, Leslie E, Giles-Corti B, Owen N (2008) Associations of neighbourhood greenness with physical and mental health: do walking, social coherence and local social interaction explain the relationships? J Epidemiol Community Health 62:e9

Sugiyama T, Cerin E, Owen N, Oyeyemi AL, Conway TL, Van Dyck D, Schipperijn J, Macfarlane DJ, Salvo D, Reis RS, Mitáš J, Sarmiento OL, Davey R, Schofield G, Orzanco-Garralda R, Sallis JF (2014) Perceived neighbourhood environmental attributes associated with adults recreational walking: IPEN adult study in 12 countries. Health Place 28:22–30

Takano T, Nakamura K, Watanabe M (2002) Urban residential environments and senior citizens' longevity in megacity areas: the importance of walkable green spaces. J Epidemiol Community Health 56:913–918

Tamosiunas A, Grazuleviciene R, Luksiene D, Dedele A, Reklaitiene R, Baceviciene M, Vencloviene J, Bernotiene G, Radisauskas R, Malinauskiene V, Milinaviciene E, Bobak M, Peasey A, Nieuwenhuijsen MJ (2014) Accessibility and use of urban green spaces, and cardiovascular health: findings from a Kaunas cohort study. Environ Health 13:20

The Marmot Review (2010) Fair society, healthy lives: strategic review of health inequalities in England post-2010. UCL IHE, London

Thiering E, Markevych I, Brüske I, Fuertes E, Kratzsch J, Sugiri D, Hoffmann B, Von Berg A, Bauer CP, Koletzko S, Berdel D, Heinrich J (2016) Associations of residential long-term air pollution exposures and satellite-derived greenness with insulin resistance in German adolescents. Environ Health Perspect 124(8):1291–1298. [Epub ahead of print]

Triguero-Mas M, Dadvand P, Cirach M, Martínez D, Medina A, Mompart A, Basagaña X, Gražulevičienė R, Nieuwenhuijsen MJ (2015) Natural outdoor environments and mental and physical health: relationships and mechanisms. Environ Int 77:35–41

Ulrich RS (1983) Aesthetic and affective response to natural environment. In: Altman I, Wohlwill JF (eds) Human behavior & environment: advances in theory & research. Plenum, New York

Ulrich RS, Simons RF, Losito BD, Fiorito E, Miles MA, Zelson M (1991) Stress recovery during exposure to natural and urban environments. J Environ Psychol 11:201–230

United Nations (2016) Sustainable development goals and related targets, New York. https://sustainabledevelopment.un.org/topics. Accessed 13 May 2016

Van Den Berg AE, Custers MHG (2011) Gardening promotes neuroendocrine and affective restoration from stress. J Health Psychol 16:3–11

Van Den Berg AE, Van Den Berg CG (2011) A comparison of children with ADHD in a natural and built setting. Child Care Health Dev 37:430–439

Van Den Berg M, Van Poppel M, Van Kamp I, Andrusaityte S, Balseviciene B, Cirach M, Danileviciute A, Ellis N, Hurst G, Masterson D, Smith G, Triguero-Mas M, Uzdanaviciute I, Wit PD, Mechelen WV, Gidlow C, Grazuleviciene R, Nieuwenhuijsen MJ, Kruize H, Maas J (2016) Visiting green space is associated with mental health and vitality: a cross-sectional study in four European cities. Health Place 38:8–15

Van Den Bosch MA, Ostergren PO, Grahn P, Skarback E, Wahrborg P (2015) Moving to serene nature may prevent poor mental health – results from a Swedish longitudinal cohort study. Int J Environ Res Public Health 12:7974–7989

Villeneuve PJ, Jerrett M, Su JG, Burnett RT, Chen H, Wheeler AJ, Goldberg MS (2012) A cohort study relating urban green space with mortality in Ontario, Canada. Environ Res 115(63):51–58. doi:10.1016/j.envres.2012.03.003

Villanueva K, Badland H, Hooper P, Koohsari MJ, Mavoa S, Davern M, Roberts R, Goldfeld S, Giles-Corti B (2015) Developing indicators of public open space to promote health and wellbeing in communities. Appl Geogr 57:112–119

Völker S, Kistemann T (2015) Developing the urban blue: comparative health responses to blue and green urban open spaces in Germany. Health Place 35:196–205

Ward Thompson C, Roe J, Aspinall P, Mitchell R, Clow A, Miller D (2012) More green space is linked to less stress in deprived communities: evidence from salivary cortisol patterns. Landsc Urban Plan 105:221–229

Ward Thompson C, Roe J, Aspinall P (2013) Woodland improvements in deprived urban communities: what impact do they have on people's activities and quality of life? Landsc Urban Plan 118:79–89

Ward Thompson C, Aspinall P, Roe J, Robertson L, Miller D (2016) Mitigating stress and supporting health in deprived urban communities: the importance of green space and the social environment. Int J Environ Res Public Health 13:440

White MP, Alcock I, Wheeler BW, Depledge MH (2013) Would you be happier living in a greener urban area? A fixed-effects analysis of panel data. Psychol Sci 24:920–928

Wilker EH, Wu C-D, Mcneely E, Mostofsky E, Spengler J, Wellenius GA, Mittleman MA (2014) Green space and mortality following ischemic stroke. Environ Res 133:42–48

WHO Regional Office for Europe (2010) Parma declaration and commitment to act, Copenhagen. http://www.euro.who.int/__data/assets/pdf_file/0011/78608/E93618.pdf. Accessed 13 May 2016

WHO Regional Office for Europe (2012) Addressing the social determinants of health: the urban dimension and the role of local government. World Health Organization, Geneva

WHO Regional Office for Europe (2013) Health 2020. A European policy framework and strategy for the 21st century, Copenhagen, http://www.euro.who.int/__data/assets/pdf_file/0011/199532/Health2020-Long.pdf?ua=1. Accessed 13 May 2016

WHO Regional Office for Europe (2014) Obesity and inequities. Guidance for addressing inequities in overweight and obesity. Copenhagen. http://www.euro.who.int/__data/assets/pdf_file/0003/247638/obesity-090514.pdf?ua=1. Accessed 13 May 2016

WHO Regional Office for Europe (2016) Urban green spaces and health. WHO Regional Office for Europe, Copenhagen

WHO Regional Office for Europe (2017) Urban green space interventions and health. A review of impacts and effectiveness. WHO Regional Office for Europe, Copenhagen

Xu Y, Dadvand P, Barrera-Gómez J, Sartini C, Marí-Dell'olmo M, Borrell C, Medina-Ramón M, Sunyer J, Basagaña X (2013) Differences on the effect of heat waves on mortality by sociode-mographic and urban landscape characteristics. J Epidemiol Community Health 67:519–525

Yang CY, Boen C, Gerken K, Li T, Schorpp K, Harris KM (2016) Social relationships and physi-ological determinants of longevity across the human life span. Proc Natl Acad Sci U S A 113:578–583

Chapter 12
Urban Green Spaces and the Potential for Health Improvement and Environmental Justice in a Changing Climate

Nadja Kabisch and Matilda Annerstedt van den Bosch

Abstract Urbanisation and climate change affect people's health and well-being in various ways. Nature-based solutions implemented as natural, sustainable solutions in cities can attenuate negative health impacts of these processes. In this chapter, urban green spaces are considered as one type of nature-based solutions that use urban ecosystem services to provide mitigation and adaptation actions and solutions to climate change and urbanisation related challenges. An overview over the relationships to urban health is presented. The city of Berlin is used as a case, to show how an unequal distribution of urban green area may be linked to an insufficient provision of ecosystem services and the related positive health outcome effect. This is discussed through the presentation of the distribution of different vulnerable population groups such as children and elderly people throughout the city area. The link to environmental justice is made and discussed in this context.

Keywords Urbanisation • Health • Well-being • Children • Elderly • Urban green space • Berlin

N. Kabisch (✉)
Department of Geography, Humboldt-Universität zu Berlin,
Unter den Linden 6, Berlin 10099, Germany

Department of Urban and Environmental Sociology, Helmholtz Centre for Environmental Research-UFZ, Permoserstrasse 15, Leipzig 04318, Germany
e-mail: nadja.kabisch@geo.hu-berlin.de

M.A. van den Bosch
School of Population and Public Health, The University of British Columbia (UBC),
314A - 2206 East Mall, Vancouver V6T 1Z3, BC, Canada

The Department of Forest and Conservation Sciences, The University of British Columbiam,
3041-2424 Main Mall, Vancouver BC V6T 1Z4, Canada
e-mail: matilda.vandenbosch@ubc.ca

© The Author(s) 2017
N. Kabisch et al. (eds.), *Nature-based Solutions to Climate Change Adaptation in Urban Areas*, Theory and Practice of Urban Sustainability Transitions,
DOI 10.1007/978-3-319-56091-5_12

12.1 Introduction

Urbanisation and climate change are increasingly-affecting the global earth surface and urban health today and create a number of challenges to urban planning. World's urban population is expected to increase by more than two-thirds by 2050, from 3.9 billion in 2014 to 6.3 billion in 2050 (United Nations, Department of Economic and Social Affairs 2014). Interlinked pressures from land conversion, soil sealing and densification of built-up areas, decrease in quantity and access to urban green and blue spaces and increase of traffic and related effects of air and noise pollution pose significant threats to human health and well-being. In addition, climate change will have a significant impact on city environments (The World Bank 2010). Main climate change effects in cities include a rise in air temperature (e.g. during heat waves), poor air quality and higher ozone concentration, as well as extreme precipitation events (European Environment Agency 2016). Urban planners and decision-makers have to deal with the challenges of urbanisation and climate change to equitably secure access to clean air and drinking water, recreational green and blue spaces and an overall healthy living environment and, with this, the provision of ecosystem services (McHale et al. 2015). Ecosystem services are various goods and benefits that biodiverse natural environments provide to people, such as nutrients, livelihoods and cultural and recreational experiences (Millenium Ecosystem Assessment 2005). They provide specific health benefits to city residents. New approaches are needed in order to efficiently adapt to and mitigate negative effects from climate change and urbanisation and to maximise opportunities for improving the health of all urban residents, independent of socioeconomic status, gender, cultural background or age. Nature-based solutions (NBS) in urban areas are one approach, which have the potential to counteract these challenges across populations. NBS to societal challenges are defined by the European Commission 2016 as "[…]Nature-based solutions to societal challenges as solutions that are inspired and supported by nature, which are cost-effective, simultaneously provide environmental, social and economic benefits and help build resilience. Such solutions bring more, and more diverse, nature and natural features and processes into cities, landscapes and seascapes, through locally adapted, resource-efficient and systemic interventions". By referring to "solutions that are inspired and supported by nature", urban green space can be implemented as components of NBS in cities. Such solutions contain natural or semi-natural areas like urban parks, vegetated roofs and facades, street trees, gardens and blue systems such as rivers, canals, lakes, wetlands or ponds as well as other types of interventions that use at least partial ecosystem functions and services to provide adaptation and mitigation actions to climate change and challenges from urbanisation (Kabisch et al. 2016a). This chapter discusses how urban green spaces can provide ecosystem services and thus act as NBS particularly to health challenges resulting from climate change and urbanisation. In this regard, the present chapter builds on the previous chapter on "Effects of Urban Green Space on Environmental Health, Equity and Resilience" by Braubach et al. (Chap. 11, this volume). Using the city of Berlin as a case, a special focus is on environmental justice with distribution of urban green spaces linked to different vulnerable population groups such as children and elderly people.

12.1.1 The Potential of Urban Green Spaces for Ecosystem Service Provision and Health Improvement

Many of the climate regulation ecosystem services counteract particularly the environmental health threats connected to urbanisation and climate change (Haase et al. 2014). Extreme weather events such as heat waves, exacerbated by the urban heat island (UHI) effect, cause premature death and illnesses (Basagaña et al. 2011; Xu et al. 2016). The UHI effect is most significant in areas of high impermeable built-up density and low share of green space (Oke 1973; Rizwan et al. 2008). Urban trees and vegetation provide climate regulation services as they reduce the UHI effect through evapotranspiration and shading and can thus help preventing heat-related morbidity and mortality (Chen et al. 2014). Also urban blue spaces can decrease heat levels and mitigate heat-related morbidity (Burkart et al. 2015). In this context, green and blue spaces may therefore be considered as examples of NBS.

There is also an interaction effect between heat and air pollution, with higher levels of pollution in hotter environments (Harlan and Ruddell 2011). Air pollution from traffic and industrial sources has increased with rising urbanisation resulting in a severe impact on human health with an estimated 600,000 premature deaths annually in Europe alone (Lelieveld et al. 2015; WHO 2016; Brauer et al. 2016). Increased exposure to poor air quality conditions can have severe health effects in the life course of individuals. (King et al. 2011; Lindström et al. 2014). The issue is particular problematic in poor areas of cities, often situated close to traffic or industry with sparse vegetation and high-quality green spaces. There is some evidence around the regulation potential of urban green space as an NBS to reduce air pollution levels in cities. However, evidence is inconsistent (for a detailed discussion and evidence on the potential of air pollution improvement of NBS, see Baró et al. Chap. 9, this volume). Some studies show significant effects (Nowak et al. 2013; Vailshery et al. 2013; Baró et al. 2014), while others show no effect (Setälä et al. 2013) or even worsened pollution levels under street tree canopies (Jin et al. 2014). However, by careful management and planning, it is likely that reduced air pollution levels can be achieved through an optimised relation between plant genotype, tree canopy density and leaf area index as an NBS (Derkzen et al. 2015; Cameron and Blanuša 2016).

Flooding is another risk factor, which is associated with climate change-related impacts and is exacerbated in dense cities with high sealing rates and less inflow. Floods can induce high economic losses because of the risk for intense infrastructure damage. Climate change projections show an increase in the risk of river floods, superficial floods as well as coastal floods due to sea level rise (European Environment Agency 2012). Next to economic effects, floods have severe effects to human health. Flooding not only poses severe direct risks to health of residents but also affects health infrastructure. The local characteristics of a city, including sealing rates and green space cover, determine the amount of damages and impacts on infrastructure and human life through flooding. Particularly, in high density districts, extreme precipitation events can cause flooding and lead to economic and infrastructural damages with risks to health of local residents. In summer 2016, an extreme precipitation event occurred in the city of Berlin, Germany. This extreme precipitation event has not been existing in the last 50 years and has led to flooding of the particular high-dense districts of Neukölln and Wedding in Berlin with damages to the transport infrastructure (Berliner Morgenpost 27.07.2016). The strategic implementation of green spaces to mitigate extreme precipitation and potential resulting floods can be accounted as an NBS based on regulating urban ecosystem services (Haase et al. 2014).

12.1.2 Unequal Distribution of Exposure to Health Threats in Urban Areas – An Issue of Environmental Justice

Many of the mentioned environmentally related health threats are unequally distributed in a city with a higher exposure to vulnerable populations in deprived areas, often living in very dense areas with high share of imperviousness, living closer to traffic, industrial sites, contaminated soil and poor accessibility to high-quality green spaces (Su et al. 2011). Health inequalities are expected to grow with ongoing urbanisation and with impacts from climate change, thus affecting people's equal chances to create healthy and prosperous lives (McMichael 2000). Apart from socioeconomically deprived populations, children and elderly populations belong to vulnerable groups with increased sensitivity to urban health risks related to climate change, such as heat stress and air pollution (Vanos 2015; Benmarhnia et al. 2015). Children are in a developing state, thus more sensitive to environmental extremes and harmful exposures. Elderly can be more vulnerable due to co-morbidity, medications and inefficient thermoregulation. Both children and elderly are restrained in their capacity to behavioural adaptation (e.g. mobility constraints). The disproportionate allocation of environmental burdens to different population groups raises concerns about environmental justice (Davis et al. 2012). Environmental justice is traditionally related to the health of low-income residents and minority groups who live in neighbourhoods with low environmental quality (for a literature review, see Downey and Hawkins 2009). For definitions of the concept of environmental justice, see Box 12.1.

Box 12.1: Definition of Environmental Justice in a European and US Context

In a European context, environmental justice has been defined as "… equal access to a clean environment and equal protection from possible environmental harm irrespective of race, income, class or any other differentiating feature of socio-economic status" (Schwarte and Adebowale 2007). The US Environmental Protection Agency defined environmental justice as "… the fair treatment and meaningful involvement of all people regardless of race, colour, national origin, or income, with respect to the development, implementation, and enforcement of environmental laws, regulations, and policies" (United States Environment Protection Agency 2017).

12.2 Links Between Urban Green Spaces, Health and Environmental Justice

12.2.1 Health Effects as Co-benefits of Nature-Based Solutions to Climate Change Mitigation and Adaptation from Urban Green Spaces

Several pathways have been proposed for explaining the link between urban natural (green and blue) areas and improved public health. Such pathways relate to, for example, the opportunities for stress recovery, physical activity and social contacts (Hartig et al. 2014). As the same factors – stress, physical inactivity and social isolation – are major risk factors for many chronic diseases (e.g. diabetes, cardiovascular and mental disorders, obesity and cancer), actors in public health urgently seek to identify strategies to reduce these risks. City living may worsen the exposure and consequences of these risk factors. For example, the higher prevalence of mental disorders in urban as compared to rural areas has, among other factors, been attributed to the relatively hectic and stressful life in cities (Peen et al. 2010). Similarly, environmental factors also contribute to physical inactivity in cities, such as high-density traffic and lack of parks and sidewalks. In this context, NBS implemented as urban natural spaces have emerged as health-promoting environments to reduce stress, encourage physical activity and provide a sense of community for increased social interactions.

Recent epidemiological studies seem to confirm the beneficial health impact of natural spaces. Demonstrated health effects in a general population are, for example, reduced mortality (Gascon et al. 2016), reduced cardiovascular morbidity (Tamosiunas et al. 2014; Donovan et al. 2015), lower blood pressure (Grazuleviciene et al. 2014) and decreased depressive symptoms (Reklaitiene et al. 2014). At the same time, improved conditions are found for pregnancy outcomes (Dadvand et al. 2012), and general physical and mental health (Annerstedt et al. 2012; Triguero-Mas et al. 2015).

12.2.2 Health, Justice and the Link to Urban Green Spaces

Urban green spaces and the benefits they provide can be disproportionately available to a subset of urban population (Ernstson 2013). Scientific literature suggests that immigrant communities in European cities often have limited access to parks and urban green spaces in their vicinity compared to nonimmigrant groups (Germann-Chiari and Seeland 2004; Comber et al. 2008; Dai 2011). Many studies have also demonstrated that health inequalities tend to decrease in greener areas (Mitchell and Popham 2008; Mitchell et al. 2015) and that deprived groups seem to benefit the most from the positive health effects of nature (Ward Thompson et al. 2012; Roe et al. 2013; Ward Thompson et al. 2016).

Children is one group with particular needs and particular health risks related to urban living. There is a risk that in today's society, with increased screen time, computerisation and less outdoor play, children will become disconnected from nature and thereby miss out on related health benefits (Louv and Hogan 2005). Research indeed demonstrates that children in areas with more green show a better cognitive and behavioural development (Amoly et al. 2014; Dadvand et al. 2015) and symptoms of various behavioural disorders are relieved in nature (Faber Taylor and Kuo 2009). In addition, street tree density and other urban greenery have been associated with less childhood obesity (Kim et al. 2014) as well as lower asthma prevalence (Lovasi et al. 2008).

The elderly population is particularly vulnerable to environmental exposures with negative impact on health. Older people's health can benefit from quality and quantity of urban green spaces (Takano et al. 2002; Barbosa et al. 2007). Proximity to green space (near home of residents) improves longevity of senior citizens (Takano et al. 2002). A study by Kawachi and Berkman (2001) showed that even the potential to be outside in a green space could increase older people's health. Sugiyama and Ward Thompson (2007) identified that neighbourhood environments are likely to contribute to older people's health by providing places as opportunity spaces to be active. They found that older people who live in a supportive environment including green spaces are likely to walk more and are equally likely to be in better health. Those studies have shown that the pure existence of urban parks motivates older people to walk and go outside, which in turn improves their health and well-being and decreases potential health costs.

12.3 Unequal Distribution of Urban Green Spaces as a Concern for Environmental Justice

Establishment and management of green spaces are examples of societal actions that cost-efficiently can improve public health and counteract health inequalities. Being a free and public asset providing benefit independent of individual resources, green spaces can positively influence health in various population groups. However, within cities, green spaces are often unequally distributed between groups of different socio-economic status, age and ethno-racial characteristics, making specific population groups more vulnerable to climate change- and urbanisation-related health impacts (Gobster 1998; Byrne and Wolch 2009). Unequal access to urban green space has, thus, become an issue of environmental justice (Kabisch and Haase 2014), and awareness of this problem has increased in order to prevent avoidable negative health impacts across the life course (Dai 2011). This means that NBS provided by green spaces may be withheld from those who need it most.

Uneven distribution of and access to urban green and blue spaces may be related to a number of interlinked factors including historic land use development, park management and design. Also in historical times, green spaces and parks were created where the rich lived. Even today, the installation and development of urban green spaces – such as parks – increases attractiveness of a neighbourhood, making it desirable for investments. In turn, raising house and rent prices can potentially lead to a displacement of those residents the green space was actually meant to be beneficial for. Such effects are called "green paradox" (Wolch et al. 2014), "eco gentrification" (Patrick and Kowalski 2011; Haffner 2015), "ecological gentrification" (Dooling 2009) or "environmental gentrification" (Checker 2011) (for an intensive discussion on the concepts, see *A. Haase*, this volume).

To ensure that all residents in a city have a minimum amount of urban green spaces in their vicinity and therefor benefit from the ecosystem services provided by them, city planning departments use threshold values to coordinate their urban green space planning and development and to safeguard current green space quantity. Some city agencies try to focus on concrete per capita threshold values (e.g. 6 m^2 per inhabitant for Berlin, 10 m^2 per inhabitant for Leipzig) or certain park or green area sizes which should be reached by a certain distance (e.g. 2 ha in 300 m or in 500 m, Handley et al. 2003). These values are still planning objectives, as they are not met in all parts of the city, at least in the case of Berlin (Kabisch and Haase 2014). Calculation of accessibility and availability of green spaces using different threshold values has already been applied and analysed in geographical information system (GIS) analyses (Kabisch et al. 2016b, c; Dai 2011; Comber et al. 2008; Barbosa et al. 2007). Kabisch et al. (2016c) assessed urban green space availability using a sample of EU cities. They identified a diverse picture across the countries. Southern European cities show below-average availability values, which may be explained by their low forest and tree cover and reflect the history of cities in Southern Europe. Comparatively, above-average availability values in Northern European cities were identified and discussed to may be a result of biophysical conditions, the presence of rich forestland in general but also of Northern European attitudes towards urban living that naturally value having forests close to home. Comber et al. (2008) showed that Hindu and Sikh groups have limited access to UGS in the city of Leicester. For Sheffield, Barbosa et al. (2007) found that 64% of Sheffield households fail to meet the recommendation of the regulatory agency English Nature (EN) that people should live no further than 300 m from their nearest green space.

However, there are also critical notes arguing that threshold values simply underestimate the actual provision of urban green spaces. Threshold values and a certain size or distance measure may say nothing about the actual accessibility for different population groups, nor do they provide real information about the quality and the safety and the use of the space. Thresholds do not consider how many people actually live within the recommended distance and therefore do not take into account the pressure on the area and the risk for crowding and overuse. For example, in cities where green spaces and waters are distributed throughout the whole city, this can result in good overall green space provision values despite low per capita values in certain dense areas. Thresholds used for defining availability or accessibility such as maximum 300 m linear distance to a green space of minimum 2 ha may also not be the most appropriate for identifying differences on a sub-district level for certain vulnerable groups. A 300 m linear distance is often longer in reality, as the linear measure, which is mostly applied in accessibility studies does not consider the actual walking route, including larger roads and other potential physical barriers. The 300 m threshold may therefore be less relevant for children and the elderly.

12.3.1   Threshold Values for Urban Green Space Provision in Berlin

Berlin is applying a 6 m^2 urban green space per capita value on a local scale to have some orientation for further green space development projects throughout the city. The threshold is met in most of Berlin's sub-districts (see Fig. 12.1). However, values are below the threshold in the very central districts with population densities of more than 14,000 inhabitants per km^2.

With regard to the percentage of specific age group distribution in the city the maps show that relatively high proportions of older individuals (more than 65 years of age) are located in the peripheral parts of the city where share of green space is particularly high (with more than 50% in some districts). In some inner city sub-districts, older individuals represent less than 10% of the sub-district population, whereas in most of the districts, the percentage of children is between 8 and 13%. There are some inner city districts such as Neukölln in the south east of the city and Mitte in the central and northern inner city parts where percentage of children is more than 17%. Here, population density is comparatively high with high sealing rates and less green spaces (see also a paper on children's health and distribution of urban green and blue spaces by Kabisch et al. 2016b).

12.4   Discussion and Conclusion

Urban green space development and maintenance as NBS for climate change adaptation and mitigation will almost certainly become increasingly important as urbanisation and climate change increase (European Environment Agency 2012). Various scenario studies and existing cases show that climate change most strongly affects those who are the most vulnerable, such as people of low income and education, children and elderly people. For those population groups, NBS implemented through, for instance, parks, street green, urban forests, pocket parks or even roof greenery could potentially function as a complementary health resource, counteracting some of the socially determined health inequalities present today in our cities (Hartig et al. 2014; Mitchell et al. 2015). In this chapter, we discussed how urban green spaces may act as an NBS to climate change and urbanisation-induced

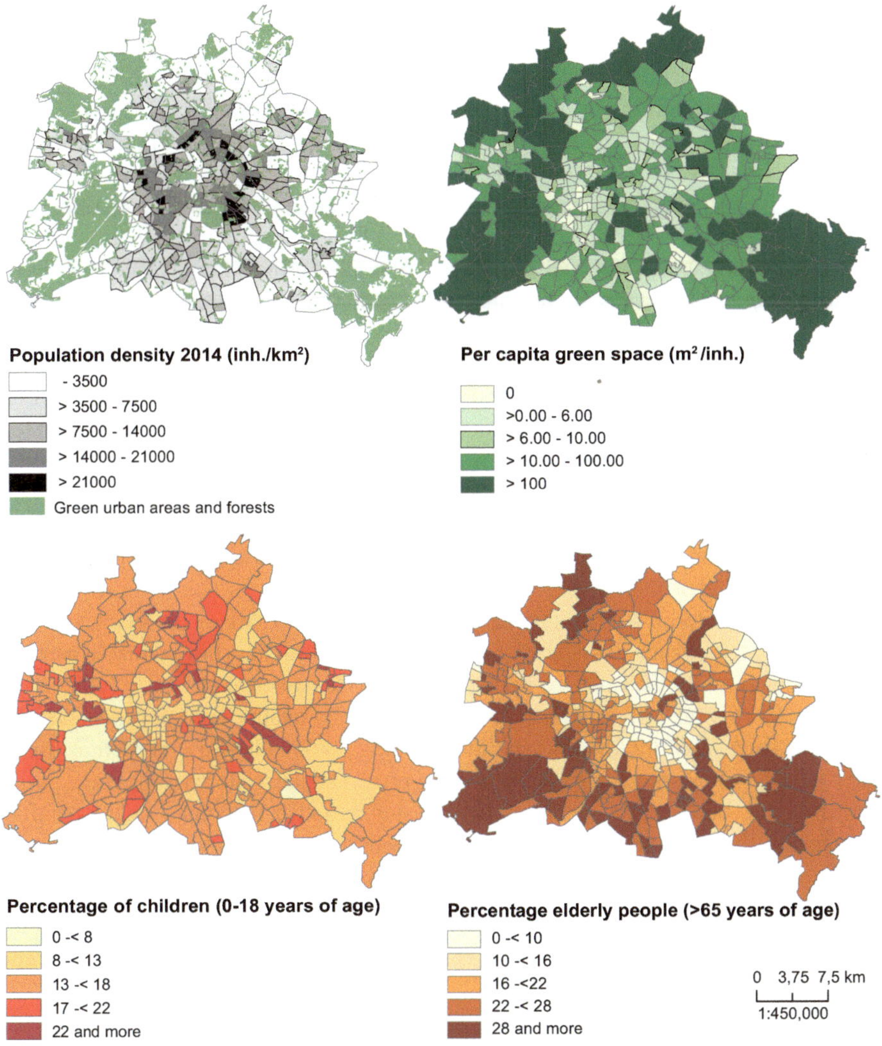

Population density 2014 (inh./km²)

- ☐ - 3500
- ☐ > 3500 - 7500
- ☐ > 7500 - 14000
- ☐ > 14000 - 21000
- ■ > 21000
- ☐ Green urban areas and forests

Per capita green space (m²/inh.)

- ☐ 0
- ☐ >0.00 - 6.00
- ☐ > 6.00 - 10.00
- ☐ > 10.00 - 100.00
- ■ > 100

Percentage of children (0-18 years of age)

- ☐ 0 -< 8
- ☐ 8 -< 13
- ☐ 13 -< 18
- ☐ 17 -< 22
- ■ 22 and more

Percentage elderly people (>65 years of age)

- ☐ 0 -< 10
- ☐ 10 -< 16
- ☐ 16 -<22
- ☐ 22 -< 28
- ■ 28 and more

0   3,75 7,5 km
1:450,000

Fig. 12.1 Population density and per capita green space in the sub-districts of Berlin 2012 (Land use data are based on the Urban Atlas 2012 (Source: http://www.eea.europa.eu/legal/copyright). Copyright holder: Directorate-General Enterprise and Industry (DG-ENTR), (Directorate-General for Regional Policy). Population data are provided by the Department of Statistics Berlin-Brandenburg, (www.statistik-berlin-brandenburg.de) and refer to 2014)

challenges and at the same time counteract health inequalities across socio-economic status and age scales. This could potentially have a substantial bearing as health inequalities is a major target for improved public health. There are no biological fundaments for differences in health between groups of different education or income, nevertheless the health gap is wide between poor and rich and continues to widen (Dai 2011; Vaughan et al. 2013).

This means that there is a need to act on social and environmental health determinants to achieve good health for all people (Martuzzi et al. 2010; Marmot et al. 2012). In order to achieve positive health outcomes it is important that urban natural spaces are available in a sufficient quantity and easily accessible to all population groups. Children, especially from less wealthy families with fewer opportunities to travel, are bound to spend much of their time in the close neighbourhood and are as such specifically vulnerable to effects of the residential environment (Koller and Mielck 2009). Therefore, an equal distribution of high-quality and safe urban natural spaces, adequate for physical activity and play, is of utmost importance in healthy urban planning. This was highlighted already in the Parma Declaration (WHO 2010) where the European member countries committed themselves "to provide each child by 2020 with access to healthy and safe environments and settings of daily life in which they can walk and cycle to kindergartens and schools, and to green spaces in which to play and undertake physical activity". Green space implementation projects as NBS should be considered as an appropriate tool for city planning and administrations to reach this commitment.

Left for future research and intensive discussions is the question of how much ecosystem service can an urban green space provide when it starts to get very frequently used, get crowded or even overused in very dense urban districts. This is sometimes the case in some of Berlin's inner city parks such as the Mauerpark or the Görlitzer Park particularly at the weekend or during holidays. Although studies showed that in general residents tend to use their nearest park most often (Neuvonen et al. 2007; Schipperijn et al. 2010), this may not hold true in cases where the quality does not meet certain standards because of the overuse and the resulting low-quality aspects such as trash, dirty toilets, vandalism and criminality. Local residents may start avoiding these parks and even use other places farther away causing negative outcomes such as traffic (Arnberger 2012). Good quality park management that adapts to local conditions and integrates the needs of local residents may improve such situations at place.

Acknowledgements This work was financially supported by GREEN SURGE, EU FP7 collaborative project, FP7-ENV.2013.6.2-5-603567.

References

Amoly E, Dadvand P, Forns J et al (2014) Green and blue spaces and behavioral development in Barcelona schoolchildren: the BREATHE project. Environ Health Perspect 122:1351–1358. doi:10.1289/ehp.1408215

Annerstedt M, Ostergren P-O, Bjork J et al (2012) Green qualities in the neighbourhood and mental health – results from a longitudinal cohort study in Southern Sweden. BMC Public Health 12:337

Arnberger A (2012) Urban densification and recreational quality of public urban green spaces—a Viennese case study. Sustainability 4:703–720. doi:10.3390/su4040703

Barbosa O, Tratalos J, Armsworth P et al (2007) Who benefits from access to green space? A case study from Sheffield, UK. Landsc Urban Plan 83:187–195. doi:10.1016/j.landurbplan.2007.04.004

Baró F, Chaparro L, Gómez-Baggethun E et al (2014) Contribution of ecosystem services to air quality and climate change mitigation policies: the case of urban forests in Barcelona, Spain. Ambio 43:466–479. doi:10.1007/s13280-014-0507-x

Basagaña X, Sartini C, Barrera-Gómez J et al (2011) Heat waves and cause-specific mortality at all ages. Epidemiology 22:765–772. doi:10.1097/EDE.0b013e31823031c5

Benmarhnia T, Deguen S, Kaufman JS, Smargiassi A (2015) Review article. Epidemiology 26:781–793. doi:10.1097/EDE.0000000000000375

Brauer M, Freedman G, Frostad J et al (2016) Ambient air pollution exposure estimation for the global burden of disease 2013. Environ Sci Technol 50:79–88. doi:10.1021/acs.est.5b03709

Burkart K, Meier F, Schneider A et al (2015) Modification of heat-related mortality in an elderly urban population by vegetation (urban green) and proximity to water (urban blue): evidence from Lisbon. Portugal Environ Health Perspect doi:10.1289/ehp.1409529

Byrne J, Wolch J (2009) Nature, race, and parks: past research and future directions for geographic research. Prog Hum Geogr 33:743–765. doi:10.1177/0309132509103156

Cameron RWF, Blanuša T (2016) Green infrastructure and ecosystem services – is the devil in the detail? Ann Bot 118:377–391. doi:10.1093/aob/mcw129

Checker M (2011) Wiped out by the "Greenwave": environmental gentrification and the paradoxical politics of urban sustainability. City Soc 23:210–229. doi:10.1111/j.1548-744X.2011.01063.x

Chen A, Yao XA, Sun R, Chen L (2014) Effect of urban green patterns on surface urban cool islands and its seasonal variations. Urban For Urban Green 13:646–654. doi:10.1016/j.ufug.2014.07.006

Comber A, Brunsdon C, Green E (2008) Using a GIS-based network analysis to determine urban greenspace accessibility for different ethnic and religious groups. Landsc Urban Plan 86:103–114. doi:10.1016/j.landurbplan.2008.01.002

Dadvand P, Sunyer J, Basagaña X et al (2012) Surrounding greenness and pregnancy outcomes in four Spanish birth cohorts. Environ Health Perspect 120:1481–1487. doi:10.1289/ehp.1205244

Dadvand P, Nieuwenhuijsen MJ, Esnaola M et al (2015) Green spaces and cognitive development in primary schoolchildren. Proc Natl Acad Sci 112:7937–7942. doi:10.1073/pnas.1503402112

Dai D (2011) Racial/ethnic and socioeconomic disparities in urban green space accessibility: where to intervene? Landsc Urban Plan 102:234–244. doi:10.1016/j.landurbplan.2011.05.002

Davis AY, Belaire JA, Farfan MA et al (2012) Green infrastructure and bird diversity across an urban socioeconomic gradient. Ecosphere 3:1–18

Derkzen ML, van Teeffelen AJ, Verburg PH (2015) REVIEW: quantifying urban ecosystem services based on high-resolution data of urban green space: an assessment for Rotterdam, the Netherlands. J Appl Ecol 52:1020–1032. doi:10.1111/1365-2664.12469

Donovan GH, Michael YL, Gatziolis D et al (2015) Is tree loss associated with cardiovascular-disease risk in the Women's health initiative? A natural experiment. Health Place 36:1–7. doi:10.1016/j.healthplace.2015.08.007

Dooling S (2009) Ecological gentrification: a research agenda exploring justice in the City. Int J Urban Reg Res 33:621–639. doi:10.1111/j.1468-2427.2009.00860.x

Downey L, Hawkins B (2009) Race, income, and enviornmental inequality in the United States. Sociol Perspect 51:759–781. doi:10.1525/sop.2008.51.4.759.RACE

Ernstson H (2013) The social production of ecosystem services: a framework for studying environmental justice and ecological complexity in urbanized landscapes. Landsc Urban Plan 109:7–17. doi:10.1016/j.landurbplan.2012.10.005

European Commission (2016) Policy topics: nature-based solutions https://ec.europa.eu/research/environment/index.cfm?pg=nbs. Accessed 11 Sept 2016

European Environment Agency (2012) Urban adaptation to climate change in Europe – challenges and opportunities for cities together with supportive national and European policies. European Environment Agency, Copenhagen

European Environment Agency (2016) Urban adaptation to climate change in Europe 2016 – transforming cities in a changing climate. Eurpean Environment Agency, Copenhagen. 978-92-9213-742-7

Faber Taylor A, Kuo FE (2009) Children with attention deficits concentrate better after walk in the park. J Atten Disord 12:402–409. doi:10.1177/1087054708323000

Gascon M, Triguero-Mas M, Martínez D et al (2016) Residential green spaces and mortality: a systematic review. Environ Int 86:60–67

Germann-Chiari C, Seeland K (2004) Are urban green spaces optimally distributed to act as places for social integration? Results of a geographical information system (GIS) approach for urban forestry research. For Policy Econ 6:3–13. doi:10.1016/S1389-9341(02)00067-9

Gobster PH (1998) Urban parks as green walls or green magnets? Interracial relations in neighborhood boundary parks. Landsc Urban Plan 41:43–55. doi:10.1016/S0169-2046(98)00045-0

Grazuleviciene R, Dedele A, Danileviciute A et al (2014) The influence of proximity to City parks on blood pressure in early pregnancy. Int J Environ Res Public Health 11:2958–2972. doi:10.3390/ijerph110302958

Haase D, Larondelle N, Andersson E et al (2014) A quantitative review of urban ecosystem service assessments: concepts, models, and implementation. Ambio 43:413–433. doi:10.1007/s13280-014-0504-0

Haffner J (2015) The dangers of eco-gentrification: what's the best way to make a city greener? Guard 5–7

Handley J, Pauleit S, Slinn P et al (2003) Accessible natural green space standards in Towns and Cities: a review and toolkit for their implementation. English Nat Res Rep, Rep. Nr. 526

Harlan SL, Ruddell DM (2011) Climate change and health in cities: impacts of heat and air pollution and potential co-benefits from mitigation and adaptation. Curr Opin Environ Sustain 3:126–134. doi:10.1016/j.cosust.2011.01.001

Hartig T, Mitchell R, de Vries S, Frumkin H (2014) Nature and health. Annu Rev Public Health 35:207–228. doi:10.1146/annurev-publhealth-032013-182443

Jin S, Guo J, Wheeler S et al (2014) Evaluation of impacts of trees on PM2.5 dispersion in urban streets. Atmos Environ 99:277–287. doi:10.1016/j.atmosenv.2014.10.002

Kabisch N, Haase D (2014) Green justice or just green? Provision of urban green spaces in Berlin, Germany. Landsc Urban Plan 122:129–139. doi:10.1016/j.landurbplan.2013.11.016

Kabisch N, Frantzeskaki N, Pauleit S et al (2016a) Nature-based solutions to climate change mitigation and adaptation in urban areas: perspectives on indicators, knowledge gaps, barriers, and opportunities for action. Ecol Soc 21:art39. doi:10.5751/ES-08373-210239

Kabisch N, Haase D, van den Bosch MA (2016b) Adding natural areas to social indicators of intraurban health inequalities among children: a case study from Berlin, Germany. Int J Environ Res Public Health 13:783. doi:10.3390/ijerph13080783

Kabisch N, Strohbach M, Haase D, Kronenberg J (2016c) Urban green space availability in European cities. Ecol Indic 70:586–596. doi:10.1016/j.ecolind.2016.02.029

Kawachi I, Berkman LF (2001) Social ties and mental health. J Urban Health 78:458–467. doi:10.1093/jurban/78.3.458

Kim JH, Lee C, Olvara NE, Ellis CD (2014) The role of landscape spatial patterns on obesity in Hispanic children residing in Inner-City neighborhoods. J Phys Act Health 11:1449–1457. doi:10.1123/jpah.2012-0503

King KE, Morenoff JD, House JS (2011) Neighborhood context and social disparities in cumulative biological risk factors. Psychosom Med 73:572–579. doi:10.1097/PSY.0b013e318227b062

Koller D, Mielck A (2009) Regional and social differences concerning overweight, participation in health check-ups and vaccination. Analysis of data from a whole birth cohort of 6-year old children in a prosperous German city. BMC Public Health 9:43. doi:10.1186/1471-2458-9-43

Lelieveld J, Evans JS, Fnais M et al (2015) The contribution of outdoor air pollution sources to premature mortality on a global scale. Nature 525:367–371. doi:10.1038/nature15371

Lindström M, Fridh M, Rosvall M (2014) Economic stress in childhood and adulthood, and poor psychological health: three life course hypotheses. Psychiatry Res 215:386–393. doi:10.1016/j.psychres.2013.11.018

Louv R, Hogan J (2005) Last child in the woods: saving our children from nature-deficit disorder. Algonquin Books of Chapel Hill, Chapel Hill

Lovasi GS, Quinn JW, Neckerman KM et al (2008) Street trees and childhood asthma. J Epidemiol Commun Heal 62:647–649

Marmot M, Allen J, Bell R et al (2012) WHO European review of social determinants of health and the health divide. Lancet 380:1011–1029. doi:10.1016/S0140-6736(12)61228-8

Martuzzi M, Mitis F, Forastiere F (2010) Inequalities, inequities, environmental justice in waste management and health. Eur J Pub Health 20:21–26. doi:10.1093/eurpub/ckp216

McHale M, Pickett S, Barbosa O et al (2015) The new global urban realm: complex, connected, diffuse, and diverse social-ecological systems. Sustainability 7:5211–5240. doi:10.3390/su7055211

McMichael A (2000) The urban environment and health in a world of increasing globalization: issues for developing countries. Bull World Health Organ 28:1117–1126

Millenium Ecosystem Assessment (2005) Ecosystems and human well-being: synthesis. Island Press, Washington, DC

Mitchell R, Popham F (2008) Effect of exposure to natural environment on health inequalities: an observational population study. Lancet 372:1655–1660. doi:10.1016/S0140-6736(08)61689-X

Mitchell RJ, Richardson EA, Shortt NK, Pearce JR (2015) Neighborhood environments and socio-economic inequalities in mental well-being. Am J Prev Med 49:80–84. doi:10.1016/j.amepre.2015.01.017

Neuvonen M, Sievänen T, Tönnes S, Koskela T (2007) Access to green areas and the frequency of visits – a case study in Helsinki. Urban For Urban Green 6:235–247. doi:10.1016/j.ufug.2007.05.003

Nowak DJ, Hirabayashi S, Bodine A, Hoehn R (2013) Modeled PM2.5 removal by trees in ten U.S. cities and associated health effects. Environ Pollut 178:395–402. doi:10.1016/j.envpol.2013.03.050

Oke TR (1973) City size and the urban heat island. Atmos Environ 7:769–779. doi:10.1016/0004-6981(73)90140-6

Patrick DJ, Kowalski A (2011) The politics of Urban sustainability: preservation, redevelopment, and landscape on the high line by submitted to Central European University Department of Sociology and Social Anthropology. In partial fulfillment of the requirements for the degree of Master of Arts, Budapest, Hungarys

Peen J, Schoevers RA, Beekman AT, Dekker J (2010) The current status of urban-rural differences in psychiatric disorders. Acta Psychiatr Scand 121:84–93. doi:10.1111/j.1600-0447.2009.01438.x

Reklaitiene R, Grazuleviciene R, Dedele A et al (2014) The relationship of green space, depressive symptoms and perceived general health in urban population. Scand J Public Health 42:669–676. doi:10.1177/1403494814544494

Rizwan AM, Dennins LYC, Liu C (2008) A review on the generation, determination and mitigation of Urban Heat Island. J Environ Sci 20:120–128. doi:10.1016/S1001-0742(08)60019-4

Roe J, Ward Thompson C, Aspinall PA et al (2013) Green space and stress: evidence from cortisol measures in deprived urban communities. Int J Environ Res Public Health 10:4086–4103. doi:10.3390/ijerph10094086

Schipperijn J, Stigsdotter UK, Randrup TB, Troelsen J (2010) Influences on the use of urban green space – a case study in Odense, Denmark. Urban For Urban Green 9:25–32. doi:10.1016/j.ufug.2009.09.002

Schwarte C, Adebowale M (2007) Environmental justice and race equality in the European union. ISBN 13 9780 954725563

Setälä H, Viippola V, Rantalainen A-L et al (2013) Does urban vegetation mitigate air pollution in northern conditions? Environ Pollut 183:104–112. doi:10.1016/j.envpol.2012.11.010

Su JG, Jerrett M, de Nazelle A, Wolch J (2011) Does exposure to air pollution in urban parks have socio-economic, racial or ethnic gradients? Environ Res 111:319–328. doi:10.1016/j.envres.2011.01.002

Sugiyama T, Ward Thompson C (2007) Older people's health, outdoor activity and supportiveness of neighbourhood environments. Landsc Urban Plan 83:168–175. doi:10.1016/j.landurbplan.2007.04.002

Takano T, Nakamura K, Watanabe M (2002) Urban residential environments and senior citizens' longevity in megacity areas: the importance of walkable green spaces. J Epidemiol Commun Heal 56:913–918. doi:10.1136/jech.56.12.913

Tamosiunas A, Grazuleviciene R, Luksiene D et al (2014) Accessibility and use of urban green spaces, and cardiovascular health: findings from a Kaunas cohort study. Environ Health 13:20. doi:10.1186/1476-069X-13-20

The World Bank (2010) Cities and climate change: an urgent agenda. World Bank, Washington, DC

Triguero-Mas M, Dadvand P, Cirach M et al (2015) Natural outdoor environments and mental and physical health: relationships and mechanisms. Environ Int 77:35–41. doi:10.1016/j.envint.2015.01.012

United Nations, Department of Economic and Social Affairs Population Division (2014) World urbanization prospects: the 2014 revision, Highlights (ST/ESA/SER.A/352). United Nations, New York

United States Environmental Protection Agency (2017) Environmental justice. https://www.epa.gov/environmentaljustice. Accessed 6 Apr 2017

Vailshery LS, Jaganmohan M, Nagendra H (2013) Effect of street trees on microclimate and air pollution in a tropical city. Urban For Urban Green 12:408–415. doi:10.1016/j.ufug.2013.03.002

Vanos JK (2015) Children's health and vulnerability in outdoor microclimates: a comprehensive review. Environ Int 76:1–15. doi:10.1016/j.envint.2014.11.016

Vaughan KB, Kaczynski AT, SAW S et al (2013) Exploring the distribution of park availability, features, and quality across Kansas City, Missouri by income and race/ethnicity: an environmental justice investigation. Ann Behav Med 45:28–38. doi:10.1007/s12160-012-9425-y

Ward Thompson C, Roe J, Aspinall P et al (2012) More green space is linked to less stress in deprived communities: evidence from salivary cortisol patterns. Landsc Urban Plan 105:221–229. doi:10.1016/j.landurbplan.2011.12.015

Ward Thompson C, Aspinall P, Roe J et al (2016) Mitigating stress and supporting health in deprived urban communities: the importance of green space and the social environment. Int J Environ Res Public Health 13:440. doi:10.3390/ijerph13040440

WHO (2010) Parma declaration on environment and health. In: Fifth ministerial conference on environment and health "Protecting children's health in a changing environment." Copenhagen

WHO (2016) Health risk assessment of air pollution – general principles. WHO Regional Office for Europe, Copenhagen

Wolch JR, Byrne J, Newell JP (2014) Urban green space, public health, and environmental justice: the challenge of making cities "just green enough". Landsc Urban Plan 125:234–244. doi:10.1016/j.landurbplan.2014.01.017

Xu Z, FitzGerald G, Guo Y et al (2016) Impact of heatwave on mortality under different heatwave definitions: a systematic review and meta-analysis. Environ Int 89–90:193–203. doi:10.1016/j.envint.2016.02.007

Chapter 13
The Contribution of Nature-Based Solutions to Socially Inclusive Urban Development– Some Reflections from a Social-environmental Perspective

Annegret Haase

Abstract Nature-based solutions have emerged to be a major approach or concept when discussing about the sustainable future four cities and are expected to represent solutions for societal problems. When looking closer at this approach, it becomes, however, obvious that the concept is loaded with too many expectations concerning the societal – and, what is more, the social – context of today's urban reality. Furthermore, nature-based solutions are not inherently socially just; when aiming at bringing together environmental sustainability and social equity/inclusion, then a range of issues have to be critically looked at. Set against this background, the paper reflects on the contribution of nature-based solutions to a socially inclusive urban development. In the focus are trade-offs and blind spots of the hitherto discussion. The paper is thought to be first and foremost a positioning paper and is based on five theses. The paper argues, among others, that nature-based solutions offer, if discussed comprehensively and seriously, a potential for creating and shaping more sustainable cities. In order to meet this objective, they should, however, be seen as more than just tools, technologies and instruments. Nature-based solutions have to be improved as a comprehensive approach, especially with respect to their societal and social embeddings and the full picture of their impacts.

Keywords Nature-based solutions • Social inclusiveness • Urban development • Social-environmental perspective

A. Haase (✉)
Department of Urban and Environmental Sociology, Helmholtz-Centre for Environmental Research – UFZ, Leipzig, Germany
e-mail: annegret.haase@ufz.de

© The Author(s) 2017
N. Kabisch et al. (eds.), *Nature-based Solutions to Climate Change Adaptation in Urban Areas*, Theory and Practice of Urban Sustainability Transitions, DOI 10.1007/978-3-319-56091-5_13

221

13.1   Introduction

In recent years, two strands of discussion have received growing attention within both social and environmental sciences. First, literature has become more focused on highlighting issues related to urbanization and urban environments; the acknowledgement that the future of our planet will be urban has also resulted in an increased interest in cities within ecology- and "green" sustainability-based studies. Second, urban policy-makers, planners and architects have "discovered" that nature and the environment can be used as a strategy or factor within urban renewal, urban rehabilitation, and smart/sustainable development agendas.

Against this backdrop, the European Commission has increasingly promoted nature-based solutions (NBS) (EC 2015) as a concept or term to describe measures or instruments that can support the maintenance, improvement and restoration of urban nature, green infrastructure and biodiversity. NBS are promoted as an approach to tackle challenges such as climate change and resource (water, energy and food) scarcity The European Commission particularly focuses on the applicability of NBS to cities – 'Urban areas and enhancing sustainable urbanization' are listed as one out of four key opportunity areas for NBS in the EC document. Compared to other terms/concepts such as green infrastructure and (urban) ecosystem services, the NBS concept has a stronger focus on linking ecological/green and economic benefits and is seen to provide a holistic approach to tackling an array of challenges as a result. NBS is not just a scientific concept but also an approach that is attractive for stakeholders, policy-makers and the business environment.

In addition, NBS are also expected to produce a range of co-benefits that contribute to quality of life in cities and are, according to the final report of the EC (2015) expected to help solve various societal challenges in cities. What is striking however, when looking at proposed NBS in the EU document, is that the listed examples either make no direct reference to societal issues or adopt a fairly undifferentiated view of urban societies and the social and power structures within cities. There is no reflection on the potential impacts of NBS on urban life, different population groups, and areas in cities. The possibility that NBS may not always be equally beneficial for all population groups (Kabisch et al. 2016) is not considered. Furthermore, it remains unclear whether NBS automatically lead to socially just and inclusive developments as well or whether they lead to tradeoffs.

Set against this background, the following section will reflect on these issues, and will particular focus on exploring the following questions:

- Which issues need to be considered when looking at the relationship between NBS and the social environment in cities?
- Which tradeoffs exist and what are their consequences?
- To what extent can NBS contribute to socially inclusive urban development?

The paper will present five theses related to the contribution of NBS to socially inclusive urban development and will build on examples from the author's research on social-environmental processes in cities during the last one and a half decades.

The focus will be on the European context but will also be relevant to the general/global scale. Prior to outlining these theses, the section below will set the context by exploring the social-environmental nexus in cities from a theoretical angle, and by outlining the foundations, ingredients and ambitions of the NBS approach.

Nature-based solutions in the following, are understood as solutions using green and/or blue infrastructure or dealing with natural resources in a responsible way in order to improve both urban quality of life in and sustainability of cities, or, how the EC put it in its report (2015, p. 5):

> Nature-based solutions are understood as living solutions inspired by, continuously supported by and using nature, which are designed to address various societal challenges in a resource efficient and adaptable manner and to provide simultaneously economic, social and environmental benefits.

Socially inclusive urban development is defined as a development considering the needs and wants of all groups of urban inhabitants as well as the different capabilities, capacities and constraints of people to benefit from goods and not to suffer from burdens.

13.2 Setting the Context: The Concept of NBS and the "Socio-environmental Nexus" in Cities

13.2.1 The Concept of NBS

NBS have been discussed as a term or concept for some years now, mainly within the fields of agriculture (management), industrial design and resilience to the impacts of climate change (Potschin et al. 2015). IUCN defines NBS for the field of nature protection in the era of climate change as

> interventions which use nature and the natural functions of healthy ecosystems to tackle some of the most pressing challenges of our time. These types of solutions help to protect the environment but also provide numerous economic and social benefits. (Cohen-Shacham et al. 2016, p. 2)

Recently, NBS have also emerged as a priority area for the EU's Horizon 2020 Research Program. The arguments why such a new approach is needed now relate to a generally growing awareness of the value of nature, the seizing of a momentum where cities have to transform and adapt to manifold challenges. With an explicit relation to economy- and technology-based solutions, the NBS approach also seeks to make NBS relevant to business and private sector actors too (pp. 5–6). According to the EC,

> NBS in cities should support the emergence of new business models decoupling growth from uneven distribution of resources and an increased reliance on local resources (p.8).

> Nature in cities should be employed for a multiple and innovative reuse of degraded urban areas (ibid.).

Moreover, NBS have the potential to contribute to human health and well-being as well as social cohesion. Consequently, recommended focus areas for action are urban regeneration and the improvement of well-being (pp. 16–17).

Generally, NBS can be seen as a positive way to employ nature more explicitly for a sustainable planning of urban areas including the involvement of the political and economic sphere. And without doubt, there are great opportunities for using green infrastructure or instruments for cost-effective, innovative and healthy solutions that, at the same time, contribute to quality of life and a successful economic performance of a city.

However, two issues must be considered critically here: First, it seems that the NBS approach as defined by the EC and also by IUCN seeks to adapt older concepts such as urban green infrastructure (UGI) or urban ecosystem services (UES) still closer to the economy-related and business-oriented sector. It employs a very business-friendly and de-politicised understanding of urban nature and sustainability and should especially invite developers and investors to take part in their creation, realization and implementation. I am fully in line with Potschin et al. (2015, p. 2) here who demand that:

> Yet, a clear link between NBS and these concepts is needed to ensure consistency and avoid redundancy or confusion.

When looking at the scientific debate, NBS are close to what While et al. (2004) in a study on UK cities coined as a "sustainability fix" to describe ways that entrepreneurial urban regimes have sought to (selectively) incorporate the green agenda and changes in rules and incentives structuring urban governance as part of an evolving geopolitics of nature and the environment in order to "greening of the urban growth machine". NBS as described in the EC report seem to fit well as a measure- or instrument-oriented concept fulfilling the idea of sustainability fix.

Second, the EC report in particular does not reflect on the fact that NBS are applied or will be implemented in cities with existing socio-economic and socio-spatial differences and inequalities; also IUCN's definition remains very vague in speaking about "numerous social benefits" that could be brought about by NBS. Subsequently, they are discussed in a de-contextualized way which risks ignoring social and justice-related trade-offs.

13.2.2   The "Socio-environmental Nexus" Within Urban and Cities-Related Research

Ongoing discussions on sustainable urbanization have resulted in an increasing body of literature and knowledge both in social and environmental sciences. What is more, the number of cross-disciplinary works has also increased. By and large, however, the interplay of social and environmental processes in urban environments remains under-researched. While green infrastructure, greening strategies and environmental goods play the role of an "add-on" in social and planning related debates on urban development, the ecological debate shows a lack of attention towards the spatial unevenness of green supply and access to green as well as social inequalities in cities as a basic factor for assessing the distribution of environmental goods and burdens and the feasibility and implementation of sustainability goals (see e.g. the conclusions of the review paper by Kabisch et al. 2016, especially p. 9).

There are a number of central matters of the so-called "socio-environmental nexus" in cities, i.e. the linkages between social and environmental processes and structures:

- an increasing importance of "the urban" in UES/UGI debates but, at the same time, a lacking attention for the existing inequalities and injustices in cities,
- an increasing employment of nature into the strategic debate on future urban development and
- a poor understanding and insufficient dealing with the trade-offs between environmental/ecological and social developments in cities.

Urban environments have become ever more important within sustainability discussion, not least due to growing urban populations, but also due to an increased appreciation of environmental benefits and services and their contributions to urban quality of life. The ecological debate draws on a range of concepts and perspectives that offer the theoretical background for a more sophisticated discussion, among others the concepts of urban green infrastructure (UGI) and urban ecosystem services (UES, see Pauleit et al. in this book). These concepts are looking at urban nature and its functions and services from a human or society-focused perspective and have grown in popularity within the research community over the past few years (see e.g. Kabisch 2015; Haase et al. 2014; Elmquist et al. 2013; Secretariat of the Convention on Biological Diversity 2012, BiodivERsA project URBES, Green Surge etc.). Kabisch et al. (2016) demand explicit consideration of socio-environmental justice and social cohesion when implementing NBS. A plethora of terms and concepts dealing with futures of sustainable or green urbanization: eco city, smart city, green city, healthy city etc. (see for an overview Jong et al. 2015) have emerged over the years; some of these concepts offer some potential to explore socio-environmental linkages.

However, what is striking within these debates is the limited attention given to basic realities of cities and urban societies concerning (a) the relationship between ES provision and political and social power structures and embeddings (see a very recent paper by Berbes-Blazquez et al. 2016) and (b) the impacts of market logics and power structures on urban policymaking, also with respect to UGI and green strategies (Checker 2011 speaks even of a post-political approach here). By and large, "urban" refers to a specific but largely undifferentiated and de-politicized context, within which ES or GI knowledge can be "applied". Or, in critical cases, nature serves as a legitimation for economically-founded upgrading:

> Environmental amenities of all kinds can act as tools for urban entrepreneurialism and gentrification. (Bryson 2013, p. 581)

There are some doubts whether concepts within the green and sustainability debate "can adequately cater for social justice" (Jong et al. 2015, p.10) but there is no deeper knowledge on this relationship. Berbes-Blazquez et al. (2016) address the relationship between ES provision and social power relations concluding that although power relations extensively impact on ES provision they are largely disregarded in the ES debate so far (p.134).

From the perspective of urban development and planning, the relationship between green spaces and the built environment, as well as the social sphere, have been a topic for a long time (see e.g. Jane Jacobs 1961 on Philadelphia parks). The history of European urban development and urban restructuring of the last decades (especially after WW II) shows many examples of how green spaces were employed to increase urban quality of life, e.g. through the enlargement and improvement of parks, the creation of new green spaces etc. By and large these measures resulted in benefits for many urban dwellers. However, when it comes to the distribution of high-quality green areas, often the better-off dwellers or areas reaped the most benefits; high-quality green (e.g. costly shaped parks) was more often created in or close to upmarket housing environments. As a result, the development of green spaces in a way contributed to the socio-spatial inequalities within the urban environment. Within last decades, urban nature has been increasingly employed as a strategic factor of urban development in a context of inner-city upgrading and the shaping of new upmarket housing schemes, among them river- and waterfronts, and the revitalization of industrial and railway areas (as examples see here Eckerd 2011; Banzhaf and McCormick 2007).

Literature on UGI, UES and NBS has dealt extensively with the positive effects of these concepts and their potential to contribute to related measures for urban green spaces and biodiversity as well as health and quality of life of urban dwellers (Kabisch 2015; Amoly et al. 2014; Krekel et al. 2016; Breuste et al. 2016). What is, however, less common is a critical analysis of the trade-offs, ambivalences and conflicts between the two spheres. It is either presumed that green developments/strategies etc. are per se beneficial for all inhabitants or social cohesion and well-being, or the social dimension is simply missing in many assessments of the impacts of green developments/UES. A debate on trade-offs, e.g. that there is an uneven distribution of and access to UES and that there are winners and losers of green strategies/projects is however slowly developing from the UES research field. Though, there is some anchor points from the perspective of socially-based urban research, e.g. the evolving debate on eco- or green gentrification. This debate is intrinsically concerned with the relationship between green strategies, general upgrading and displacement (Dooling 2009; Quastel 2009; Banzhaf and McCormick 2007: "clean up and clear out"). The debate focuses on revitalization of urban brownfield sites, creation of new neighbourhoods/new built housing projects and upgrading of neighbourhoods, or, as Curran and Hamilton (2012, p. 1027) state:

... environmental improvements result in the displacement of working-class residents as cleanup and reuse of undesirable land uses make a neighbourhood more attractive and drive up real estate prices.

The NYC High Line, a linear park built in Manhattan on an elevated section of a disused railroad spur 2006–2011, represents still the most prominent and well-known example for those processes but meanwhile, the debate has internationalized including research in Europe, China and other contexts. This debate, however, is rooted mainly in the social-science or critical urban context and is not addressed in the UGI or UES communities so far. It represents, however, also a kind of "add-on" of the general gentrification debate; I am in line here also with Bryson (2013, p. 578) who concludes that.

> … the urban natural environment plays an important and understudied role in shaping gentrification processes in contemporary cities.

The issues outlined above demonstrate the complexity of the "socio-environmental" nexus in cities and the many challenges confronting researchers. A general obstacle to overcome seems of the lack of two-sided/multi-disciplinary/inclusive debates; the debate on UES and UGI, for example, largely disregards any literature on gentrification and segregation in cities, while the debate on eco-gentrification is largely grounded in the social science perspective and does not differentiate the role and character of "green" or "eco". Eco-gentrification is an "add-on" and seems to be not well-connected to the main streams of debate. Furthermore, there is also a disregard for the political and economic realities of cities when dealing with socio-environmental issues and a lack of understanding of the variegated contexts of cities and path dependencies. In this regard, the NBS concept could be seen in a negative light, as it propagates generalized and de-politicized arguments. On the other hand, NBS could be seen to offer a new approach for discussing the role of nature in sustainable urban development, and therefore a concept that cannot be ignored.

In summary: When reflecting on the conceptual aims of NBS, there are some doubts whether the socio-environmental realities and premises of real-world cities (across Europe) are adequately addressed and considered. The following section will introduce and expand on five theses that should serve to enhance the discussion and qualify the debate on the potential opportunities and limitations of NBS for sustainable urbanization. The aim of the reflections is not simply to criticize NBS as such but also contribute to discussions about their benefits and their trade-offs. The underlying assumption is that if we aim to maximize the potential of NBS, must also be aware of their limits.

The following section outlines these ideas in more detail and explores ways of improving the discussion on NBS use and application.

13.3   Nature-Based Solutions to Support Socially Inclusive Development – Five Theses

Thesis 1
Nature-based solutions are not inherently socially inclusive or just. Under certain conditions, they might even work as triggers for segregation and displacement or be employed as deliberate strategies for selective upgrading.

It is a popular assumption that greening or green strategies generally, or per se, support social justice and social cohesion. This assumption is also made by the EC when introducing the approach of NBS. But the reality in cities is much more complicated, and any debate on the impacts of NBS should consider the full range of possible benefits and potential impacts a strategy, measure or development might have.

Cities are characterized by social and spatial differentiation. Goods and burdens are unevenly distributed over the territory of a city, urban dwellers have uneven access to goods and are unevenly exposed to burdens; living conditions of urban dwellers with respect to housing, services, green etc. differ considerably.

While green developments such as new parks, enlargement of green areas etc. by and large bring about benefits for many urban dwellers, it is more often the more well-off citizens who reap the most benefits, since they can afford to live in areas close to parks, water- or riverfronts or far away from traffic axes or industrial areas. In recent years, urban regeneration projects (of urban brownfield sites as mentioned by the EC report) and new housing projects (as part of re-urbanization or re-densification strategies) in particular, have employed green and blue qualities and related services as elements of upgrading. There are many examples of this from large cities across the European continent and Northern America; the emerging debate on eco- or green gentrification has addressed the relationship between environmental qualities and upgrading and displacement deliberately and directly. Examples are new housing areas including many waterfront and riverside developments as well as new, unconventional parks, in short, projects that privilege high-profile developments over the general provision of green and blue services for a larger population (Millington 2015; Banzhaf and Walsh 2006).

Another example is urban gardening projects, a type of NBS explicitly mentioned by the EC report on NBS, too (2015). Many urban gardening projects undoubtedly aim to follow sustainable and community-supportive goals. However, as Steinberg (2015, p. 20) concludes in her study on community gardens in NYC, LA and Vancouver,

> …under urban neoliberalism, sustainability is all too easily appropriated as a tool for economic, rather than environmental and human, development ….

Community gardens, even when they emerge as spaces for collective and political actualization against neoliberal forces, may, under certain conditions, contribute to the premises for a rising attractiveness of a residential area and thus – in most cases involuntarily – help to initiate or intensify upgrading and displacement in the affected areas (Baier 2011, pp.184–85).

As a consequence, urban areas that are marketed as eco-friendly and green (and/ or blue) become less affordable for lower income groups (see e.g. Kramer 2013; Eckerd 2011). Smart or green growth thus might go hand in hand with social exclusion or even displacement ("clearing up and clearing out"). To be here: The upgrade in itself is not the problem but rather the oftentimes selective character and the uneven distribution of the resulting benefits (high-quality green housing, access to high-quality green space) and burdens (displacement, exclusion from high-end green spaces) or, to put it differently, the lack of moderately priced or social housing within those development schemes. NBS cannot resolve the problem of socio-spatial inequality in cities; under circumstances, they can have very positive co-effects for poorer dwellers, too (Curran and Hamilton 2012; McKendry and Janos 2015). But much depends on a deliberate consideration of existing inequalities and the potential unequal chances of gaining access to the benefits of green projects or policies. New governance modes and larger-scale participation as demanded by Kabisch et al. (2016) might be a step into the right direction to overcome this challenge, although the political and power aspect that is inherent within inequality issues should deserves more attention.

Thesis 2
In order to meet the ambition of combined effects of NBS, existing trade-offs between ecological 'solutions' and their social environment, embedding and impacts have to be seriously examined and discussed.
Thesis 1 dealt with the myth of "unconditionally good green solutions". The second thesis sets existing trade-offs into the focus. As Jane Jacobs stated for Philadelphia parks in 1961 (!), green areas may contribute to improvement in good areas and to further degradation in bad areas. Subsequently, NBS should not be looked at in an isolated but in an integrated manner. For a balanced view, existing trade-offs or even conflicts between the impacts of NBS for urban space, life and people also have to be considered. This would be a topic for a whole paper on its own; however, some examples to illustrate this issue are outlined below.

Green developments, in many cases, contribute to an improved quality of life for urban dwellers. There are, however, also cases where such interventions e.g. an improved access to green, may not increase social coherence but instead go hand in hand with rising prices and displacements. Thesis 1 above addressed incidences of displacement in US or European cities through top-end green housing projects. Here, although causal effects are hard to identify, one can see a link between the 'greening' and market-oriented upgrading. Existing literature has discussed possible solutions to avoid such trade-offs: One proposed option is to carefully analyze social and socio-spatial effects of new housing developments or improvements; the scholarly debate discusses several approaches here; some argue for the integration of social indicators in ex-ante measurements or monitoring systems (see Pearsall 2012 who proposes the inclusion of the vulnerability analysis and indicators into existing urban sustainability planning or environmental quality reviews). Other authors plead for "just green enough" policies that seek to maintain industrial character and working class neighbourhood while improving quality of life without upscale prestige developments or

> interrogate how urban sustainability can be used to open up a space for diversity and democracy in the neoliberal city. (Curran and Hamilton 2012, p. 1027)

Similar discussions have emerged in relation to the introduction of environmentally-friendly technologies such as thermal insulation of residential buildings which under some circumstances also may lead to increasing rents and displacement (Großmann et al. 2014).

The described trade-offs operate at different scales; so, it may happen that a positive effect on one scale (e.g. a new green space that contributes to quality of life for many inhabitants) may also result negative effect at another scale (rising housing costs in the direct environment of the new green space may lead to displacement of lower-income households or to exclusionary displacement). Trade-offs, moreover, might operate in very different ways: they may be deliberate ingredients of policies (see the employment of green for purposes of upgrading) or just appear as unintended side-effects of policies.

In order to not provide solutions including few and excluding many, the NBS discussion should at least include a debate on exiting or potential trade-offs between environmental and social developments. The focus should be on identifying these trade-offs and examining how they can be avoided or minimized or at least mitigated in a way that meets both the ecological and social dimensions of sustainability. In planning contexts, trade-offs should be considered in light of general goals of quality of life and equity.

Thesis 3
Nature-based solutions, as currently discussed, do not sufficiently address political and social power structures and embeddings.
One of the most critical shortcomings of the concept of NBS as introduced by the EC (2015) is the de-politicized way it is discussed. As a result, the approach runs the risk of shunning politics, de-linking sustainability from politics/power and, as a result, contributing to a post-political or de-politicized view of green developments. Solutions are discussed against the context of a 'peaceful and good will based lab', showing little understanding of the conflict-saturated realities of modern societies (see also Brand 2016: 25).

Policies in cities are characterized by the interplay of different actors in the field of policies, market and civic society. Decision-making is more or less democratic, top-down or bottom-up, inclusion and participation more or less organized and channeled. Strategies and the ability to assert own interests depend on political power, financial resources and ownership. The respective resources are, in most cases, unevenly distributed; different actors and their interests have very different chances to realize their wants and needs.

The main argument is here that the scientific and practical debate should analyze more explicitly the relations between NBS and their social and political contexts. There are some debates that have dealt with the link between environment-based urban development and the political or power sphere which may deserve some attention in the NBS debate.

In the debate on green gentrification, some scholars have examined the post-political character of green upgrading, stating that it is

> … becoming a mode of post-political governance that shuns politics and de-links sustainability from justice. Thereby, it disables meaningful resistance. (Checker 2011, p. 212)

Helpful for a more critical debate are approaches which, for example, analyze the political ecologies of urban regeneration as proposed by Quastel (2009, p. 694):

Political ecologies of gentrification involve tracing the powers of government planners, real estate developers, consumers, and social organizations as they act in relation to urban ecologies and discourses of the environment. Tracing the effects of such discourses on gentrification, and how gentrification utilizes such discourses, contributes to showing how environmental discourses and policies involve issues of distribution, power, and inequality.

In order to move the NBS discussion forward, political and power structures require further consideration. A study by Berbes-Blazquez et al. (2016) is one of very few publications which look at the relationship between ES provision and social power relations. It concludes that there is a need to analyze how power relations underpin policymaking and governance that also determine access, use and management of ES (p. 138).

How could NBS be better embedded into political and power systems? How can they be used for a deliberately inclusive policy-making? A first step for all involved actors would be to recognize the political and power context in which any solution will be settled. A second step would be to apply participative and integrative modes of governance and participation to be sure that the varying wants and needs of the population are considered (see also Kabisch et al. 2016). A third step would be to screen proposed solutions (first and foremost those that involve market players and their interests) with respect to justice issues and to make sure that disadvantaged groups will be more likely to benefit from the implemented measures; here, the expertise of both scientists and practitioners, but also by civic society activists and NGOs is required. In this way, a solution would immediately contribute to justice and social coherence. Here, the creation of incentives for market actors to act in a socially sustainable way could be a step forward. In short: NBS or urban greening strategies are not "beyond political and power structures" but form part of those; for a comprehensive understanding of their impacts and functioning, they should be discussed against this background, and hence, be politicized or re-politicized (see also McKendry and Janos 2015, p. 56). If not, NBS run "the danger – unintentionally – for preparing the epistemic-political terrain for a greening of capitalism" (Brand 2016, p. 27) that might safeguard or create liveability in some places and for some people but will miss the general aim to provide solutions for structural problems that are rooted in the logics and mechanisms of the capitalist system and today's global inequalities and injustices created by it.

Thesis 4

The concept of nature-based solutions so far do not consider socio-spatial differences as well as different levels of in- and exclusion of people in cities, different everyday life routines as well as differing needs and wants of a heterogeneous and diversifying urban society.

The concept of NBS does not specify its addressees, especially when it comes to the "urban population" or "urban dwellers". This is not just problematic with respect to existing socio-spatial inequalities. It is also denies or ignores the existence of a wide range of perceptions, needs and wants concerning green and urban nature among urban dwellers. A great challenge for the debate on NBS is to adequately consider today's ever more differentiated and diverse urban societies. How does one cope with different, diverging or even opposing wants and needs? How can different views on urban nature and green developments be integrated? Who defines what benefits, qualities, good and healthy urban life are at all? How does one cope with the uneven integration of people into participative and opinion-building processes?

In the literature on UES, e.g., there are some studies that point to different perceptions, wants and needs of urban dwellers (e.g. Botzat et al. 2016; Gentin 2011) as well as their different opportunities to make use of or benefit from urban nature (Jones et al. 2009). There is a need to link the NBS or "green" debate with (social science based) debates on urban diversity, inequalities, capabilities, and (in)justice. Existing studies on UGI, UES and NBS have so far a bias with regard to the views and perceptions of middle-class inhabitants and the needs and wants of specific (environment-minded) lifestyle groups; this largely by-passes the views of various precarious population groups including the unemployed, poor families, alternative subcultures, migrants, other minorities (see Berbes-Blazques et al. 2016). The eco-gentrification debate, again, considers this fact, stating that ecological projects or solutions often designated for a bourgeois aesthetic (Dooling 2009) and that urban gardening and farms picture a niche or "glossy representations of sustainability" (Steinberg 2015). By and large, the views of low-income groups or people not taking part in participatory events tend to be less considered when discussing on green developments, the potentials of UES or NBS as solutions for a sustainable neighbourhood development.

If we really want to understand the potential of NBS for contributing to social cohesion, communication, empowerment, reduction of inequities etc., we must acknowledge the wide range of potentially opposing perceptions, wants and needs. The development of a more differentiated view and, as explored by Low (2013), the consideration of matters of distributional (Is there green space for all and a fair allocation?), procedural (Is there a fair access to green space?) and interactional justice (Does the green space allow for all to interact safely?) when creating and implementing NBS such as parks, playgrounds or other types of open space would be a large progress for the whole debate on NBS and would it make more than just a middle-class discussion. Solutions that directly address the needs of disadvantaged groups would offer an incentive for those people to get involved and would directly contribute to social inclusiveness. Particularly now, in a situation where most larger European cities are seeing new waves of in-migration and immigration including international migrants and (non-European) refugees, the question of how different parts of urban society look at urban nature and how this nature can be employed for different needs and wants, long-term or even temporary, seems to be crucial. It remains to be seen whether the NBS concept has the capacity to develop such a differentiated view.

Thesis 5
NBS as they are discussed run the risk to get overloaded regarding their capacities to respond to societal questions and to be easily transferable.

NBS may be a prerequisite for more sustainable urban development including social inclusiveness. But they are not enough in themselves for reaching (automatically) socially inclusive solutions. At the same time, it seems to be challenging to bring together environmental, social and economic priorities in a balanced way. This is reported e.g. by a study by McKendry and Janos (2015, p. 55) on neighbourhoods in Seattle and Chicago that apply urban environmentalism as a key aspect for their postindustrial development strategy and where, so far,

> ... efforts to promote economic growth, social and environmental equity, and environmental protection sit together uneasily.

Pretending this, it is crucial to get a clearer picture of (a) the opportunities and impacts that NBS may have in a more complex setting of urban regeneration (Rodriguez-Labajos and Martínez-Alier 2013) and (b) how different priorities such as economic, environmental and social ones might work together in a coadjutant and cross-fertilizing way. To put it differently: The debate on if and how greening strategies should be deliberately employed for urban restructuring and upgrading has to be complemented by a debate on to what extent urban decision-makers are ready to integrate ideas of social justice and inclusiveness into these greening strategies (see also McKendry and Janos 2015, p. 57). Clearly, NBS offer proper potential for promoting sustainability in a wider sense but they should be seen as part of a more complex-based solution, not as 'the solution' itself. NBS will, in most cases, rather selectively contribute to more sustainability and human well-being; they will be relevant for specific groups, activities, uses and contexts. Reflecting on their limitations as well as the potential allows for a much clearer picture of their outcomes/impacts.

Limits exist surely also with respect to transferability. 'Solutions' that work in one place might have very different effects elsewhere. Expensive greening strategies on former brownfield sites, for example, work only in contexts where the resulting costs can be paid by state and/or private actors. Projects related to urban gardening may be successful given the presence of a motivated and active local community, but will fail if there is apathy and disinterest. The majority of existing examples of successful NBS stem largely from rich cities or, at least, not from poor cities or cities in less developed regions of the world; this is not an incident. Poor cities, in many cases, have other priorities and often do not have the money that is needed for the realization or implementation of NBS; private actors, while easily to be convinced at wealthy places, will be more reluctant there. Therefore, it seems to be crucial to look at the context in which solutions should be implemented, first and foremost on political and power structures and ways of decision-making and in−/ exclusion of certain groups of inhabitants and/or actors. The same holds true for path-dependencies: as it is also underlined by Berbes-Blazquez et al. (2016, p.139) as a general matter for ES provision and management, research and practice debate have to more thoroughly recognize historical ties between social power relations and greening strategies as well as their employment into urban planning, housing construction and green space development.

13.4 In Lieu of a Conclusion …

The reflections presented in the five theses above demonstrated that the relationship between NBS and the social environment in cities is a complex one and must be considered with respect to their various synergies and trade-offs by both the academic debate and policy-makers and practitioners in cities. NBS offer, if discussed comprehensively and seriously, a potential for creating and shaping more sustainable cities, also in terms of social sustainability and inclusiveness. In order to meet this objective however, they should be seen as more than just tools, technologies and instruments to be applied within urban contexts. Although nature or nature-based solutions are not the only solution to societal challenges in cities, they may become part of a more strategic alliance of environmentally and socially healthy premises and aligned measures for shaping urban space, housing and infrastructure/amenities and aligned measures for shaping urban space, housing and infrastructure/amenities and can serve for more than a "narrow and insufficient corridor of ecological modernization" (Brand 2016, p. 26). To realize such a vision, NBS have to be improved as an approach, especially with respect to their social embeddings and the full picture of their impacts. The NBS concept may shift and change, but can be enriched and completed by new perspectives and ingredients. Even from a critical perspective, NBS in their current conception cannot be seen as incompatible or irreconcilable with social inclusiveness. A balanced view which incorporates different debates, lines of argument and includes a consideration of both pros and cons is however vital. Furthermore, a more holistic view, combining economic, social, environmental and technical elements and components of 'solutions', which does not pit the single dimensions against each other but sees them as ingredients of a larger and inseparable unity, is needed. There is much potential for creating a more comprehensive and in-depth crossover between social and environmental/ecological debates and it remains to be seen whether a convergence between strands of thought will take place and how these discussions will evolve.

References

Amoly E et al (2014) Green and blue spaces and behavioral development in Barcelona schoolchildren: the BREATHE project. Environ Health Perspect 122(12):1351–1358

Baier A (2011) Urbane Landwirtschaft und Stadtteilentwicklung. Die Nachbarschaftsgärten in Leipzig. In: Müller C (ed) Urban Gardening. Über die Rückkehr der Gärten in die Stadt. oekom, München, pp 173–189

Banzhaf HS, McCormick E (2007) Moving beyond cleanup: identifying the Crucibles of Environmental Gentrification. In: Andrew Young School of Policy Studies, research paper series, working paper 07/29. http://aysps.gsu.edu/publications/2007/index.htm (9.6.2016)

Banzhaf HS, Walsh RP (2006) Do people vote with their feet? An empirical test of environmental gentrification. Discussion paper RFF DP 06–10

Berbes-Blazquez M, Gonzales JA, Pasqual U (2016) Towards an ecosystem services approach that addresses social power relations. Curr Opin Environ Sustain 19:134–143

Botzat A, Fischer LK, Kowarik I (2016) Unexploited opportunities in understanding liveable and biodiverse cities. A review on urban biodiversity perception and valuation. Unpublished manuscript

Brand U (2016) "Trasformation" as the New Critical Orthodoxy. GAIA 25(1):23–27

Breuste J, Pauleit S, Haase D, Sauerwein M (2016) Stadtökosysteme. Springer, Heidelberg

Bryson J (2013) The nature of gentrification. Geogr Compass 7/8(2013):578–587

Checker M (2011) Wiped out by the "Greenwave": environmental gentrification and the paradoxical politics of urban sustainability. City Soc 23(2):210–229

Cohen-Shacham E, Walters G, Janzen C, Maginnis S (eds) (2016) Nature-based solutions to address global societal challenges. IUCN, Gland. xiii + 97pp

Curran W, Hamilton T (2012) Just green enough: contesting environmental gentrification in Greenpoint, Brooklyn. Local Environ 17(9):1027–1042

de Jong M et al (2015) Sustainable – smart – resilient – low carbon – eco – knowledge cities; making sense of a multitude of concepts promoting sustainable urbanization. J Clean Prod. http://dx.doi.org/10.1016/j.jclepro.2015.02.004

Dooling S (2009) Ecological gentrification: a research agenda exploring justice in the city. IJURR 33(3):621–639

Eckerd A (2011) Cleaning up without clearing out? A spatial assessment of environmental gentrification. Urban Aff Rev 47(1):31–59

Elmquist T et al (2013) Urbanization, biodiversity and ecosystem services: challenges and opportunities. Springer

European Commission (EC) (2015) Towards an EC research and innovation policy agenda of nature-based solutions and re-naturing cities. European Commission (EC), Brussels

Gentin S (2011) Outdoor recreation and ethnicity in Europe – a review. Urban For Urban Green 10:153–161

Großmann K et al (2014) Energetische Sanierung: Sozialräumliche Strukturen von Städten berücksichtigen. GAIA 23(4):309–312

Haase D, Haase A, Rink D (2014) Conceptualizing the nexus between urban shrinkage and ecosystem services. Landsc Urban Plan 132:159–169

Jacobs J (1961) The death and life of great American cities. Penguin Books, New York

Jones A, Brainard J, Batman IJ, Lovett AA (2009) Equity of access to public parks in Birmingham, England. Environ Res J 3:237–256

Kabisch N (2015) Ecosystem service implementation and governance challenges in urban green space planning – the case of Berlin, Germany. Land Use Policy 42:557–567

Kabisch N, Frantzeskaki N, Pauleit S, Naumann S, Davis M, Artmann M, Haase D, Knapp S, Korn H, Stadler J, Zaunberger K, Bonn A (2016) Nature-based solutions to climate change mitigation and adaptation in urban areas: perspectives on indicators, knowledge gaps, barriers, and opportunities for action. Ecol Soc 21(2):39. http://dx.doi.org/10.5751/ES-08373-210239

Kramer A (2013) Divergent affordability: Transit access and housing in North American cities. In: A thesis presented to the University of Waterloo in fulfillment of the thesis requirement for the degree of Doctor of Philosophy in Planning, Waterloo, ON, Canada

Krekel C, Kolbe J, Wüstemann H (2016) The greener, the happier? The effect of urban land use on residential well-being. Environ Econ 121:117–127

Low S (2013) Public space and diversity: distributive, procedural and interactional justice for parks. In: Young G, Stevenson D (eds) The Ashgate research companion to planning and culture. Ashgate Publishing, Surrey, pp 295–310

McKendry C, Janos N (2015) Greening the industrial city: equity, environment, and economic growth in Seattle and Chicago. Int Environ Agreements 15:45–60. doi:10.1007/s10784-014-9267-0

Millington N (2015) From urban scar to 'park in the sky': terrain vague, urban design, and the remaking of New York City's High Line Park. Environ Plan A 47:1–15

Pearsall H (2012) Moving out or moving in? Resilience to environmental gentrification in New York City. Local Environ 17(9):1013–1026

Potschin M, Kretsch C, Haines-Young R, Furman E, Berry P, Baró F (2015) Nature-based solutions. In: Potschin M, Jax K (eds) OpenNESS ecosystem service reference book. EC FP7 Grant Agreement no. 308428. Available via: www.openness-project.eu/library/reference-book

Quastel N (2009) The political ecologies of gentrification. Urban Geogr 30(7):694–725

Rodriguez-Labajos B, Martínez-Alier J (2013) The economics of ecosystems and biodiversity: recent instances for debate. Conserv Soc 11:326–342

Secretariat of the Convention on Biological Diversity (2012) Cities and biodiversity outlook. Montreal, 64 pp. https://www.cbd.int/doc/health/cbo-action-policy-en.pdf

Steinberg GW (2015) Cultivating resistance? Urban sustainability, neoliberalism, and community gardens. Thesis submitted in partial fulfillment of the requirements for the Degree of Master of Arts, SIMON FRASER UNIVERSITY Spring 2015

While A, Jonas AEG, Gibbs D (2004) The environment and the entrepreneurial city: searching for the urban 'sustainability fix' in Manchester and Leeds. IJURR 28(3):549–569

Chapter 14
Urban Gardens as Multifunctional Nature-Based Solutions for Societal Goals in a Changing Climate

Ines Cabral, Sandra Costa, Ulrike Weiland, and Aletta Bonn

Abstract Urban gardens can contribute to climate mitigation and adaptation through a range of provisioning, regulating, and cultural ecosystem services as multifunctional nature-based solutions in a city. Besides providing food, urban gardens contribute to water regulation through unsealed soils, to improved air circulation and cooling through plant transpiration and shading, offering microclimate oases to many users, such as gardeners, visitors, and immediate neighbors. In combination with other green and blue infrastructures, urban gardens can thereby help to mitigate and adapt to the urban heat island effect. They also provide important habitat for

I. Cabral (✉)
German Centre for Integrative Biodiversity Research (iDiv) Halle-Jena-Leipzig,
Deutscher Platz 5e, 04103 Leipzig, Germany

Martin Luther Universität Halle Wittenberg, Institute of Biology,
Wolfgang-Langenbeck-Str. 4, 06120 Halle, Germany

Department of Ecosystem Services, Helmholtz-Centre for Environmental Research – UFZ,
Permoser Str. 15, 04318 Leipzig, Germany
e-mail: ines.cabral@idiv.de

S. Costa
Birmingham City University,
The Parkside Building, 5 Cardigan Street, B4 7BD, Birmingham, UK
e-mail: sandra.costa@bcu.ac.uk

U. Weiland
University of Leipzig, Institute of Geography, Johannisallee 19a, 04103 Leipzig, Germany
e-mail: uweiland@uni-leipzig.de

A. Bonn
Department of Ecosystem Services, Helmholtz-Centre for Environmental Research – UFZ,
Permoster Str. 15, 04318 Leipzig, Germany

Friedrich-Schiller-Universität Jena, Institute of Ecology,
Dornburger Str. 159, 07743 Jena, Germany

German Centre for Integrative Biodiversity Research (iDiv) Halle-Jena-Leipzig,
Deutscher Platz 5e, 04103 Leipzig, Germany
e-mail: aletta.bonn@idiv.de

© The Author(s) 2017
N. Kabisch et al. (eds.), *Nature-based Solutions to Climate Change Adaptation in Urban Areas*, Theory and Practice of Urban Sustainability Transitions,
DOI 10.1007/978-3-319-56091-5_14

237

wildlife and genetic diversity. Urban gardens create opportunities for leisure and recreation and thereby promote health and well-being, as well as a sense of place, cultural identity, and social cohesion – important factors for societies to adapt to change. Exploring case studies across Europe, we discuss differences between garden types and their contribution to achieving sustainability goals for city communities.

Keywords Urban gardens • Community gardens • Allotment gardens • Climate change adaptation and mitigation • Ecosystem services • Climate change and food security • Nature-based solutions

14.1   Introduction

Climate change involves complex environmental, political, and socio economic interactions that cannot be addressed in isolation of holistic societal and human well-being concerns, including social cohesion, equity, and social justice interests (IPPC 2001; Adger and Barnett 2009; Haase, Chap. 13, Kabisch et al., Chap. 12 and Braubach et al., Chap. 11, this volume). With a changing climate, policy makers and advisors face challenges on how to mitigate and adapt while taking into consideration societal goals. One of the ways to deal with societal goals in the context of climate change is through nature-based solutions, including a wide range of green and blue infrastructure measures.

This chapter focuses on urban gardens, particularly allotments and community gardens, as one type of green infrastructure. Allotment gardens are mostly larger estates divided into plots that are allocated under rental payments to a single person or a family for non commercial cultivation of fruits, vegetables, and ornamental plants and recreational purposes. They are normally ruled and managed by local authorities, associations, or private or public organizations. In contrast, community gardens are single pieces of land that are gardened and managed collectively by a group of people. Community gardens, may have a permanent or temporary character, and are often characterized by informal claiming of urban voids with the purpose of local community development (Adams et al. 2013).

In this chapter, we explore how allotment and community gardens can serve as multifunctional nature-based solutions to achieve both climate-related and societal goals. Throughout the chapter, case studies illustrate how ecosystem services are provided by urban gardens in cities such as Lisbon, Leipzig, Manchester, and Poznan, all set in different socio-ecological contexts across Europe from northern to southern regions and from eastern to western regions. We first outline a brief history of urban gardens, then reflect on similarities and differences of the case studies, and discuss the multi functional dimensions of urban gardens from provision of genetic diversity, places for recreation promoting human well-being to drivers for social cohesion in a changing climate. In conclusion, we present prospects for the future of urban gardens in Europe.

14.2 History of Urban Gardens

Urban gardens have played an important role in cities ever since cities exist (Bell and Fox-Kämper 2016). The first European allotment garden was located in Kappeln, Germany, and dates back to 1814. The allotment garden movement became prominent after 1861, when the first allotment association in Leipzig was created by the *Schrebergarten movement,* spreading subsequently across central Europe in the next century.

Historically, the primary goal of allotment gardens was to mitigate poverty among factory workers during the industrial revolution by providing food. Another specific goal, at least by Dr. Schreber, was to provide opportunities for recreation, especially for children. As such, societal goals are imminently entwined with allotments. In fact, most allotments in Central and Northern Europe provide playgrounds in the common areas and often have a clubhouse for cultural events (Cabral and Weiland 2016), promoting recreation and social cohesion. Today, the *Office International du Coin de Terre et des Jardins famil-iaux* (http://jardins-familiaux.org/) accounts for two million allotment gardens across Europe.

During the First and Second World Wars, England, (Speak et al. 2015), Germany and Sweden (Barthel et al. 2010) relied on allotment gardens to provide 10% of each country's food needs. During the 1970s oil crisis, European allot-ments proved to be equally important, mitigating unemployment and austerity and acting as a reliable way of producing food (Adam-Bradford and Veenhuizen 2016). In fact, the recent economic and financial recession in Southern European countries has caused a resurgence of urban gardens, e.g., in Spain during the 2000s (Camps-Calvet et al. 2015), Portugal after 2000, and later in Greece in 2010, to address issues of food security as well as climate change through a greater emphasis on self-sufficiency. The rise of urban gardens and other urban agriculture initiatives is also associated with a desire to reduce food miles, another avenue to contribute to reduced carbon emissions and climate mitigation in cities. The renewed interest in urban gardens prompted a European research action to assess challenges and opportunities posed by these spaces (COST action TU1201; Bell et al. 2016; Box 14.1).

Box 14.1: European Urban Gardens

The COST Action TU1201 (http://www.urbanallotments.eu) is a research network involving 31 European countries and New Zealand as a partner country, bringing together experts from various fields established in academia, municipalities, and urban gardening associations. To understand the contribution of allotment and community gardens in Europe in achieving urban sustainability with regard to social and ecological aspects, and economic resilience in a changing future, and their role in urban design and urban policy, the research consortium explored extensive case studies (Bell et al. 2016).

Allotment and community gardens vary across Europe in terms of historical, cultural, political, and planning contexts from well-established institutions in Northern European countries to rather recent developments in Southern Europe. Access to such spaces may be prevented by pressures for urban development, the lack of maintenance or investment in existing and new sites, as well as difficulties in contractual arrangements. Consequently, new community urban gardens are a response to the inadequacy of conventional schemes, imbalance between demand and offer, and slow adaptation of the planning systems to the recent socio economic and political crises and changes (Caputo et al. 2016). They are also an expression of group manifestations, green activism, or similar socio political engagements (Shepard 2013; Hardman and Larkham 2014) and entrepreneurship. Situated in specific urban and social contexts (Ioannou et al. 2016), emergent community gardens tend to be absent of rigid capacity, allow for temporary occupation and mobility, and enable new urban socio-spatial experimentations and practices, thus transcending the conventional concept of allotment (Caputo et al. 2016).

The diversity of urban gardens therefore offers significant opportunities to meet different societal and urban challenges through providing ecosystem services in proximity with neighborhoods as part of the cities' green infrastructure in a changing climate.

Spatially, allotment gardens are distributed across all Europe with a focus in Central and Northern Europe (Bell et al. 2016). Allotments are commonly found along train tracks, water canals, or adjacent to previous industrial areas, as these were formerly marginal lands (now often protected through city planning laws for their noise and flood risk). Today, European allotments are under threat in several cities (such as Warsaw, Poznan, Basel, Riga, and Vienna) due to real estate pressure (Costa et al. 2016). In Southern Europe, however, they have been growing in number and size as part of several city planning strategies.

Community gardens provide another form of gardening spreading in Europe for the last 20 years as a complement to allotment gardens. Community gardens are filling urban voids, by squatting within brownfields in large- and medium-sized cities. These are usually initiatives of younger gardeners, who use these spaces mostly for social purposes, such as recreation, education, physical and mental health services (Genter et al. 2015; Wood et al. 2015), and thus for social cohesion.

The history of community gardens is quite recent and followed a New York trend during the financial crisis of 1970. In Europe, this form of gardening has been intertwined with the guerrilla gardening and transition town movements. While this form of gardening is at times based on illegal occupation or squatting, in some cities it has become legal for interim use. There are several examples of municipalities and local governments that encourage the creation of community gardens (also known as guerrilla, intercultural, or neighborhood gardens). Some of them are Barcelona (Langemeyer 2015), Milan (Silvestri 2014), Athens (Anthopoulou 2012), Berlin (Appel et al. 2011), and Leipzig (Weiland 2015). These gardens represent a new social movement aiming to augment resilience in socio-ecological systems (Ioannou et al. 2016).

14.3 Urban Gardens Across Europe: The Importance of Socio economic Context and Urban Planning

The case studies in this chapter illustrate the main trends in Europe with regard to urban gardens as nature-based solution for urban integration, municipal policy, and ecosystem services, with Leipzig and Lisbon representing northern and southern geophysical and urban planning-related differences. Leipzig has a long tradition of allotment gardens (Box 14.2) and a large number of successful and legal community gardens, the oldest being 20 years old. These community gardens promote environmental education as well as for social cohesion (Box 14.3) and therefore are important contributing factors in cities for climate adaptation. Lisbon, on the other hand, has a very recent, yet successful, history of allotment gardens (Box 14.4). It has, however, few community gardens, due to restrictive policies by the city (e.g., former community garden Horta do Monte, Graça). While in Leipzig accessibility is restricted in allotment sites located on private land and this law is enforced by the allotment club (especially during winter time), allotments located on municipal land are enforced to open their common areas, i.e., playgrounds and restaurants, to the public, during spring and summer time. In Lisbon, the strategy for promoting recreation in allotments has led to the integration of these spaces into existing urban parks to create ecological corridors (Cabral and Weiland 2016), although this has incurred some privacy loss and some acts of vandalism.

14.4 Urban Gardens as Banks of Genetic Diversity

Urban gardens can promote habitat for diverse plant species, both ornamental species and cultivated species, while these can sometimes include non native and invasive species (Smith et al. 2006; Bigirimana et al. 2012; Jaganmohan et al. 2012) which may spread with a changing climate. However, recent studies have shown that allotments in Poznan and Manchester can host many native species, especially when many plots are abandoned (Speak et al. 2015; Borysiak et al. 2017, Box 14.5), a result we could also show for Leipzig (Cabral et al. 2017).

Box 14.2: Urban Gardens in Leipzig: A Long History of Allotment Gardening

The city of Leipzig has a long-lasting tradition of allotment gardening initiated by the Schreber movement in the late nineteenth century. Nowadays, there are 270 allotment sites (Fig. 14.1) allocated in 1229 ha in a total of 39,000 plots, which equals to around 4% of the city area. This represents one of the highest densities of urban gardens among European cities (23m² per citizen). In addition to the significant areas of riparian forest and many domestic gardens, the allotments contribute to microclimate regulation (Strohbach and Haase 2012). Due to strict allotment codes, the distribution of large trees is limited to communal spaces in allotments, and thereby the garden's contribution to climate mitigation measures is limited (Cabral et al. 2017). Nevertheless, the large area of unsealed surfaces (Fig. 14.2) allows for local cooling through evapotranspiration and run off regulation as contribution to climate adaptation goals. While up to 1989 food provision was a main goal and publicly promoted to combat food shortages, the emphasis of gardening has now shifted toward recreational services, also with the development of community gardens (Cabral and Weiland 2016; Box 14.3). Therefore, urban gardens are regarded as an important asset for the city with plans to energize these spaces by interlinking them with urban parks. Allotment competitions and an exhibition by the Leipzig Botanical Garden are used to promote gardening techniques that are beneficial to biodiversity and ecosystem services and to raise awareness of sustainability among gardeners and the public (www.gartenwerkstatt-leipzig.de).

Fig. 14.1 Map of Leipzig allotment gardens (in *black*) and community gardens (numbered 1-8) (Credit: Roland Kraemer, UFZ/iDiv)

Box 14.2 (continued)

Fig. 14.2 Kleingarten
Naturfreunde (Credit:
Cabral)

**Box 14.3: Community Garden Initiatives in Leipzig: Interplay of Spatial
Policy and Cultural Context as Impulse for Social Innovation**

Leipzig's community garden initiatives (CGIs) represent a broad variety as in
many European cities, such as intercultural gardens, community-supported
gardens, a start-up business (Fig. 14.3), and gardens run by environmental
NGOs for environmental education. The initiatives are influenced by both
availability of space and socio-cultural contexts. With regard to spatial policy,
Leipzig suffered deindustrialization and lost almost one third of its population
until 2010 (Haase et al. 2014), and the city had to cope with many inner urban
voids that can now be used by CGIs for interim use. In terms of cultural context, many community garden actors belong to younger urbanites aiming at
sustainable urban development, urban transition, and/or political change –
while differing in details and rigidity. Their common background led more
than 120 CGIs nationwide to sign an Urban Gardening Manifest. Therein,
urban gardens are defined as common goods counteracting privatization and
commercialization, as spaces for cultural, social, and cultural variety of neigh-
borly cooperation, as bridges between cities and rural agriculture, and as
places of environmental education and common learning as well as places of
silence and endowed time (Müller and Überall 2014). Importantly, urban gardens are explicitly acknowledged in the manifest for their contribution to a
better climate, quality of life, and environmental justice.

Box 14.3 (continued)

Fig. 14.3 Community gardens: Annalinde (Credit: Cabral) and community-supported agriculture, start-up "ernte mich" (Credit: Cabral and Weiland)

This interplay of space availability and cultural movement therefore provides opportunities for social innovation to grow and as an experimental space to explore ecological alternatives (Müller and Überall 2014) under global change.

Box 14.4: Urban Gardens in Lisbon: Providing Food and Occupational Activities

In 2011, the Greenway Network Strategy for Lisbon established the construction of several new allotments. Most of these allotments (Fig. 14.4) are located in poorly maintained municipal green areas, thus assuring its maintenance at a lower cost for the city aiming to support deprived communities. Presently, Lisbon's gardening area accounts for 84 ha and represents 1.5 m^2 per capita or 1% of the total city area (Cabral and Weiland 2016). As such allotment gardens alone do not exert a significant contribution to the city climate mitigation and adaptation, their main function is to serve food provision, e.g., for immigrants and struggling families, as well as cultural services, including recreation and a sense of place, e.g., for retired and unemployed people, and education opportunities, e.g., for school children (Cabral and Weiland 2016). The strategy of connecting these spaces to existing parks and therefore maintaining the existing permeable areas also contributes to improved green infrastructures by establishing ecological corridors in the city.

Box 14.4 (continued)

Fig. 14.4   Vale de Chelas Horticultural park (credit: Costa)

Box 14.5: Urban Gardens in Manchester and Poznan: Provision of Habitats for Biodiversity and Ecosystem Services

Speak et al. (2015) compared the biodiversity and ecosystem services provided by allotments in Poznan (Poland) and Manchester (UK). Together with another study by Borysiak et al. (2017), on 11 representative allotment estates in Poznan, 357 spontaneous species were found, among which 72% were native. This is probably due to the fact that some allotment estates can host a high number of spontaneous plants which is correlated with a high number of abandoned plots. Thus the authors could show that allotments make a significant contribution to biodiversity conservation unlike many previous studies have documented.

The authors also found remarkable differences in ecosystem service provision between sites and between countries. While in Poznan the plots are large and host large trees, which contribute to (micro)climate regulation, Manchester allotments have fewer trees since they allocate more space to food cultivation, while recreational use also differs, responding to societal preferences and possibly socio economic needs. Their study also shows that urban gardens can serve a multitude of purposes and may – especially at the individual gardener level – serve as adaptation potential to respond to changing climate and urbanization pressure.

Since many landrace cultivars in Europe, i.e., traditional crop varieties adapted to a specific geographical area, are threatened by extinction as they have often been substituted by commercial strains, some authors argue that the maintenance of these landraces in urban gardens can convert these spaces into a gene banks (Barthel et al. 2010, 2013). In this way, urban gardens can make a significant contribution to climate change adaptation, as these varieties are more likely to adapt to changing and extreme conditions and may therefore contribute to resilience in an urban environment.

14.5   Urban Gardens as Places for Promotion of Health and Human Well-Being

Allotment gardens can make significant contributions to human health and well-being, with main benefits arising from the contact with nature and the social opportunities provided by such places, which thus provide a stress-relieving refuge, enable self-development, and contribute to healthier lifestyles (Genter et al. 2015). There is a general understanding of the perceived benefits for health promotion arising from engaging in community garden activities (Armstrong 2000). Hawkins et al. (2011) found lower levels of stress among gardeners, ages 50–88, than their peers who performed indoor exercises. Part of the benefits seem to arise from the activity of gardening itself. Van Den Berg and Custers (2011) found that gardening can promote relief from acute stress. In their study, allotment gardeners were submitted to a stressful Stroop task and then assigned to 30 min of outdoor gardening while measuring salivary cortisol levels and self-reported mood; decrease in salivary cortisol and thus levels of stress and positive moods were fully restored after gardening. Another study in the UK, comparing the mental well-being of allotment gardeners with non-gardeners, found that the allotment gardeners' self-esteem, mood, and general health were significantly improved as a result of just one allotment session (Wood et al. 2015). These studies begin to demonstrate not only the perceived impact of allotments and community gardens in health and well-being but also its actual impact. Understanding the significance of the contribution of urban gardens as a nature-based solution toward well-being is opportune in the current context, where there is an increasing political interest in public health, in well-being agendas, and in the impacts of the environment on mental and physical health.

14.6   Urban Gardens as Drivers for Social Cohesion

One of the characteristics that can be studied regarding allotments and community gardens is their social relevance for the people involved but also for the surrounding communities. Adaptation to change (e.g., to climate change) is always a challenge which depends on individual and community knowledge, attitudes to risk, and cultural predisposition (Adger et al. 2009), among other factors. Urban gardens can help build social capacity to implement change (Adger 2003; Smit and Wandel 2006) by providing environmental education, intergenerational learning, and understanding of natural processes, cycles, and processes of climate change itself. According to Barthel and Isendahl (2013), they have the capacity to respond to needs of socio-ecological resilience. Firth et al. (2011) have shown that the social benefits associated with community gardens are broad and include increasing social cohesion and the ability of sharing common values, aims and behaviors, social support, and social connections developed through social bonds and networks. There are many examples of collective action (Adger 2003) around the creation of community gardens to increase resilience in communities. Thus, these have the ability to increase resilience and adaptive capacity of social structures (Folke et al. 2002) in times of change and crises (e.g., producing fresh food for self-consumption and for local communities) and to provide a place for socialization.

While the importance of urban gardening is recognized as a driver for community building and social cohesion, the ways in which it happens may not always be immediate and direct (Rodrigues et al. in review). In the early stages of development, such benefits do not necessarily lengthen beyond the site and the people involved to the realms of the neighborhood or the city (Veen 2015; Firth et al. 2011). Moreover, the integration of community gardens within existing social structures and in bringing together people from different socio economic backgrounds may face some challenges (Veen 2015). However, social relations in gardens which have a main focus on the social benefits of gardening have larger effects beyond the garden itself (Veen 2015).

In contexts of migrant and refugee communities flows, urban gardens have the potential to integrate diverse ethnic and social groups, and even though there might be vulnerable issues to solve through the process of engagement development, there are good examples of how these promote social integration and cohesion. For example, community gardens in Glasgow cultivate collective practices that promote an "equality-of-participation in place and community making," accentuating relations between people, organizational processes, and institutions and the opportunity to develop and use public spaces in Glasgow (Crossan et al. 2016). Community gardens enable better social networks and organizational capacity within the communities they are located, but this appears to be further enhanced in lower income and minority neighborhoods (Armstrong 2000). Due to climate change, natural catastrophes can lead to migration movements more often. Since urban gardens can promote integration in societies and promote knowledge exchange and cultural tolerance, they can ultimately avoid social instability during crisis and distressful situations and serve as models and tools to adapt to climate change.

14.7   Future Opportunities for Urban Gardening

Allotment and community gardens are not the only forms of urban gardening. In fact, a new hybrid form of gardening, called zero-acreage farming (Zfarm), is growing in several cities around the world. It includes rooftop gardens (Fig. 14.5), rooftop greenhouses, indoor farms (Fig. 14.6), windowsill gardens, and balcony gardens. While rooftop gardens have proved to be efficient in Southern European cities like Bologna (Orsini et al. 2015) or Barcelona and Milan (Sanyé-Mengual et al. 2015), roof greenhouses on the other hand are more suitable for Northern European cities, by allowing an extension of the growing season in cold regions. On a smaller scale, gardening balconies with edible species can also provide local food for households in any latitude.

Fig. 14.5 Rooftop garden in Bologna (Credit: Orsini)

Fig. 14.6 Indoor farm at Pasona's Tokyo building (Credit: De Zeen)

But gardening in buildings can do more than just provide food. Roof gardens, for example, can also contribute to climate and water regulating services by offsetting the effects of hard surfaces and impervious surfaces caused by buildings. In fact, shading a roof surface can lower its temperature, and soil beds can absorb rainwater, reducing surface run off (see also Enzi et al., Chap. 10, this volume). Additionally, roof gardens' soil beds reinforce the thermal insulation of a building rooftop, leading to lower energy consumption and thus lower carbon emissions. There is also another advantage in gardening in an elevated site: a rooftop has potentially more solar exposure within a city than a ground located garden, thus enhancing plant growth.

From a social perspective, gardening on a common area such as a residential building rooftop, if shared by tenants, provides an opportunity for meeting neighbors and sharing knowledge on gardening. From an ecological perspective, rooftops can attract pollinators and birds and create a more biodiverse ecosystem in the city skyline (Orsini et al. 2014).

Within public spaces such as parks, and pedestrian streets, promoting edible species can become a source of food with financial advantages for both the municipality and its citizens. In fact, some cities have implemented this goal successfully, such as in the case of Berlin and Leipzig (*Essbare Stadt* project) or Hague (*Urbania Hoeve*) where both municipalities provide maps containing edible trees for free harvesting. At an even bigger scale, the city of Munich has transformed in 2005 a former airport into an exhibition area, called *Die Plantage,* which contains mostly edible species (Philips 2013). In a remarkable case, the city of Todmorden (Box 14.6) has seen a guerrilla gardening movement turn the whole city into an edible garden, as citizens

---

Box 14.6: The Incredible Edible Project: A Community Movement Turned into an International Reference

The Incredible Edible project is a community gardening initiative which was started by a small local group in Todmorden, England (http://www.incredible-edible-todmorden.co.uk). The town has a population of about 17,000 and was once an important textile production town, while it now suffers from one of Britain's highest unemployment rates. In 2008, the community decided to restore the local food production system and started growing and promoting local food, aiming to change behavior toward the environment and to create a more resilient city (Paull 2013). This is realized by growing food to share in public spaces, running networking and training workshops, and organizing events to communicate and exchange knowledge about gardening (Fig. 14.7). The success of the initiative has brought nearly 120 Incredible Edible groups in the UK and more than 700 worldwide to the network. Thus, the Incredible Edible Network has clearly a societal agenda (Adams et al. 2013) and has begun to directly influence decision-makers at national level such as urbanists by being shortlisted for urbanism design competitions (Paull 2013) and showcasing the potential of nature-based solutions to environmental and societal goals.

Box 14.6 (continued)

Fig. 14.7 The green route map and a vegetable patch along the canal in Todmorden (Credit: Paull 2013)

took up the task of planting edible species in every empty space in the city. Todmorden has become a popular reference after establishing the incredible edible network which extends now to hundreds of cities in Europe and around the world (Paull 2013).

14.8   Conclusions

Allotments and community gardens are not the largest contributors for climate mitigation at the city level, as parks and private gardens can contribute more due to larger amounts of biomass in large trees. Their role in microclimate regulation is heightened when gardens are interlinked with other green spaces, thus enhancing their performance. Additionally, they play important roles for water regulation in providing unsealed surfaces for regulating run off and providing open space to escape the urban heat island. They also provide habitat for native flora and may serve as gene banks for adapted land cultivars. Furthermore, they are important societal meeting places that can contribute to recreation, health, and well-being as well as social cohesion.

New urban gardens, or community gardens, are a response to the current needs, not only for food provision but also for socio political expressions and manifestations.

Grassroots movements have proven that gardening is an important cultural contribution for future generations and that the future of cities needs to address this demand by providing more space and more autonomy for better governance. The densification of urban environments necessitates the use of innovative nature-based solutions that respond and adapt to physical, socio economic challenges and embrace new and creative typologies of spaces and approaches of urban gardening that can be undertaken, for example, in existing or new developed buildings. These are all important factors to facilitate climate change adaptation in urban communities.

References

Adam-Bradford A, van Veenhuizen R (2016) Role of urban agriculture in disasters and emergencies. In: de Zeeuw H, Drechsel P (eds) Cities and agriculture: developing resilient urban food systems. Routledge, Oxon, pp 387–410

Adams D, Scott AJ, Hardman M (2013) Guerrilla warfare in the planning system: revolution or progress towards sustainability? Geogr Ann 95(4):375–387

Adger WN (2003) Social capital, collective action, and adaptation to climate change. Econ Geogr 79:387–404

Adger WN, Barnett J (2009) Four reasons for concern about adaptation to climate change. Environ Plan A 41:2800–2805

Adger WN, Dessai S, Goulden M (2009) Are there social limits to adaptation to climate change. Climat Change 93(3):335–354

Anthopoulou T (2012) Urban agriculture. Social inclusion and sustainable city. Case study of two municipal gardens in northern Greece. Panteion University, Athens. (research report in greek)

Appel I, Grebe C, Spitthoever M (2011) Aktuelle Gartenitinitiaven. Kleingärten und neue Gärten in deutschen Großstädten. [Recent gardening initiatives: allotments and new gardens in big german cities]. Kassel University Press, Kassel

Armstrong D (2000) A survey of community gardens in upstate New York: implications for health promotion and community development. Health & Place 6:319–327

Barthel S, Folke C, Colding J (2010) Socio-ecological memory in urban gardens -retaining the capacity for management of ecosystem services. Glob Environ Chang 20(2):255–265

Barthel S, Isendahl C (2013) Urban gardens, agriculture and water management: sources of resilience for long term food security in cities. Ecol Econ 86(224):234

Bell S, Fox-Kämper R (2016) A history of urban gardens in Europe. In: Bell S et al (eds) Urban allotment gardens in Europe. Routledge, London/New York, pp 8–32

Bell S, Fox-Kämper R, Keshavarz N, Benson M, Caputo S, Noori S, Voigt A (2016) Urban allotment gardens in Europe. Routledge, London/New York

Bigirimana J, Bogaert J, De Cannière C, Bigendako M-J, Parmentier I (2012) Domestic garden plant diversity in Bujumbura, Burundi: role of the socio-economical status of the neighborhood and alien species invasion risk. Landsc Urban Plan 107:118–126

Borysiak J, Mizgajski A, Speak A (2017) Floral biodiversity of allotment gardens and its contribution to urban green infrastructure. Urban Ecosys. 20:323–335

Cabral I, Weiland U (2016) Urban gardening in Lisbon and Leipzig: a comparative study on governance. TU1201. In: Tappert, S., Drilling, M. (eds) (2016): Growing in Cities. Interdisciplinary Perspectives on Urban Gardening. Conference Proceedings. Basel: University of Applied Sciences: 66-79. Available on http://www.sozialestadtentwicklung.ch/tagungen/growing_cities.pdf

Cabral I, Keim J, Engelmann R, Kraemer R, Siebert J, Bonn A (2017) Assessing ecosystem services of allotments and community gardens: a Leipzig, Germany case study. Urban Urban Green 23(4):44–53

Camps-Calvet M, Langemeyer J, Calvet-Mir L, Gómez-Baggethun E, March H (2015) Sowing resilience and contestation in times of crises: the case of urban gardening movements in Barcelona. Partecipazione e Conflitto 8(2):417–442

Caputo S, Schwab E, Tsiambaos K, Benson M, Bonnavaud H, Demircan N, Pourias J (2016) Emergent approaches to urban gardening. In: Bell S et al (eds) Urban allotment gardens in Europe. Routledge, London/New York, pp 229–253

Costa S, Fox-Kamper R, Good R, Sentic I (2016) The position of urban allotment gardens within the urban fabric. In: Bell S et al (eds) Urban allotment gardens in Europe. Routledge, London/New York, pp 201–228

Crossan J, Cumbers A, McMaster R, Shaw D (2016) Contesting neoliberal urbanism in Glasgow's community gardens: the practice of DIY citizenship. Antipode 48:937–955

Firth C, Maye D, Pearson D (2011) Developing "community" in community gardens. Local Environ 16(6):555–568

Folke C, Carpenter S, Emqvist T, Gunderson L, Holling CS, Walker B (2002) Resilience and sustainable development: building adaptive capacity in a world of transformations. Ambio 31:437–440

Genter C, Roberts A, Richardson J, Sheaff M (2015) The contribution of allotment gardening to health and wellbeing: a systematic review of the literature. Br J Occup Ther 78(10):593–605

Haase A, Rink D, Grossmann K, Bernt M, Mykhenko V (2014) Conceptualizing urban shrinkage. Environ Plann A46 7:1519–1534

Hardman M, Larkham PJ (2014) Informal urban agriculture: the secret lives of Guerrilla gardeners. Springer, Cham

Hawkins JL, Thirlaway KJ, Back K, Clayton DA (2011) Allotment gardening and other leisure activities for stress reduction and healthy ageing. Hort Technol 21(5):577–585

Ioannou B, Morán N, Sondermann M, Certomà C, Hardman M (2016) Grassroots gardening movements. In: Bell S et al (eds) Urban allotment gardens in Europe. Routledge, London/New York, pp 62–90

IPCC (2001) In: Metz B, Davidson O, Swart R, Pan J (eds) Climate Change 2001: mitigation – contribution of Working Group III to the third assessment report of the Intergovernmental Panel on Climate Change (IPCC). Cambridge University Press, Cambridge

Jaganmohan M, Vailshery LS, Gopal D, Nagendra H (2012) Plant diversity and distribution in urban domestic gardens and apartments in Bangalore, India. Urban Ecosyst 15:911–925

Langemeyer J (2015) Urban ecosystem services: the value of green spaces in cities. PhD thesis, Stockholm University

Müller C, Überall D (2014) Urban gardening Manifest. Die Stadt ist unser Garten-Manifest. Available at http://www.urbangardeningmanifest.de. Accessed 25 Sept 2016

Orsini F, Gasperi D, Marchetti L, Piovene C, Draghetti S, Ramazzotti S, Bazzocchi G, Gianquinto G (2014) Exploring the production capacity of rooftop gardens (RTGs) in urban agriculture: the potential impact on food and nutrition security, biodiversity and other ecosystem services in the city of Bologna. Food Secur 6:781–792

Paull J (2013) "Please Pick Me": how incredible edible Todmorden is repurposing the commons for open source food and agricultural biodiversity. In: Franzo J et al (eds) Diversifying foods and diets: using agricultural biodiversity to improve nutrition and health. Earthscan/Routledge, Oxford, pp 336–345

Philips A (2013) Designing urban agriculture. Wiley, Hoboken

Rodrigues FM, Costa S, Andrade C, Muiños G, Mouro C (under review) Patterns and perceptions in urban gardening: the reimagining of new landscape forms. In Coles R, Costa S (eds) Food growing in the city: exploring the productive urban landscape as new paradigm for inclusive and productive approaches to the design and planning of urban open spaces. Special Issue: Landscape and Urban Planning

Sanyé-Mengual E, Orsini F, Gianquinto G, Oliver-Solà J, Montero IJ, Rieradevall J (2015) Rooftop farming: an opportunity towards urban sustainability? In: Conference paper for ISIE2015: 8th biannal conference of the international society of industrial ecology. University of Surrey, Guilford, UK

Shepard B (2013) Community gardens, creative community organizing and environmental activism. In: Gray M, Coates J, Hetherington T (eds) Environmental social work. Routledge, New York, pp 121–134

Silvestri G (2014) Sustainability transitions in Milan (Italy) with urban agriculture. Research case study report, DRIFT, Erasmus University of Rotterdam, Rotterdam

Smit B, Wandel J (2006) Adaptation, adaptive capacity and vulnerability. Glob Environ Chang–Urb Policy Dimens 16(3):282–292

Smith RM, Thompson K, Hodgson JG, Warren PH, Gaston KJ (2006) Urban domestic gardens (IX): composition and richness of the vascular plant flora, and implications for native biodiversity. Biol Conserv 129:312–322

Speak AF, Mizgajski A, Borysiak J (2015) Allotment gardens and parks: provision of ecosystems services with an emphasis on biodiversity. Urban For Urban Green 14(4):772–781

Strohbach M, Haase D (2012) Above-ground carbon storage by urban trees in Leipzig, Germany: analysis of patterns in an European city. Landsc Urban Plan 104(2012):95–104

Tappert S, Drilling M (eds) (2016) Growing in cities. Interdisciplinary perspectives on urban gardening. Conference Proceedings. University of Applied Sciences, Basel, pp 66–79. Available on http://www.sozialestadtentwicklung.ch/tagungen/growing_cities.pdf

Van Den Berg AE, Custers MHG (2011) Gardening promotes neuroendocrine and affective restoration from stress. J Health Psychol 16(1):3–11

Veen EJ (2015) Community gardens in urban areas: a critical reflection on the extent to which they strengthen social cohesion and provide alternative food. Wageningen University, Wageningen

Weiland U (2015) Gemeinschaftsgaerten in Leipzig. Landschaften in Deutschland Online. URL: http://landschaften-in-deutschland.de/themen/78_B_150-urbanes-gaertnern/. Accessed 10 Oct 16

Wood CJ, Pretty J, Griffin M (2015) A case-control study of the health and well-being benefits of allotment gardening. J Public Health UK 38:e336–e344

Part IV
Policy, Governance and Planning Implications for Nature-Based Solutions

Chapter 15
Mainstreaming Nature-Based Solutions for Climate Change Adaptation in Urban Governance and Planning

Christine Wamsler, Stephan Pauleit, Teresa Zölch, Sophie Schetke, and André Mascarenhas

Abstract The concept of mainstreaming climate change adaptation to foster sustainable urban development and resilience is receiving increasing interest. In particular, the need to mainstream ecosystem- or nature-based solutions into urban governance and planning is widely advocated by both academic and governmental bodies.

Adaptation mainstreaming is the inclusion of climate risk considerations in sector policy and practice. It is motivated by the need to challenge common ideas, attitudes, or activities and change dominant paradigms at multiple levels of governance. It seeks to increase sustainability and resilience by expanding the focus – from preventing or resisting climate hazards – to a broader systems framework in

C. Wamsler (✉)
Lund University Centre for Sustainability Studies, Lund, Sweden

Centre for Natural Disaster Science, Uppsala, Sweden
e-mail: christine.wamsler@lucsus.lu.se

S. Pauleit • T. Zölch
Centre for Urban Ecology and Climate Adaptation (ZSK) and Chair for Strategic Landscape Planning and Management, Technical University of Munich, Munich, Germany
e-mail: pauleit@wzw.tum.de; teresa.zoelch@tum.de

S. Schetke
Institute for Geodesy and Geoinformation, University of Bonn, Bonn, Germany
e-mail: schetke@uni-bonn.de

A. Mascarenhas
Center for Environmental and Sustainability Research (CENSE),
Universidade NOVA de Lisboa, Lisbon, Portugal

Geography Department, Humboldt-Universität zu Berlin, Berlin, Germany
e-mail: andre.mascarenhas@fct.unl.pt; andre.mascarenhas@hu-berlin.de

© The Author(s) 2017   257
N. Kabisch et al. (eds.), *Nature-based Solutions to Climate Change Adaptation in Urban Areas*, Theory and Practice of Urban Sustainability Transitions,
DOI 10.1007/978-3-319-56091-5_15

which we learn to live and cope with an ever-changing, and sometimes risky, environment. It aims to address the root causes of risk (including power structures) and failed approaches to sustainable development.

This chapter begins with an introduction to the concept of adaptation mainstreaming. It then presents an integrated framework that illustrates potential mainstreaming measures and strategies at different levels of governance, and discusses their application in urban planning practice with a focus on nature-based solutions. Case studies from Germany and Portugal illustrate the text. Four key principles for successful adaptation mainstreaming are highlighted. First, at the local level, adaptation mainstreaming requires the active consideration and combination of four approaches/measures to reduce climate risk on the ground. Second, to ensure their sustainable implementation, mainstreaming strategies must be implemented at the local, institutional and interinstitutional level. Third, the different measures and strategies only lead to sustainable change in combination. Finally, experience in mainstreaming other cross-cutting issues (notably climate change mitigation) can create synergies and support progress.

However, in practice there is still a long way to go. Current approaches often remain characterised by individual actions and the creation of separate, bolted-on structures and mechanisms.

Keywords Mainstreaming • Risk reduction • Disaster risk reduction • Climate policy integration • Environmental policy integration • Resilience • Climate change adaptation • Climate change mitigation • Germany • Portugal • Sweden • UK • Central America • Brazil

15.1 Introduction

Climate change and related hazards are an increasing threat to urban development, which, in turn, is increasing vulnerability to these hazards by inconsiderate development choices (OECD 2009; UNISDR 2008). This trend is expected to continue. Climate change is projected to magnify the frequency and intensity of weather-related hazards, which already account for the majority of annual losses from disasters (IPCC 2012, 2014; UNISDR 2010).

Consequently, resilience has emerged as a central concept in international and national development policy together with the concept of mainstreaming (Pervin et al. 2013; Wilkinson et al. 2014). Today, increasing resilience through the mainstreaming (or integration) of climate risk considerations into sector work is a unifying goal for the domains of climate change adaptation and disaster risk reduction (Andrade et al. 2011; Wamsler 2014). Resilience, i.e., "[T]the ability of a system, community or society exposed to hazards to resist, absorb, accommodate to and recover from their effects (…)" (UNISDR 2009, 2015), is also a feature of related global agendas, including the Sendai Framework for Disaster Risk Reduction 2015–2030, Sustainable Development Goals (SDGs) and the Paris Climate Agreement.

Against this background, the objective of this chapter is to present the key principles that show how adaptation mainstreaming can be achieved in urban governance and planning,[1] and discuss its role in fostering nature-based solutions and resilience. It should be noted that the term 'climate change adaptation'[2] denotes here an approach that integrates risk reduction and adaptation considerations.

15.2   What Is Adaptation Mainstreaming?

Adaptation mainstreaming refers to the inclusion of adaptation considerations into all sector policy and practice in order to reduce climate risk. The concept has two principal origins. One strand has developed from risk reduction mainstreaming, which has been strongly supported since the World Conference on Disaster Risk Reduction in Kobe, Japan in 2005 (UNISDR 2005), and which itself builds on mainstreaming experience in cross-cutting domains such as HIV/Aids and gender (Daly 2005; Holden 2004; Mazey 2002). The second has its roots in environmental policy integration (United Nations 1987; Lenschow 2002; Van Asselt et al. 2015), and more specifically climate policy mainstreaming, which has been promoted since around 1997 (Collier 1997). The initial objective of climate policy mainstreaming was to integrate the objective of reducing greenhouse gas emissions into other sectoral policies. Over time, the focus has gradually broadened and currently it also explicitly includes adaptation considerations (Berkhout et al. 2015; Wamsler and Pauleit 2016).

15.3   Why Is Adaptation Mainstreaming Relevant?

Adaption mainstreaming is particularly relevant in the context of nature-based solutions. Although climate adaptation in general, and nature-based solutions in particular are widely advocated (Daily and Matson 2009; Daily et al. 2009; Gaffin et al. 2012; Ojea 2015; Pasquini and Cowling 2014), they have not been systematically implemented (Andrade et al. 2011; IPCC 2014; Sitas et al. 2014; Vignola et al. 2009; Wamsler et al. 2014; Wamsler 2015a). Many local authorities and other urban stakeholders are unsure about what they can do to change this situation. The problem remains: how can climate adaptation be systematically mainstreamed into urban governance and planning, and ultimately increase resilience?

To answer this question, this chapter provides an integrated framework for adaptation mainstreaming. It builds upon frameworks that have been developed for mainstreaming environmental policy, climate adaptation, disaster risk reduction and

---

[1] Planning can be seen as a key sector, or avenue, for adaptation (IPCC 2014; Measham et al. 2011) and draws attention to respective governance arrangements (Agrawal 2008).

[2] Note that the terms "climate change adaptation", "climate adaptation" and "adaptation" are used as synonyms in this chapter.

other cross-cutting domains. The framework was empirically developed and tested between 2003 and 2016 in close collaboration with governmental and non-governmental organizations, in both developed and so-called developing countries (e.g., Germany, Sweden, the United Kingdom, Central America, Brazil), a process that has involved theoretical developments, analyses and the elaboration of practical guidelines. For more detailed information on the methodology, related case studies etc., see Wamsler (2006/2007, 2007, 2009, 2014, 2015a, b); Wamsler et al. (2014); and Wamsler and Pauleit (2016).

15.4 The Framework: Adaptation Mainstreaming at the Local Level

At the local level, adaptation mainstreaming requires the consideration and combination of four approaches to comprehensively reduce climate risk. All four approaches aim to reduce risk, but in different ways. They can be illustrated by the example of a potential landslide. The first approach aims to reduce (current and future) hazard exposure, which can be achieved by moving away from the landslide hazard or by reducing exposure on-site. The second approach aims to reduce vulnerability of the landslide-exposed area. Here, the aim is to create an environment that can withstand hazard impacts, without losing any of the community's main functions. The third aims to ensure an effective response post-landslide. Here the goal is to prepare response mechanisms and structures, before potential hazard impacts. Finally, the fourth approach aims to ensure effective recovery. Here, the aim is to prepare recovery mechanisms and structures, again, before potential hazard impacts.

The specific activities that are associated with each approach vary as a function of the hazard. Nevertheless, the principles do not change. For example, if the hazard is a flood rather than a landslide, the four approaches can be applied: reduce exposure, reduce vulnerability, prepare an effective response and prepare an effective recovery. Whenever possible, both multi-hazard and multi-purpose measures (i.e., measures that address both adaptation and other municipal objectives) should be implemented.

Nature-based solutions can be applied in the context of all four approaches: For example, exposure to flood can be reduced through beach nourishment, restoring or managing mangroves, or improved water management on the outskirts of urban areas. To reduce exposure to landslides, slopes can be stabilized through planting or the use of retention walls, which can combine elements of grey and green infrastructure. Careful urban planning, for example in the form of protected natural environments, can help to distance residential areas or critical infrastructure from a hazard, or at least ensure that settlements do not develop into hazard-prone areas.

With respect to reducing vulnerability, creating redundancy through nature-based solutions is an important element. Green infrastructure can help to reduce vulnerability by reducing dependency on only one system e.g. for heating, cooling, transportation or drainage. In addition, in the case of flood risk, measures for vulnerability reduction include the creation of buffer zones, retention ponds or increased permeable surfaces, for instance through the promotion of green roofs or urban agriculture. In the case of heat, measures include drought-resistant plants and improved insulation (e.g. through green walls).

In the context of response preparedness, typical measures include early warning systems and preparations for temporary refuge. In this context, nature-based solutions include well-designed green areas that can provide space for temporary shelter or protection (e.g., use of elevated green platforms during flash floods). Another example is the preparation of cooling mechanisms or structures. These include mobile planting systems or fountains, which can be used during heatwaves.

Recovery preparedness measures through nature-based solutions can include the use of materials or green infrastructure elements that can be easily recovered or replaced, along with preparations for post-disaster assistance. Examples are designated green areas that can be used for accommodation during reconstruction, preparations for the clearing or re-use of rubble (including green materials), or the provision of health and psychological support. The latter can include the support for greening private lots, being a multi-purpose measure with positive impacts on health and well-being. Other preparedness measures are awareness-raising campaigns and guidance on what to do after certain hazards, which can be linked to nature-based solutions (e.g. environmental learning parks). Although the contribution of nature-based solutions is often more indirect and contingent on other factors when it comes to preparedness measures, they are equally important.

It is crucial to understand and, ultimately, implement all four approaches, since local resilience is a function of inclusiveness and flexibility, rather than the effectiveness of a single approach (or measure) (Inderberg et al. 2015; Wamsler and Brink 2014; Wamsler and Pauleit 2016). Inclusiveness refers to the use of not just one or two, but all of the four risk-reducing approaches. Hence, whenever possible, measures that combine two, three or possibly all four approaches within a single activity should be preferred (e.g., in the creation of a climate park; cf., Box 15.1). Flexibility relates to the number and diversity of activities implemented for each approach, which must include both grey and green infrastructure together with socio-economic and political/institutional measures. This is crucial not only to identify multi-purpose solutions, but to also address the root causes of risk and, in turn, achieve sustainable change.

Appropriate activities for each approach must be identified for each individual context. Consideration must here be given to: (i) urban–rural linkages; (ii) urban characteristics (i.e., the urban fabric, the environment, economic and governance systems); and (iii) inter-area differences[3] (Wamsler, 2014; Wamsler and Brink 2016a). The latter can be characterized in terms of inclusion–exclusion, integration–marginalization, wealth–poverty, equality–inequality, and formality–informality.[4]

---

[3] A framework for analyzing urban–rural differences and linkages, and how they relate to weather and non-weather-related hazards is presented in Wamsler (2014) and Wamsler and Brink (2016a).

[4] A framework for a systematic vulnerability and capacity assessment that can identify appropriate risk-reduction measures is presented in Wamsler (2014).

Box 15.1 Magdalenenpark – A Climate Park for Munich, Germany
In 2013–2014, the NGO 'BUND Naturschutz' and the City of Munich, in cooperation with a team of landscape architects from the Technical University of Munich developed the design for a climate park. What is a climate park? It is an urban green space that: (i) is adapted to climate change; (ii) positively influences the microclimate; and (iii) stimulates citizens' adaptive awareness and behaviour (Brandl et al. 2014). Consistent with the four risk-reducing approaches introduced in this chapter, holistic adaptation measures were developed to address risk, and the microclimate modelling tool EnviMet was used to evaluate the design. This highlighted several features. First, increased vegetation improves thermal comfort and can reduce exposure to heat (Approach 1). Second, a variety of plants (e.g., drought-resistant species) increases the diversity of flora and reduces its vulnerability to heat stress (Approach 2). Third, it is important to select plants and materials (e.g., for benches) that can easily recover from heat or precipitation events (Approach 4). Finally, the park provides nearby residents from large housing estates with a place of refuge and temporary shelter during strong winds, precipitation or heat (Approach 3). The design is complemented by a pedagogical concept, related to all four risk-reducing approaches (Strategy VI). This includes providing information on adaptive behaviour, climate change and climate variability, and demonstrating the influence of different types of surfaces and vegetation covers. It also provides an opportunity to study the phenology of seasonal vegetation (e.g. bulbs that flower in early spring along hedges and in woodland, flowering cherries and meadows in summer, and fruit harvesting in autumn). The climate park is planned for an existing open green space in the western suburb of Neuaubing, but its implementation requires further negotiation with the landowner. Land ownership issues are a common challenge and highlight the importance of linking on-the-ground measures with mainstreaming strategies at institutional and interinstitutional level to ensure their sustainable implementation as described in this chapter.

All of the above approaches/measures and considerations only apply to mainstreaming at the local or operational level. However, also organizations themselves need to change (procedures, organizational structures, policies and regulations, etc.), rather than simply 'mainstreaming' change into selected on-the-ground activities (Persson and Klein 2009; Van Asselt et al. 2015; Wamsler and Pauleit 2016).

15.5 The Framework: Adaptation Mainstreaming at the (Inter)Institutional Levels

Mainstreaming needs to take place at the local (operational), institutional and the interinstitutional[5] level in order to achieve sustainable change and to unite top-down and bottom-up efforts that together create a holistic and distributed governance system for climate adaptation (Fig. 15.1). To assure the sustainable implementation of the local approaches/measures presented in the previous section, there must also be changes at the institutional and interinstitutional levels in order to:

- Institutionalize adaptation so that its integration at local level becomes standard procedure, which also includes the creation of mechanisms and structures for monitoring and learning[6];

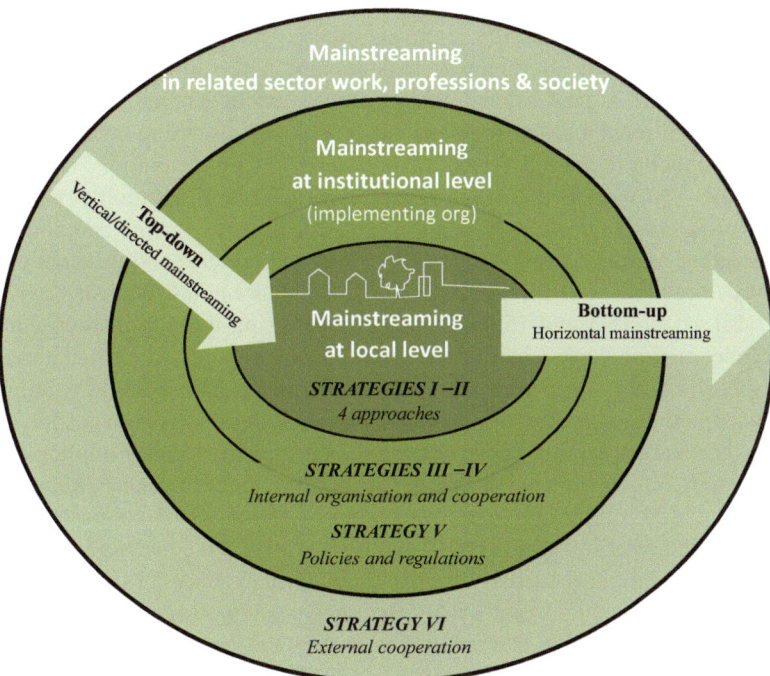

Fig. 15.1 Mainstreaming framework

---

[5] Here, the institutional level refers to the implementing organization. The interinstitutional level refers to the interlinkages between the implementing organization and other actors, to mainstream adaptation into related sector work, professions and society as a whole (Fig. 15.1).

[6] The issue of learning is crucial to build resilience. The four key features of resilience are anticipation, recognition, adaptation and learning (Becker 2014), which are an inherent part of the framework presented here.

- Ensure that organizations themselves can continue to function during climate change impacts;
- Cooperate in creating a multilevel governance system for climate adaptation that also includes citizens and, where possible;
- Drive improved education on adaptation mainstreaming and science–policy integration.

In sum, six mainstreaming strategies operate at the three levels (local, institutional and interinstitutional).[7] They are illustrated in Fig. 15.1 and Boxes 15.1, 15.2, 15.3 and 15.4. The first two (Strategies I–II) focus on the local or household level and relate to the way the four risk-reducing approaches (outlined above) are included in on-the-ground initiatives (either integrated or added on). Three strategies focus on the institutional level and address issues of internal organization (Strategy III) and cooperation (Strategy IV), together with policies and regulations (Strategy V). For example, it is crucial to ensure that organizational procedures and routines foster, rather than hamper (cf. Uittenbroek 2016), the implementation of nature-based solutions. The sixth strategy (Strategy VI) focuses on the interinstitutional level, external cooperation with other organizations (local, regional, national and international) and citizens. It addresses sector work, professional training and society in general.

At the institutional and interinstitutional level, mainstreaming involves both targeted and implicit integration. At the institutional level, municipalities' mainstreaming strategies can include the development of standalone adaptation strategies, the creation of interdepartmental working groups for climate change adaptation, and the integration of adaptation objectives into sectoral policies and instruments (e.g. green structure plans), comprehensive or detailed planning (cf. Wamsler 2015a; cf., Boxes 15.1, 15.2, 15.3 and 15.4). Strategies addressing the interinstitutional level can include municipal participation in regional innovation platforms that aim to create new business and cooperation models for financing nature-based solutions for climate adaptation, or the creation of city-citizen collaboration (cf. Wamsler 2015a). Unlike other mainstreaming frameworks that address targeted and implicit integration separately (cf. Uittenbroek 2016), the framework that is presented here includes both since related actions and processes are fluent and mutually supportive.

Politics and power must be explicitly addressed at all three mainstreaming levels, as they are a potential root cause of risk and represent avenues (or barriers) to transformation (cf., Boxes 15.1 and 15.4). The analysis of power (relations) is thus a precondition to sustainable change and to understand how transformation can be achieved or be hindered (Daly 2005; Digeser 1992; Partzsch 2015). This should include an evaluation of shared power (*power with*, cooperation and learning), the exercise of power (*power to*, resistance and empowerment) and *power over* others (coercion and manipulation) in potential mainstreaming approaches and strategies (cf. Allen 1998; Verloo 2005). Issues related to power need to be fundamentally challenged whenever and wherever modernity and globalization lead societies down an unsustainable road (Manuel-Navarrete 2010).

---

[7] For detailed definitions of the six strategies see Wamsler and Pauleit (2016).

Box 15.2 Urban River Restoration – The 'Isarplan' Project in Munich, Germany

The river Isar is the most important green corridor in the city of Munich. Since the beginning of the 19th century this pre-alpine river has been increasingly regulated, both to reduce the risk of flooding and for power generation. However, flood risk and legal requirements related to flood protection necessitated its fundamental redesign. The Isarplan project was implemented in 2000–2011, and the main objectives were: (i) to improve flood control; (ii) to restore the river's ecological functions; and (iii) to improve recreational opportunities for the city's population. The city of Munich and the Regional Office for Water Management were responsible for its planning and implementation. The obvious solution was to continue to elevate the river's dams. However, the project provided an opportunity to explore a novel approach to ecological restoration, which would meet all three objectives. Consequently, the river bed was widened into surrounding flood plains, existing embankments were removed, and a naturalistic system of riverbed rock ramps was implemented that allowed fish to move upstream. (Oppermann 2005; Pauleit 2005; Pauleit and Kollmann 2015).

The Isarplan project is a good example of mainstreaming nature-based solutions for climate adaptation in urban planning at local and institutional levels. At the local level, the restoration of natural river banks and the widening of its channel reduced exposure to flood risk (Approach 1, Strategy I). The expansion of flood plains within the city created buffer zones that reduced vulnerability (Approach 2, Strategy I). Nature-based solutions are now also being planned for adjacent neighbourhoods, e.g., more greened routes that reduce risk and improve multipurpose recreation facilities (Approaches 2–4, Strategies I–II). Finally, water quality has been improved to the point where it is now possible to swim in the river. This was achieved by modernising sewage treatment plants in upstream municipalities. At the institutional level, an interdepartmental working group was responsible for coordinating the project; this group provided support and fostered a multi-benefit approach (Strategies III–IV). In addition, the project was designed by an interdisciplinary group of engineers, landscape architects, city planners and biologists, both internal and external to the city administration (Strategy VI). Today, the river Isar has been successfully transformed—not only into an appealing green space—but also into a support for comprehensive flood protection and management.

Box 15.3 Mainstreaming Climate Adaptation into Settlement Development in Bonn, Germany

The City of Bonn uses a four-step approach to advance mainstreaming of nature-based solutions for climate change adaptation in the development of new residential areas (Helbig and Gädker 2015):

1. Cross-departmental information events on climate change (involving urban planning, urban drainage, environment, urban greening and health departments) and follow-up events to discuss options, strategies and trade-offs with regional actors (politicians, scientists and public utility companies) (Strategy VI).
2. Educational events moderated by scientists that broaden the knowledge base of city administrators and council members in terms of climate modeling, exceptional rain events and heat stress (Strategy IV).
3. The launch of an integrated climate protection concept, which includes a section on adaptation measures (science, management and administration) (Strategy V).
4. Mainstreaming of climate adaptation into municipal working routines and local settlement development (Strategies II and III).

These four steps include a broad range of actors and decision-makers from sectors that are both internal and external to the city administration. Whilst the first two steps aim to build a shared knowledge base about the impacts of climate change, the third and fourth aim to sustainably mainstream climate change adaptation at the local and institutional levels (internal cooperation and policies). In order to enrich the discussion and provide insights into the options for implementing nature-based solutions in residential areas, the municipality has also asked the University of Bonn (the Department of Urban Planning and Real Estate Development) to develop prototype scenarios for a potential infill site (Kötter et al. 2014). These scenarios, which were developed in the context of student seminars, focus on vulnerability reduction for local residents (Approach 2), while other types of risk-reducing approaches and their anchoring at institutional levels are excluded.

Box 15.4 The Lisbon Metropolitan Area's Ecological Network

The Lisbon Metropolitan Area (LMA) in Portugal is considered to be highly vulnerable to global climate change (Giorgi and Lionello 2008). As the region with the biggest population concentration, climate adaptation is a pressing issue for spatial and environmental planning.

The 2002 LMA Regional Spatial Plan defined a Metropolitan Ecological Network (MEN), which created a knowledge base of nature-based solutions for adaptation. The MEN defines areas and ecological corridors, which are organised into three hierarchical levels according to their importance for the environmental structure. Each level includes strategic guidelines and concrete climate change adaptation measures (mainly vulnerability reduction; Approach 2), as well as management requirements that ensure mainstreaming at the institutional level (e.g., policy and regulations - Strategy V, municipal responsibilities and financial resources - Strategy III). A noteworthy feature of the MEN is that it provides an integrated and consistent overview of designated nature conservation areas (Natura 2000 sites, natural parks, nature reserves, protected landscapes) and other ecologically-relevant areas. The lowest hierarchical level includes compact or fragmented urban areas that carry out important functions. The aim is to integrate these areas, designated as urban green zones, into municipal adaptation planning (Strategy V).

Whilst the MEN is expected to contribute to adaptation mainstreaming at the local level, it mainly addresses changes at the institutional and interinstitutional levels of governance. It can also be seen as a multi-purpose measure; although it is principally aimed at nature conservation, at the same time it supports ecosystem services that are relevant for climate change adaptation.

Its implementation highlights the potential for a governance framework with an expanded focus—from preventing or resisting climate hazards, to broader systems thinking. In 2013–2014, a participatory exercise was conducted in the LMA with stakeholders from local authorities, the national environmental authority and academia, who identified the MEN, and urban green space in general, as important drivers for sustainable planning (Mascarenhas et al. 2016).

Numerous challenges remain for mainstreaming nature-based solutions into planning practice in LMA. In addition to a fragmented institutional structure and power struggles, there is still insufficient knowledge about existing ecosystem services, how they affect human well-being and how they are, in turn, affected by planning decisions.

15.6   From Theory to Practice: Gaps and Synergies

In practice, mainstreaming can create synergy effects by promoting innovation in sector-specific policies, linking and aligning sector-specific and adaptation objectives, and encouraging more efficient use of human, physical and financial resources (Lafferty and Hovden 2003; Adelle and Russel 2013; Rauken et al. 2014; Runhaar et al. 2014; Dewulf et al. 2015; Persson et al. 2015). The framework presented here can support such developments, guide an organization's mainstreaming process and hold decision-makers accountable for any promises they may make in their endeavours. It provides an overall context for action and can be applied in combination with other frameworks that target particular mainstreaming levels, strategies or approaches, such as financial integration or internal risk management (CDKN 2015; Pervin et al. 2013; UNDP 2010; UNDP-UNEP 2011).

The mainstreaming framework presented here formed the cornerstone for the development of three operational guidelines that translate the mainstreaming theory into practice. The guidelines were developed for different urban stakeholders, including non-governmental organizations (Wamsler 2006/2007, 2009) and local authorities (Wamsler 2015b; Wamsler and Brink 2016b). The latter aim to assist municipal officials and local politicians to both assess and progress the mainstreaming of climate change adaptation into planning and governance mechanisms.

The first step in this process is to categorize existing and planned activities relevant for climate adaptation according to the mainstreaming level and the strategy they address. Questions include: What kind of activities are implemented at the local, institutional or interinstitutional level? Do activities at the institutional level address the integration of climate adaptation into the overall planning vision, comprehensive planning, detailed planning, related planning instruments, the internal organizational structure, and human and financial assets? In the second and third steps, existing and planned activities are assessed against established benchmarks. This includes an assessment of the use of the four risk-reducing approaches and related synergy creation, whether a single or multi-hazard approach is adopted, and whether physical, socio-economic, environmental and political/ institutional measures are implemented at different levels. The outcome of this assessment provides a clear overview of current progress and helps to identify gaps or ways to improve (for further details, see the operational guidelines presented in Wamsler 2015b; Wamsler and Brink 2016b).

Both the mainstreaming theory and the related guidelines have proven to be very useful, and their application has helped cities in the identification of various gaps. These gaps also highlight how mainstreaming and resilience are interlinked. Here, we focus on three specific examples: (i) the lack of systematic mainstreaming of nature-based solutions; (ii) the focus on municipal self-reliance or governing; and (iii) fragmented climate policy mainstreaming.

The first gap, the lack of systematic mainstreaming of nature-based solutions (which requires a combination of the presented approaches and strategies), is illustrated by the following statement by a member of staff from a German municipality:

"We deal with the issue of adaptation in a very broad or general sense, and the differentiation between constructive and other types of adaptation measures [i.e., nature-based solutions] is, in practice, not yet a topic. We are not there yet [...] We have a smörgasbord of ideas, we still don't have an overview. This will come with further conceptualization (Wamsler 2015a). These words provide a link back to the starting point of this chapter. Specifically, they refer to the need for a better understanding and the systematic implementation of nature-based solutions to increase resilience (e.g. to overcome the current focus on grey infrastructure solutions in on-the-ground initiatives) (Wamsler et al. 2016).

The second gap is a lack of cooperation with other stakeholders to support nature-based solutions. This is illustrated by the following statement by a member of staff from a German municipality, "The first step now is [...] we are focusing on what we can do [...] what we can implement ourselves. And then we will look, after that, further afield, and try to involve others [e.g. private actors or citizens]" (Wamsler 2015a). This statement illustrates the widespread phenomenon of municipal self-reliance in climate adaptation. At the same time the importance of involving other stakeholders is generally acknowledged, especially in the context of nature-based solutions, "We are more and more dependent [...] on everybody, every citizen must become active and get engaged, because the city cannot handle it [climate hazards] by itself anymore. The city depends on citizens' support" (Wamsler 2015a).

The third gap is fragmented climate policy mainstreaming, also regarding nature-based solutions. Specifically, it concerns the failure to integrate climate mitigation and climate adaptation mainstreaming. This is illustrated by the following statement from a staff member from a German municipality responsible for climate adaptation, "I am not sure who is responsible for climate mitigation [...] and related mainstreaming processes. It is dealt with separately" (Wamsler 2015a). This fragmented approach can hamper progress, as sustainable urban development requires integrated planning policy and practice. In addition, research has shown that adaptation mainstreaming can be spurred by an organization's experience with mainstreaming climate change mitigation (Wamsler and Pauleit 2016).

15.7 Conclusions

The mainstreaming of nature-based solutions for climate change adaptation can support incremental and transformative changes that address the root causes of risk and lead to sustainable development. Its implementation requires the consideration of four key principles. First, at the local level, all four approaches/ measures aimed at reducing climate risk on the ground have to be addressed, with nature-based solutions offering a broad range of applications. Second, mainstreaming strategies must be implemented at the local, institutional and interinstitutional levels in order to ensure the sustainable implementation of on-the-ground measures, and challenge current practice and procedures at multiple levels of governance. Third, measures and strategies can only lead to sustainable change in combination. Finally, previous experience in mainstreaming other cross-cutting issues (in particular climate change mitigation) can help to create synergies and progress adaptation mainstreaming.

Adaptation mainstreaming is a potentially effective way to foster urban resilience. In fact, the concept of mainstreaming is inherently linked to the concept of resilience, as it aims to challenge familiar ideas, attitudes, or activities and change dominant paradigms at multiple levels of governance. It can help to increase resilience by expanding the focus – from preventing or resisting hazards – to a broader systems framework in which the different stakeholders learn to live and cope with an ever-changing, and sometimes risky, environment. It can also lead to a more inclusive planning and risk governance system, which translates into the ability to change in response to altered circumstances and to carry on functioning even when individual parts fail. This can be achieved by including and linking physical, social, economic, environmental and political/ institutional aspects, different risk-reduction approaches and sector work, together with climate adaptation and mitigation considerations. Finally, the issue of power structures must be considered in any current (or potential) mainstreaming approaches or strategies.

As it stands today, practice remains characterised by individual measures, the creation of bolted-on structures, and related actions that are often seen purely as managerial governance exercises. A more systematic approach to mainstreaming nature-based solutions (as presented in this chapter), which also explicitly considers power structures, is urgently needed in order to ensure that root causes of risk and any avenues (or barriers) to transformation are addressed.

References

Adelle C, Russel D (2013) Climate policy integration: a case of déjà vu? Environ Policy Gov 23:1–12

Agrawal A (2008) The role of local institutions in adaptation to climate change. Paper prepared for the Social Dimensions of Climate Change, Social Development Department, The World Bank, Washington

Allen A (1998) Rethinking power. Hypatia 13(1):21–40

Andrade A, Córdoba R, Dave R, Girot P, Herrera-F B, Munroe R, Oglethorpe J, Paaby P, Pramova E, Watson J, Vergara W (2011) Draft principles and guidelines for integrating ecosystem-based approaches to adaptation in project and policy design. *Policy brief, CATIE* no. 46

Becker P (2014) Sustainability science: managing risk and resilience for sustainable development. Elsevier, Amsterdam and Oxford

Berkhout F, Bouwer LM, Bayer J, Bouzid M, Cabeza M, Hanger S, Hof A, Hunter P, Meller L, Patt A, Pfluger B, Rayner T, Reichardt K, van Teeffelen A (2015) European policy responses to climate change: progress on mainstreaming emissions reduction and adaptation. Reg Environ Chang 15:949–959

Brandl A, Keller R, Pauleit S (2014) *Magdalenenpark – Ein Klimapark für München*. Final project report. München, Germany

CDKN (2015) Mainstreaming climate compatible development. Chapter 4: resourcing climate compatible development

Collier U (1997) Sustainability, subsidiarity and deregulation: new directions in EU environmental policy. Environ Pollut 6(2):1–23

Daily GC, Matson P (2009) Ecosystem services: from theory to implementation. Proc Natl Acad Sci U. S. A. 105:9455–9456

Daily GC, Polasky S, Goldstein J, Kareiva PM, Mooney HA, Pejchar L, Ricketts TH, Salzman J, Shallenberger R (2009) Ecosystem services in decision making: time to deliver. Front Ecol Environ 7:21–28

Daly M (2005) Gender mainstreaming in theory and practice. Soc Policy 12(3):433–450

Dewulf A, Meijerink S, Runhaar H (2015) Editorial for the special issue on the governance of adaptation to climate change as a multi-level, multi-sector and multi-actor challenge: a European comparative perspective. J Water Climate Change 6(1):1–8

Digeser P (1992) The fourth face of power. J Politics 54(4):977–1007

Gaffin SR, Rosenzweig C, Kong A (2012) Adapting to climate change through urban green infrastructure. Nat Clim Chang 2(10):704–704

Giorgi F, Lionello P (2008) Climate change projections for the Mediterranean region. Glob Planet Chang 63:90–104

Helbig J, Gädker J (2015) Initiierung und Entwicklung von Klimaanpassungsaktivitäten in der Stadt Bonn, In: (Difu), S.K.K.K. beim D.I. für U.G. (Ed.), Klimaschutz & Klimaanpassung. Themenhefte, Köln, p 46–53

Holden S (2004) Mainstreaming HIV/AIDS in development and humanitarian programmers. Oxfam GB, Oxford

Inderberg TH, Eriksen S, O'Brien K, Sygna K (2015) Climate change adaptation and development: transforming paradigms and practices. Routledge, New York

IPCC (2012) Managing the risks of extreme events and disasters to advance climate change adaptation. SREX report. Cambridge University Press, Cambridge, UK

IPCC (2014) Climate change 2014: impacts, adaptation, and vulnerability. Cambridge University Press, Cambridge, UK

Kötter T, Schetke S, Katzschner L (2014) Das quartier "Gymnicher Hof," MSc seminar on urban renewal. Msc-programm geodesy and geoinformation. University of Bonn, Bonn

Lafferty W, Hovden E (2003) Environmental policy integration: towards an analytical framework. Environ Pollut 12(3):1–22

Lenschow A (ed) (2002) Environmental policy integration: greening sectorial policies in Europe. Earthscan, London, UK

Manuel-Navarrete D (2010) Power, realism, and the ideal of human emancipation in a climate of change. Wiley Interdiscip Rev Clim Chang 1(6):781–785

Mascarenhas A, Ramos TB, Haase D, Santos R (2016) Participatory selection of ecosystem services for spatial planning: Insights from the Lisbon Metropolitan Area, Portugal. Ecosyst Serv 18:87–99

Mazey S (2002) Gender mainstreaming strategies in the EU. Fem Leg Stud 10:227–240

Measham TG, Preston BL, Smith TF, Brooke C, Gorddard R, Withycombe G, Morrison C (2011) Adapting to climate change through local municipal planning: barriers and challenges. Mitig Adapt Strateg Glob Chang 16:889–909

OECD (2009) Integrating climate change adaptation into development co-operation: policy guidance. OECD Publishing, Paris, France p 197

Ojea E (2015) Challenges for mainstreaming ecosystem-based adaptation into the international climate agenda. Curr Opin Environ Sustain 14:41–48

Oppermann B (2005) Redesign of the River Isar in Munich, Germany Getting coherent quality for green structures through competitive process design? In: Werquin AC, Duhem B, Lindholm G, Oppermann B, Pauleit S, Tjallingii S (eds) Green structure and urban planning (Vol COST Action C11). European Commission, Brussels

Partzsch, L (2015) Kein Wandel ohne Macht – Nachhaltigkeitsforschung braucht ein mehrdimensionales Machtverständnis. GAIA – Ecol Perspect Sci Soc 24(1):48–56 (9)

Pasquini L, Cowling R (2014) Opportunities and challenges for mainstreaming ecosystem-based adaptation in local government: evidence from the Western Cape, South Africa. Environ Dev Sustain 17:1–20

Pauleit S (2005) Munich. In: Werquin AC, Duhem B, Lindholm G, Oppermann B, Pauleit S, Tjallingii S (eds) Green structure and urban planning (Vol COST Action C11). European Commission, Brussels

Pauleit S, Kollmann J (2015) Die Isarrenaturierung in München. Hochwasserschutz, Ökologie und Erholung integrativ? In: *DGGL Jahrbuch* 2015, pp. 34–39

Persson Å, Klein RJT (2009) Mainstreaming adaptation to climate change into official development assistance: Building on environmental policy integration theory. In: Harris P (ed) Climate change and foreign policy: case studies from East to West. Routledge, London, pp 162–177

Persson Å, Eckerberg K, Nilsson M (2015) Institutionalization or wither away: 25 years of environmental policy integration in Swedish energy and agricultural policy under shifting governance models in Sweden. Environ Plann C 47:1–18

Pervin M, Sultana S, Phirum A, Camara IF, Nzau VM, Phonnasane V, Khounsy P, Kaur N, Anderson S (2013) A framework for mainstreaming climate resilience into development planning. IIED Working Paper, Climate Change

Rauken T, Mydske PK, Winsvold M (2014) Mainstreaming climate change adaptation at the local level. Local Environ 20(4):408–423

Runhaar H, Driessen PJ, Uittenbroek C (2014) Towards a systematic framework for the analysis of environmental policy integration. Environmental Policy and Governance 24(4):233–246

Sitas N, Prozesky H, Esler K, Reyers B (2014) Exploring the gap between ecosystem service research and management in development planning. Sustainability 6:3802–3824

UNDP (2010) Mainstreaming disaster risk reduction into development at the national level: a practical framework

UNDP-UNEP (2011) Mainstreaming climate change adaptation into development planning: a guide for practitioners. Environment for the MDGs. UNDP-UNEP Poverty-Environment Initiative

UNISDR (2005) Hyogo framework for action 2005–2015: building resilience of nations and communities to disasters. *World conference on disaster reduction*, Kobe, Hyogo, Japan

UNISDR (2008) Links between disaster risk reduction, development and climate change: A briefing for Sweden's Commission on Climate Change and Development. p. 5

UNISDR (2009) Terminology on disaster risk reduction. UNISDR Online Glossary. https://www.unisdr.org/we/inform/terminology#letter-r

UNISDR (2010) Briefing note 03: strengthening climate change adaptation through effective disaster risk reduction. p. 10

UNISDR (2015) Sendai framework for disaster risk reduction 2015–2030. *World conference on disaster reduction*, Sendai, Japan

United Nations (1987) Our common future – Brundtland report. Oxford University Press, Oxford

Uittenbroek CJ (2016) From policy document to implementation: organizational routines as possible barriers to mainstreaming climate adaptation. J Environ Policy Plan 18(2):161–176

Van Asselt H, Rayner T, Persson Å (2015) Climate policy integration. In: Bäckstrand K, Lövbrand E (eds) Research handbook on climate governance. Edward Elgar, Cheltenham, UK, chapter 34, 24:388–399

Verloo M (2005) Displacement and empowerment: Reflections on the concept and practice of the council of Europe approach to gender mainstreaming and gender equality. Soc Policy 12(3):344–365

Vignola R, Locatelli B, Martinez C, Imbach P (2009) Ecosystem-based adaptation to climate change: what role for policy-makers, society and scientists? Mitig Adapt Strateg Glob Chang 14:691–696

Wamsler C (2007) Managing Urban Disaster Risk: Analysis and Adaptation Frameworks for Integrated Settlement Development Programming for the Urban Poor, Dec. 2007, Doctoral thesis, Lund: Lund University

Wamsler C (2006/2007) Operational Framework for Integrating Risk Reduction for Aid Organisations Working in Human Settlement Development [English and Spanish version], Benfield Hazard Research Centre (BHRC) *Disaster Studies Working Paper* No. 14, London: BHCR

Wamsler C (2009) Operational Framework for Integrating Risk Reduction and Climate Change Adaptation into Urban Development, Brookes World Poverty Institute (BWPI), *Working Paper Series* No. 101, Manchester: BWPI

Wamsler C (2014) Cities, disaster risk and adaptation. Routledge, London

Wamsler C, Brink E (2014) Moving beyond short-term coping and adaptation. Environ Urban 26:1–26, special issue on 'Towards Resilience and Transformation for Cities'.

Wamsler C, Luederitz C, Brink E (2014) Local levers for change: mainstreaming ecosystem-based adaptation into municipal planning to foster sustainability transitions. Glob Environ Chang 29:189–201

Wamsler C (2015a) Mainstreaming ecosystem-based adaptation: transformation toward sustainability in urban governance and planning. Ecol Soc 20(2):30

Wamsler C (2015b) Guideline for integrating climate change adaptation into municipal planning and governance. Working Paper 31. Disaster Studies and Management Working Paper Series, University College London (UCL) Hazard Centre, London.

Wamsler C, Brink E (2016a) The urban domino effect: a conceptualization of cities' interconnectedness of risk. Int J Disaster Resilience Built Environ 7(2):80–113

Wamsler C, Brink E (2016b) Promoting nature-based solutions: guideline for integrating ecosystem-based adaptation into municipal planning and governance, Disaster Studies and Management Working Paper 32. UCL, London. Accessible online. http://www.ucl.ac.uk/hazardcentre/resources/working_papers/working_papers_folder/UCLDisasterStudies ManagementWorkingPaper32

Wamsler C, Niven L, Beery T, Bramryd T, Ekelund N, Jönsson I, Osmani A, Palo T, Stålhammar S (2016) Operationalizing ecosystem-based adaptation: harnessing ecosystem services to buffer communities against climate change. Ecol Soc 21(1):31

Wamsler C, Pauleit S (2016) Making headway in climate policy mainstreaming and ecosystem-based adaptation: two pioneering countries, different pathways, one goal. Springer, Climatic Change, Netherlands

Wilkinson E, Carabine E, Peters K, Brickell E, Scott A, Allinson C, Jones L, Bahadur A (2014) Existing knowledge integrating disaster risk reduction, environment and climate change into development practice. ODI Advancing Integration Series Working Paper. Overseas Development Institute, London

Chapter 16
Partnerships for Nature-Based Solutions in Urban Areas – Showcasing Successful Examples

Chantal van Ham and Helen Klimmek

Abstract Increasing the uptake of nature-based solutions (NBS) requires greater collaboration amongst different policy areas, sectors and stakeholders. This chapter showcases examples of multi-stakeholder partnerships, private sector leadership, and citizen engagement, which have supported the development or implementation of NBS in urban areas. It aims to complement the theoretical contributions of the previous chapters of this book by providing real-world insights into how such partnerships can promote climate resilience and nature conservation, as well as the lessons that can be learned from them. It thereby hopes to spark ideas for future research and the development of collaborative, multi-stakeholder partnerships for NBS.

Keywords Multidisciplinary partnerships • natural capital • citizen engagement • nature-based solutions

16.1 Introduction

Recent research (Kabisch et al. 2016) has shown that there is a need to forge new networks and develop trans-disciplinary and inclusive partnerships and governance approaches in order to foster the uptake of nature-based solutions (NBS) in response to climate-related challenges. Producing stronger evidence on NBS for climate change adaptation and mitigation, and raising awareness of their benefits to society, are also key priorities for policy and practice.

Partnerships are collaborative arrangements which are important for implementing sustainability agendas due to two distinct and defining characteristics: (a) They can create and catalyse synergies between different parts of society by pooling

---

This chapter is written from IUCNs perspective as a global membership organisation uniquely composed of state and government agencies, NGOs, scientific institutions and business associations.

C. van Ham (✉) • H. Klimmek
IUCN European Regional Office, Brussels, Belgium
e-mail: Chantal.vanham@iucn.org; Helen.klimmek@iucn.org

© The Author(s) 2017   275
N. Kabisch et al. (eds.), *Nature-based Solutions to Climate Change Adaptation in Urban Areas*, Theory and Practice of Urban Sustainability Transitions,
DOI 10.1007/978-3-319-56091-5_16

together resources and skills, knowledge, institutional and governance capacities and (b) they are flexible and versatile in the roles they adopt, as partners match and complement their competencies and capacities to undertake a task or aim to achieve a common target (Frantzeskaki et al. 2014).

For IUCN, partnerships are a key driving force for successful conservation action. The socio-economic and environmental challenges confronting society today are complex and far from clear-cut. Bringing together diverse stakeholders such as governments, NGOs, scientists, businesses, local communities and indigenous peoples groups, can help to address these challenges in a comprehensive and inclusive way.

The complexity of urban environments underlines the importance of multi-disciplinary and multi-scale partnerships in cities. Cities represent a new class of ecosystems shaped by the dynamic interactions between ecological and social systems (CBD 2012). Urban citizens depend on ecosystems both within and beyond cities for a wide variety of goods and services (e.g. food, water, energy, climate regulation), and while cities are increasingly recognised for their role in conservation (CBD 2012), urbanisation also presents a major environmental challenge, for example by driving habitat conversion (McDonald et al. 2013).

Nature offers great untapped potential for improving the quality of life of urban citizens and finding solutions to challenges such as rising temperatures (the urban heat island effect) or flooding (CBD 2012). The challenge lies in developing and adopting urban planning and management approaches that ensure the delivery of regulating, provisioning, supporting, and cultural ecosystem services, while also promoting the sustainable use of resources.

Ideally, stakeholders from different policy areas and sectors should come together to develop holistic approaches to managing natural capital – the world's stocks of natural assets which includes geology, soil, air, water and all living things. In reality however, collaboration between sectors and stakeholders is often hindered by a lack of exchange and cooperation, presenting a barrier to effective policymaking (Science for Environment Policy 2016) and the implementation of successful conservation initiatives.

Multidisciplinary and cross-cutting concepts such as NBS have the potential to facilitate cooperation between sectors and contribute to a more holistic approach to tackling socio-economic and environmental challenges. From IUCNs perspective, NBS are interventions which use nature, and the ecosystem services they provide, to address societal challenges such as climate change. Well-functioning ecosystems that deliver services needed by society are at the core of these types of solutions, which include, for instance, the creation or restoration of large ecosystems; investing in natural infrastructure and watershed management for water, food and energy security and climate change adaptation; ecosystem-based mitigation oriented solutions, such as the conservation and sustainable management of forests; and using ecologically engineered solutions, such as intertidal habitats or oyster reefs to protect shorelines and reduce sea-level rise impact and coastal inundation.

The aim of this chapter is to profile a broad range of partnerships led by the private sector, local communities and local/regional governments, which have restored, conserved and managed ecosystems to the benefit of people and the environment. The following sections include reflections on key outcomes and lessons that can be learned from each of these examples, which can provide the basis for further research into the implementation of NBS. The examples contained in this chapter were sourced through IUCNs global knowledge network and connections, supplemented by existing literature, and were chosen based on their suitability for highlighting success factors of, and challenges to partnerships for NBS. The examples were also selected to reflect diversity in terms of the partners involved and their geographical location.

16.2 The Private Sector – A Valuable Partner for Implementing NBS

The private sector (i.e. for-profit businesses) is a key partner to engage with in the process of meeting global biodiversity conservation targets. While the private sector can have negative impacts on biodiversity, it also has the potential to offer innovative solutions to urban challenges. Businesses can provide insights and perspectives which are complementary to those from governments and civil society. In particular, their knowledge of markets, management experience, and ability to harness advanced research and development to deliver solutions, can be valuable assets in the context of implementing NBS (IUCN 2012a).

Many businesses are increasingly realising that their future depends (directly or indirectly) on natural resources and that solely relying on man-made infrastructure is not enough (Ozment et al. 2015). Man-made storm surge barriers, for example, can help protect harbours, but can also seriously increase surge levels in surrounding areas. River ecosystems throughout Europe have been severely impacted by engineering projects for flood protection, navigation, water supply and hydroelectricity; it is estimated that less than 20% of Europe's rivers and floodplains are in their natural state (RESTORE 2016). A combination of measures is needed to tackle flood and climate change related challenges effectively. This must include land-use management and nature-based measures which embrace natural systems as a means of enhancing our well-being and reducing risk (Munich RE 2015). Growing oyster reefs, for example, can help to reduce coastal erosion and protect businesses from storm surges, while also filtering contaminated seawater and supporting local fisheries (RESTORE 2016).

The importance and value of engaging with the private sector is aptly demonstrated in relation to climate change. According to a survey conducted by the Economist Intelligence Unit in 2014, 90% of business leaders believe that they have a role in building resilience and preparing cities for the impact of climate change (Kongrukgreatiyos 2014). This has already translated into action – in 2014, the private sector was the largest source of climate finance, devoting $243 billion to climate-related investments (Buchner et al. 2015). Partnerships between businesses, cities, civil society organisations, scientists and other urban stakeholders are crucial to showcasing the value of natural capital as the foundation for economic prosperity and human well-being, and help to bring about changes in business practices and leverage contributions from the private sector.

During the Conference of the Parties of the United Nations Framework Convention on Climate Change (COP of UNFCCC) in Paris in 2015, the World Business Council for Sustainable Development (WBCSD) launched Natural Infrastructure for Business – an online platform to increase awareness of business opportunities for investing in ecosystems, or natural infrastructure, and scale up action. The ultimate objective of the initiative is that by 2020 companies systematically assess natural infrastructure options when investing in new sites or projects, thereby contributing to the protection, restoration and creation of new ecosystems. The online platform contains case studies from different industries leveraging various ecosystem services and decision-making tools, including a cost-benefit analysis tool. Some of these case studies are highlighted below.

16.2.1   Examples of Private Sector Led Partnerships for NBS

16.2.1.1   Volkswagen Restores Nature to Secure Reliable Water Supply

One example from the Natural Infrastructure for Business database is the Volkswagen initiative in the Puebla-Tlaxcala Valley in Mexico (WBCSD 2010). This project was initiated after years of deforestation from illegal logging and livestock farming, had led to increased water runoff and a reduction in capture and storage in the groundwater table. Realising that a reliable water supply was critical to ensuring the future of the company's production efforts, Volkswagen de Mexico, in partnership with the Comisión Nacional de Áreas Naturales Protegidas and the Secretary of the Environment for Mexico, invested in a project to plant trees, dig pits, and earthen banks to enable more than 1,300,000 cubic meters of additional water per year to be fed into the ground reserves in the source region, which is significantly more groundwater than Volkswagen in México itself consumes every year (WBCSD 2010). In the long term, these measures will help to ensure the provision of fresh water for the growing city of Puebla, while securing a reliable water supply for the stability of the company's production plant in the region (WBCSD 2010). The additional biomass will also help sequester carbon dioxide and improve living conditions for the native fauna (WBCSD 2010), demonstrating the multiple benefits restoration efforts can bring to both people and the natural environment.

16.2.1.2   Rehabilitation of Quarries Provides Multiple Benefits for Nature, People and Business

Another example showcased in the WBCSD Natural Infrastructure for Business platform (Rushworth and Warau 2015) is a project in Bellegarde in the South of France, initiated by LafargeHolcim, a global leader in the building materials industry. This project focuses on stormwater management and flood prevention through targeted quarry rehabilitation and management programmes that provide stormwater catchments and create wetland habitats (Rushworth and Warau 2015). The sand and gravel quarry of Bellegarde has been in operation since 1970 and LafargeHolcim has worked with the local municipality to develop flood prevention infrastructure and create wetlands. The extracted quarry areas have been converted into stormwater reservoirs with a capacity of 2.5 million cubic meters, reducing the risk of flooding for local communities. Rehabilitation measures included the creation of shoreline areas and gently sloped riverbanks with varied contours, which have created diverse natural habitats such as ponds, resting places, and small islands that are favourable to many species. Research has shown that wetlands created from quarries in France have become a habitat for 132 species of birds (more than 48 percent of the French avifauna), 17% of the flora (1001 vascular plant species), and 63% of dragonflies found in France. Quarries have also become important refuge areas for many pro-tected species. In addition to improving biodiversity in the area, the measures have also resulted in water quality and recreational benefits. Provided that there is access to sufficient land area to accommodate quarrying activities, the approach adopted by LafargeHolcim could be replicated in other areas and result in similar benefits.

16.2.2 Reflections and Lessons Learned

These examples demonstrate the potential of NBS to address multiple needs; spe-cific interventions in, for example, reed beds, wetlands and forests can provide significant benefits to species' populations, while also improving water quality and quantity. Sharing these types of best practices via an online platform, such as the WBCSD Natural Infrastructure for Business platform, can help to promote investments in natural infrastructure, and provide the basis for developing similar initiatives, adapted to local contexts.

Based on IUCNs experience, a key criterion for successful partnerships with the business sector is a shared understanding of landscape, land use, ecosystem relation-ships, benefits of investment in natural capital, key policies, development strategies and legal frameworks, and rights and responsibilities over resources. Acting in partnership also means being clear on the values of different stakeholders, the needs of the natural environment and local communities. Business actors may for example want to know the quantified impacts of water shortage on their business operations, whereas conservation actors may want to assess the actual impacts of water pollution on biodiversity. These aspects should be kept in mind when developing partnerships with the private sector.

16.3 Citizen Participation and Leadership

From IUCN's perspective, recognising and respecting the rights of people who live close to and rely on nature is a central component of effective and inclusive conservation action.

Though citizen participation in environmental decision-making can bring its own sets of challenges (Irvin and Stansbury 2004), it can also support sustainable development (Abbott 2013) and is often promoted by governments based on the assumption that citizen participation can help make governance more democratic and effective (Irvin et al. 2004). There are numerous examples from around the world where people have come together to restore the landscape's ecological functions and enhance well-being within and around cities, for example by growing organic food, building nature-friendly spaces or restoring rivers and creeks (Herzog 2013; URBES 2014a). Such examples illustrate the potential of citizens to bring about meaningful social and environmental change.

While policymakers and urban planners recognise the value of engaging with local communities, engaging citizens in urban planning and management decisions is not always easy. Municipality-promoted participation processes require political support and backing, as well as mechanisms and policies that promote inclusive governance practices (Greensurge 2015). Funding is also needed to ensure high participation levels and sound participatory engagement processes.

The section below highlights instances of citizen leadership which have helped to integrate local concerns into environmental management plans and foster the delivery of a range of ecosystem services and benefits.

16.3.1 Examples of Partnerships Building on Citizen Participation

16.3.1.1 The Miyun Watershed (Beijing) – Illustrating the Value of Engaging Local Stakeholders

The Miyun watershed is generally understood to comprise the six sub-catchments of the Chao He and Bai He Rivers, which together feed the Miyun reservoir. Located north of Beijing, the catchment covers an area of 15,788 km2. In total, around one million people live in the watershed area and the reservoir supplies between 60–80% of urban drinking water needs; an estimated 17 million people rely on it for their drinking water. This makes the watershed one of the most important water protection areas in the world (Li and Emerton 2012).

In the past 30–40 years, several attempts had been made to reforest the Miyun landscape in response to worsening water crises (Li and Emerton 2012). Conifers (*Pinophyta*) and other species were planted to compensate for the disappearance of the original broadleaf forest, and strict controls on logging, land and forest use were implemented (Li and Emerton 2012). Due to a lack of active management however, the newly planted trees did not achieve a healthy ecological status – around three-quarters of the trees were categorised as 'sub-healthy' or 'unhealthy'. The strict controls also economically disadvantaged the local communities whose livelihoods had previously been associated with forest products (Li and Emerton 2012).

Against this backdrop, IUCN initiated a project in the Miyun watershed in 2007, which introduced a new set of forest management tools and brought about a shift from a strictly protective and very conservative regime, to one based on sustainable use and active management by local communities. The policy advocacy activities undertaken as part of the project focused on showcasing the multiple benefits of a multi-functional forest landscape. This reassured the Chinese government of the local community's capacity to responsibly manage the area's forests, and helped to bring about a formal agreement to recognise different forest management and use regimes, harmonising the technical information held by government foresters with local knowledge and interests (Li and Emerton 2012; IUCN 2012b).

By bringing together many diverse stakeholders and sectors at different levels, the project effectively developed a more integrated form of landscape management and restored the Miyun landscape in a way that recognises the multiple needs and functions of the watershed. With this approach, the initiative brought about a regeneration of natural forest and improvements in forest structure, quality and function (Li and Emerton 2012).

16.3.1.2 The Harava Survery Tool – An Innovative Mechanism to Support Citizen Participation in Vitoria-Gasteiz

As outlined above, citizen engagement requires significant investment and planning. However, there are a range of innovative tools that can support the process. Harava, for example, is an interactive map-based survey tool for smart planning, which enables organisations to conduct structured surveys with spatial data to inform decision-making, by collecting insights, ideas, and feedback from citizens who have practical knowledge and understanding of their surroundings.

Following an agreement between the city of Vitoria-Gasteiz (Spain) and Tecnalia, and with support from partners (SYKE Finnish Environment Institute), Harava was used to develop an urban management plan for Vitoria Gasteiz in 2013 (Ayuntamiento de Vitoria-Gasteiz, nd). The platform provided citizens with the opportunity to actively participate in the urban planning process by allowing them to take part in a survey covering a range of topics related to favourite and most frequented public areas, and more general views related to urban development and social inclusiveness.

Three hundred citizens participated in the two-month consultation process, providing information about 2497 spatial elements within the city (Herranz-Pasual et al. 2014). The consultation captured information such as how often and for what purpose citizens visit the city centre and rural areas. The tool also allowed citizens to convey their views on areas for improvement e.g. the need more trade and economic activities, particularly small businesses, to help make the city a more liveable place, as well as the need for better bike and pedestrian connections and public transport (Herranz-Pasual et al. 2014).

By engaging citizens in these types of participatory processes, governments have the opportunity to obtain information they might otherwise not have access to and on this basis, adapt spatial planning and management approaches to make cities more liveable and appealing to the people who live in them. Given that liveability is closely linked to the existence of green spaces (Beatley 2012) this type of collaboration has the potential to provide a sound basis for developing innovative solutions to urban challenges and implementing NBS, which benefit both local people and the environment.

16.3.1.3 Berlin – The Power of Citizen Engagement and Leadership

The Vitoria-Gasteiz example illustrates how governments can employ innovative mechanisms to engage citizens and ensure that city planning incorporates the needs and wishes of its people. But there are also instances where social activism has been the driver of conservation action.

In Berlin, Germany, the decision to protect the disused Tempelhof airfield from housing development and convert it into one of the city's most popular parks came as a result of a citizen-initiated referendum in May 2014 and a series of open community meetings, citizen working groups and consultations through an online platform. Citizen engagement in determining the future of the Tempelhof site started through public meetings, forums and lectures before the airport was closed in 2008. A web dialogue drew 68,000 users and 2500 idea contributors, and surveys were distributed to 6000 local households and to 1000 households in Berlin (Burgess 2014). Moderated focus groups were established to engage migrants groups, often marginalised in consultation processes. Following the closure of the airport, consultations and large-scale public events continued to take place. In 2009, 3500 people attended a "Call for Ideas", and more than 2000 people visited an Open House event showcasing concept ideas for developing the site (Burgess 2014). Finally, it was the "100% Tempelhofer Feld" civil society group, who pushed for the referendum that determined the future of the site (Burgess 2014).

Today, Tempelhof Park is one of the most popular parks in Berlin, hosting a variety of recreation facilities. Sealed areas such as former runways are used for cycling and running, while some lawn areas have become nature conservation zones and other zones have been designated for urban gardening, educational activities or recreational activities such as barbecuing (Burgess 2014). The 100% Tempelhofer Feld group continues to actively work to protect the natural areas and cultural heritage of the park and ensure continued open public access for the future.

The transformation of the Tempelhof Park in Berlin from a disused airfield to a lively urban park illustrates how strong social engagement can result in the creation or maintenance of green areas which support biodiversity conservation, contributing to urban resilience and providing cultural and recreational ecosystem services (URBES 2014b).

16.3.2 Reflections and Lessons Learned

The above examples illustrate how involving local users of natural resources in planning and decision-making can help to support the implementation of more effective environmental management regimes which benefit both people and nature.

Citizen engagement in urban planning and ecosystem management can be time consuming and costly, and requires the development of trust between stakeholder groups and flexibility to accommodate changes in planning and processes (Li and Emerton 2012). When done successfully however, citizen participation and engagement can support urban planning by helping to uncover the needs and wishes of local people, thereby providing the basis for increasing the livability of urban spaces, which has the potential to benefit both people and nature. Citizen engagement can also provide an entry-point to identifying potential NBS which could address the key societal challenges identified by urban citizens in a holistic manner that is respectful of community needs and aspirations.

16.4 Integrated Urban and Regional Planning for NBS

Governments are increasingly searching for cost-effective and holistic ways of addressing environmental challenges, which not only reliably deliver their immediate intended impacts, such as space for recreation and reduced air pollution, but also bring additional benefits to society, such as improved health and well-being.

Policymakers in cities and at the sub-national level can lead the way in making the transition towards increasing resilience and integrating ecological concerns within urban planning and decision-making. Instead of an infrastructure agenda in which nature is a problem, a cost, and a political risk, nature can become part of the solution.

Cities around Europe have already shown a commitment to integrating nature into their urban planning and management, and thereby demonstrated awareness of the importance of protecting natural capital. The Regional Climate Plan for Paris, for example, highlights the importance of protecting ecosystems in order to adapt to and mitigate climate change (Conseil Regional D'Île-De-France 2011). The city also recognises that forest management practices which optimise their capacity for adaptation and resilience can lead to multiple benefits for biodiversity, people and the city (Natureparif 2015).

Additional examples of cities that have integrated nature within their planning and strategies are outlined below.

16.4.1 Examples for the Integration of Nature in Urban Planning

16.4.1.1 Gibsons' Eco-Asset Strategy for Climate Adaptation and Resilience

Mapping and assessing ecosystems and their services is essential to ensure that their values are taken into account in decision-making and integrated across policies and sectors. The town of Gibsons, north of Vancouver in Canada, is pioneering a strategy that could contribute to the efforts of municipalities in Canada and elsewhere to improve climate resilience. Gibsons' "Eco-Asset strategy" focuses on identifying existing natural assets such as green space, forests, topsoil, aquifers and creeks that provide municipal services such as storm water management; measuring the value of the municipal services provided by these assets; and making this information operational by integrating it into municipal asset management. This is proving to be an effective financial and municipal management approach that complements strategies to maintain, replace and build both traditional engineered assets such as roads and storm sewers and engineered 'green assets' such as rain gardens, parks and bio-swales. Integrating natural assets into decision-making can support municipal climate change adaptation and resilience building efforts in a cost-effective way. Gibsons' aquifer, for example, which is part of the municipal asset management strategy, requires about $28,000 annually in monitoring costs. This is a cost-effective NBS to water security compared to the much higher operational costs of a filtration and treatment plant.

16.4.1.2 Philadelphia's Natural Solution for Stormwater Management

Philadelphia is another city in North America which has integrated nature into city planning (Qin et al. 2015). Already in the 19th century, the city had acquired approximately 3600 hectares of natural areas to help filter and regulate its potable water, and the land remains protected as parkland (Gartner et al. 2014). Confronted by frequent sewer overflows during storms, Philadelphia recently conducted a cost-benefit analysis of green infrastructure options—such as tree planting, permeable pavement and green roofs—and conventional grey options, such as storage tunnels (UNEP 2014). The economic benefits associated with green infrastructure ranged from $1.94 billion to $4.45 billion, compared to just $0.06 billion to $0.14 billion from grey infrastructure (UNEP 2014). In 2011, the city adopted the "Green City, Clean Waters" plan to reduce stormwater pollution by greening public spaces and creating a living landscape that slows, filters and consumes rainfall. City officials expect to reduce stormwater and sewage pollution entering the waterways by 85% when the project is completed (Qin et al. 2015).

16.4.1.3 Nature Flood Management in the UK – "Slow the Flow"

In the UK, a natural flood management scheme played a prominent role in preventing floods in a small town in North Yorkshire, Pickering, in December 2015. An upstream flood storage reservoir was installed, 40,000 trees planted and heather moorland restored to soak up incoming water (Harrabin 2016). Based on the success of this "Slow the Flow" scheme, options for developing a 25-year plan which looks at the management of river catchment areas to improve flood resilience for the environment are now being explored (Harrabin 2016). A major study by Forest Research, an arm of the Forestry Commission in the UK, recently found that planting trees in hills and along watercourses could significantly reduce flooding, soil erosion and water pollution and highlighted the need to "increase incentives for woodland planting by making these better reflect the full range of water and other benefits" (Nisbet et al. 2011). Quantifying water benefits and evaluating how woodland can be best integrated with agriculture and urban activities for water and wider environmental benefits, while minimising any water trade-offs, is a critical step in order to garner support from local stakeholders such as landowners and farmers (Nisbet et al. 2011).

16.4.2 Reflections and Lessons Learned

Integrating nature within planning and policies can have clear benefits for citizens, not only by improving water quality or climate resilience, but also by saving money. Green spaces in cities can also add value to commercial and private property and can contribute to a city's tax revenues as well as attract more visitors and private sector investment (CBD 2012).

Urban planning often fails to fully recognise the connection between cities and their natural surroundings. Natural infrastructure must play a more influential role in the planning and design of cities and urban regions, but this is often hampered by budgetary constraints. The examples above demonstrate that making the protection of natural assets and enhancement of ecosystem functions a prominent part of decision-making can offer cost-effective solutions to a range of challenges.

A major challenge to upscaling the implementation of NBS is the lack of a solid evidence base showcasing the benefits of NBS over traditional approaches to climate change adaptation. As a result, policy-makers tend to favour the implementation of traditional engineering solutions for climate adaptation, instead of investing in NBS (Rizvi et al. 2015). More concrete data on the cost-effectiveness of nature-based approaches and field evidence is required to showcase the solutions ecosystems have to offer (Rizvi et al. 2015) and the benefits they can bring.

16.5   Conclusion

This chapter has demonstrated that multi-stakeholder partnerships for NBS can lead to substantial social, economic and environmental benefits and can support adaptation to climate change (see Table 16.1 below).

The chapter also highlighted a number of lessons that can inform future partnerships for NBS:

Table 16.1   Summary of NBS partnerships and resulting benefits discussed in this chapter

Example	Type of partnership	Benefits
Natural Infrastructure for Business	Online platform for members (businesses) of the World Business Council for Sustainable Development	Increased awareness of the business opportunities for investing in ecosystems
Volkswagen, ecosystem restoration initiative in Puebla-Tlaxcala Valley, Mexico	Volkswagen, National protected areas Commission and Secretary of the Environment, Mexico	Secure drinking water supply and water supply for Volkswagen's production plant, carbon sequestration, biodiversity
Rehabilitation of quarry in Bellegarde, France	LafargeHolcim, French National Musueum of National History, local municipality Bellegarde	Stormwater management, flood prevention, biodiversity, water quality, recreational benefits
Miyun watershed, Beijng	IUCN (with support from DGIS–Netherlands Ministry of Foreign Affairs), local communities, local government	Strengthened livelihoods, sustainable forest use, regeneration of natural forest and improvements in forest structure, quality and function
Vitoria-Gasteiz citizen participation	Municipality of Vitoria-Gasteiz, citizens, Tecnalia, SYKE Finnish Environment Institute	Access to information to make the city more liveable, sustainable urban development
Citizen engagement in Tempelhof Park, Berlin	City of Berlin, citizen working groups	Public access to Tempelhof Park, providing natural space and cultural and educational opportunities
Gibsons Eco-asset Strategy	Municipality of Gibsons, scientific partners and engineers	Mapping and assessing ecosystems and their services within the municipality and integration of the value of nature into municipal asset management
Philadelphia's stormwater management	Philadelphia Water Department, private developers, US Environmental Protection Agency, universities, citizens	Cost-benefit analysis to integrate natural solutions into city planning for reduced storm water and sewage pollution
Pickering "Slow the flow"	Forestry Commission, Town of Pickering	Improved flood resilience, assessment of water trade-offs with agriculture and urban development

- Multidisciplinary and inclusive partnerships can foster the uptake of NBS in response to climate-related challenges. They can create and catalyse synergies between different parts of society by pooling together resources skills and knowledge.
- Involving citizens in urban decision making can help to make cities more liveable, identify opportunities for implementing NBS, and create trust, ownership and stewardship.
- Innovative tools (e.g. Harava) can help to incorporate different stakeholder views within urban planning and policymaking and have the potential to support the development of NBS.
- Creating trust and learning to understand each other's language better can help to form the basis for joint action.
- The development of replicable business models which quantify the values of nature at local and landscape level and present a reliable return on investment can help to gain private sector support and leverage public investment in NBS.
- There is a need to develop a more solid evidence base on the multiple benefits, and particularly the cost-effectiveness of nature-based approaches to gain more wide-spread support for NBS at city level. Experts on measuring the qualitative and quantitative economic and social benefits and services provided by ecosystems, can help to create visibility for the value of a city's natural assets and promote the uptake of NBS in urban planning and management.

Sharing these types of examples and lessons can serve as a strong foundation for promoting NBS and can help to inspire future partnerships for, and investments in NBS.

References

Abbott J (2013) Sharing the city: community participation in urban management. Routledge. Ayunamiento de Vitoria-Gasteiz (nd) Plataforma Harava-Encuesta Online. Informe de Resultados ¡Ayúdanos a ver Vitoria-Gasteiz! 1ª Fase del Proceso Participativo para el Avance de la Revisión del PGOU de Vitoria-Gasteiz. Available at: http://www.vitoria-gasteiz.org/wb021/http/contenidosEstaticos/adjuntos/es/47/16/54716.pdf Accessed 18 May 2016

Beatley T (2012) Green urbanism: learning from European cities. Island Press, Washington, DC

Buchner BK, Trabacchi C, Mazza F, Abramskiehn D, Wang D (2015) Global Landscape of Climate Finance. Climate Policy Initiative

Burgess K (2014) Community participation in parks development: two examples from Berlin [Blog] the nature of cities. Available at: http://www.thenatureofcities.com/ 2014/12/10/community-participation-in-parks-development-two-examples-from-berlin/ Accessed 18 May 2016

Conseil Regional D'Île-De-France (2011) Plan Regional Pour Le Climat D'Île-De-France. Available at: https://www.iledefrance.fr/sites/default/files/mariane/RAPCR43-11RAP.pdf

Convention on Biological Diversity (2012) Cities and biodiversity outlook. Montreal. Available at: http://www.cbobook.org

Frantzeskaki N, Wittmayer J, Loorbach D (2014) The role of partnerships in 'realizing' urban sustainability in Rotterdam's City Ports Area, the Netherlands. J Clean Prod 65:406–417

Gartner T, Mehan GT III, Mulligan J, Roberson JA, Stangel P, Qin Y (2014) Protecting forested watersheds is smart economics for water utilities. J Am Water Works Assoc 106(9):54–64

Greensurge (2015) Report of Case Study City Portraits. Available at: http://greensurge.eu/filer/GREEN_SURGE_Report_of_City_Portraits.pdf

Harrabin R (2016) Pickering leaky dams flood prevention scheme 'a success'. BBC. Available at: http://www.bbc.com/news/uk-england-york-north-yorkshire-36029197 Accessed 11 May 2016

Herzog C (2013) People Take Over Nature in Cities with their Own Hands. Available at: http://www.thenatureofcities.com/2013/11/10/people-take-over-nature-in-cities-with-their-own-hands/

Herranz-Pasual K, Oaniagua C, Abajo B, and Feliu E (2014) How do you imagine Vitoria-Gasteiz in 2025 – Experiences of citizen participation in urban planning using the Harava tool [Blog] Available at: http://sateenvarjolla.blogspot.be/2014/03/how-do-you-imagine-vitoria-gasteiz-in.html Accessed 18 May 2016

Irvin RA, Stansbury J (2004) Citizen participation in decision making: is it worth the effort? Public Adm Rev 64(1):55–65

IUCN (2012a) IUCN business engagement strategy Available at: http://www.iucn.org/theme/business-and-biodiversity/resources/working-business-and-biodiversity-programme

IUCN (2012b) Livelihoods and landscapes strategy: results and resolutions. Gland: IUCN. Available at: https://portals.iucn.org/library/node/10221

Kabisch N, Frantzeskaki N, Pauleit S, Naumann S, Davis M, Artmann M, Haase D, Knapp S, Korn H, Stadler J, Zaunberger K, Bonn A (2016) Nature-based solutions to climate change mitigation and adaptation in urban areas: perspectives on indicators, knowledge gaps, barriers, and opportunities for action. Ecol Soc 2(2)

Kongrukgreatiyos K (2014) The private sector's role in climate change resilience [blog] the Rockerfeller foundation Available at: https://www.rockefellerfoundation.org/blog/private-sectors-role-climate-change/ Accessed 10 May 2016

Li J, Emerton L (2012) Moving closer to nature: lessons for landscapes and livelihoods from the Miyun landscape, China. IUCN, Gland, Switzerland

McDonald RI, Marcotullio PJ, Güneralp B (2013) Urbanization and global trends in biodiversity and ecosystem services. In: Urbanization, biodiversity and ecosystem services: challenges and opportunities. Springer, Dordrecht, pp 31–52

Munich RE (2015) Insurance solutions for industry. Topics Risk Solutions, Issue 3. Available at: http://www.munichre.com/us/weather-resilience-and-protection/rise-weather/productions-publications/publications/topics-risksolutions/index.html

Natureparif (2015) Climate: Nature-based solutions for climate change mitigation and adaptation in Paris Region–Overview of Propositions for discussion at the 21st session of the Conference of the Parties on Climate Change (COP21). Paris: Natureparif. Available at: https://www.researchgate.net/publication/284030171_Climate_Nature-based_solutions_for_climate_change_mitigation_and_adaptation_in_Paris_Region. Accessed 18 May 2016

Nisbet T, Silgram M, Shah N, Morrow K, Broadmeadow S (2011) Woodland for water: woodland measures for meeting water framework directive objectives. Forest Research Monograph, 4. Available at: http://www.forestry.gov.uk/pdf/FRMG004_Woodland4Water.pdf/$FILE/FRMG004_Woodland4Water.pdf

Ozment S, DiFrancesco K, Gartner T (2015) The role of natural infrastructure in the water, energy and food nexus, Nexus Dialolgue Synthesis Papers. Switzerland: IUCN

Qin Y, Gartner T, Otto B (2015) Cities can save money by investing in natural infrastructure for water [Blog] world resources institute. Available at: http://www.wri.org/blog/2015/10/cities-can-save-money-investing-natural-infrastructure-water?utm_campaign=wridigest&utm_source=wridigest-2015-10-27&utm_medium=email&utm_content=learnmore Accessed 10 May 2016

RESTORE–Rivers: Engaging, Supporting and Transferring Knowledge for Restoration in Europe (2016). Available at: http://ec.europa.eu/environment/life/project/Projects/index.cfm?fuseaction=search.dspPage&n_proj_id=3780&docType=pdf Accessed 10 May 2016

Rizvi AR, Baig S, Verdone M (2015) Ecosystems based adaptation: knowledge gaps in making an economic case for investing in nature based solutions for climate change. Gland: IUCN. Available at: https://portals.iucn.org/library/node/45156

Rushworth J, Warau M (2015) World Business Council for Sustainable Development natural infrastructure case study. Water Management and Flood Prevention in France, http://www.naturalinfrastructureforbusiness.org/wp-content/uploads/2015/11/LafargeHolcim_NI4BizCaseStudy_WaterManagementFloodPrevention.pdf

Science for Environment Policy (2016) Research for environmental policymaking: how to prioritise, communicate and measure impact. Thematic Issue 54. Issue produced for the European Commission DG Environment by the Science Communication Unit, UWE, Bristol. Available at: http://ec.europa.eu/science-environment-policy Accessed 18 May 2016

UNEP (2014) Green Infrastructure Guide for Water Management: Ecosystem-based management approaches for water-related infrastructure projects. Available at: http://www.medspring.eu/sites/default/files/Green-infrastructure-Guide-UNEP.pdf Accessed 18 May 2016

URBES (Urban Biodiversity and Ecosystem Services Project) (2014a) Factsheet 7: Urban agriculture: landscapes connecting people, food and biodiversity. Available at: https://www.iucn.org/sites/dev/files/import/downloads/urbes_factsheet_07_web_2.pdf

URBES (Urban Biodiversity and Ecosystem Services Project) (2014b) Factsheet 5: Urban resilience and sustainability, two sides of the same coin? Available at: http://www.mistraurbanfutures.org/sites/default/files/urbes_factsheet_05_web.pdf

World Business Council for Sustainable Development (2010) Volkswagen–Replenishing groundwater through reforestation in Mexico. Responding to the Biodiversity Challenge Business contributions to the Convention on Biological Diversity. Available at: http://www.wbcsd.org/web/nagoya/RespondingtotheBiodiversityChallenge.pdf. Accessed 18 May 2016

Chapter 17
The Challenge of Innovation Diffusion: Nature-Based Solutions in Poland

Jakub Kronenberg, Tomasz Bergier, and Karolina Maliszewska

Abstract Nature-based solutions (NBS) are currently seen and discussed as innovations, including within the European Commission. We assume that this should result in their broader popularity and implementation in EU countries. We analyse the diffusion of NBS in Poland, a post-socialist country, in the case of which less has been written on NBS and urban green and blue infrastructure than in West European countries. In spite of the above assumption, we indicate that the rate of NBS acceptance in Poland is relatively low and their visibility is limited. Our study uses Amoeba, a tool for understanding, mapping and planning for innovation diffusion and cultural change processes to understand the reasons for this situation and to seek the methods of its improvement. We focus on two case studies, green roofs and ecological corridors, and analyse the roles played by different stakeholders, their attitudes towards these innovations and their influence on NBS diffusion in Poland, as well as the interactions between them.

Keywords Diffusion of sustainability innovations • Stakeholder analysis • Urban green infrastructure • Green roofs • Ecological corridors

J. Kronenberg (✉)
Faculty of Economics and Sociology, University of Lodz, P.O.W. 3/5, 90-255 Lodz, Poland
e-mail: kronenbe@uni.lodz.pl

T. Bergier
Department of Environmental Protection and Management, AGH University of Science and Technology, Mickiewicza 30, 30-059 Krakow, Poland
e-mail: tbergier@agh.edu.pl

K. Maliszewska
Sendzimir Foundation, Chocimska 12/4, 00-791 Warsaw, Poland
e-mail: karolina.maliszewska@sendzimir.org.pl

© The Author(s) 2017   291
N. Kabisch et al. (eds.), *Nature-based Solutions to Climate Change Adaptation in Urban Areas*, Theory and Practice of Urban Sustainability Transitions,
DOI 10.1007/978-3-319-56091-5_17

17.1   Introduction

Nature-based solutions (NBS) are promoted by the European Union as an innovation meant to solve many societal problems. As a supposedly new idea, an innovative solution to outstanding problems, they are being promoted by the European Commission's Directorate-General for Research and Innovation which is responsible for defining and implementing European Research and Innovation policy. The European Commission and its experts (2015) argue that NBS fit well into the dominant discourse on 'sustainable and green growth' that NBS are cost-effective and that they offer a business opportunity for European companies to take the lead in this area in international markets.

If EU authorities see NBS as a window of opportunity not only to protect the environment but also – or perhaps principally – to improve business prospects and the position of the EU in international markets, then we can assume that this approach will be further reflected in national policies and on-the-ground management in EU countries. However so far, the EU discourse on NBS seems to have attracted relatively little attention in Poland (Kronenberg 2016). Indeed, Poland, which is one of the new post-socialist EU members, with an economy which has undergone a radical transformation, often reveals many differences in how new concepts and political ideas spread, compared to the relatively better known Western democratic countries (Kronenberg and Bergier 2012).

In this chapter we aim to address the following research questions: why the concept of NBS is so slowly accepted in Poland, what factors and drivers control the process of its diffusion, and what are the challenges and opportunities to promote it further? To realise these goals, we use Amoeba – a tool developed by Alan AtKisson to analyse the dynamics of cultural changes leading to the widespread acceptance of the innovation, especially those connected with sustainable development (AtKisson 2009). We explain the broader context of NBS in Poland and apply Amoeba to two examples – green roofs and ecological corridors. Finally, based on our analysis, we draw broader conclusions regarding further opportunities to promote NBS in Poland.

17.2   Method

While analysing the dynamics of innovation diffusion, it is crucial to understand the roles played by the different stakeholders, their interests and reasons why they promote or hinder an innovation, as well as interactions between them. To successfully transform an innovation to the mainstream, it has to be accepted by the public (mainstreamers); however the mechanisms leading to this shift are very complex, and some social groups have the crucial role in this process (e.g. leaders, celebrities, early adopters). To describe and study the process of NBS acceptance in Poland, we decided to use the method called Amoeba (AtKisson 2009), which was designed for such purposes.

This method uses the metaphor of an amoeba to describe and understand the process of innovation diffusion. An amoeba extends a pseudopodium ('false foot') to reach food, the rest of its body is dragged into that direction and it consumes the food item after completely surrounding it. Innovation acceptance by the society follows an analogous pattern: a food item represents an innovation, a pseudopodium – an innovator, who initiates the move of the whole society towards the innovation. However, the innovator alone is too weak and distant to do so; thus the role of change agents and transformers is so important – they mount the innovation into the society and have the power to make the movement more massive. Meanwhile, there are groups who slow down or block the process. The stakeholders and their roles are described in Table 17.1 and shown in Fig. 17.1. The sum of their activities controls the dynamics of innovation diffusion and eventually decides whether an innovation enters the mainstream or is rejected (or stays in a niche). The goal of Amoeba is to better understand the dynamics of these processes, the power balance and interactions between stakeholders. The method could be used to analyse any innovation, and any stakeholder group or their role has no positive or negative connotation (for instance, a sustainability activist could be a reactionary in a case of nuclear energy).

We decided to use Amoeba to analyse the dynamics of NBS acceptance in Poland because it provides a clear structure and makes it possible to map all stakeholders, but also to comprehensively analyse and describe their roles and influence on innovation diffusion. Based on the results of such an analysis, it is also possible to suggest the means of innovation promotion, as well as to identify the crucial

Table 17.1 Key stakeholders represented in Amoeba and their roles

Name	Description
Innovator	The source of new ideas (e.g. an inventor, a thinker)
Change agents	Translating an innovation into an idea that can sell. Although they remain outside of the mainstream, they know how to communicate with the mainstream (e.g. consultants and marketing specialists)
Transformers	Early adopters of an innovation. They are keen to adopt new ideas and want to promote positive change. However, they would not accept to do so at the expense of their own credibility, position and influence. Hence they only adopt innovations that they feel the mainstream would ultimately adopt
Mainstreamers	Representatives of the majority who are neither for nor against change. They adopt an innovation when they see that 'everybody else' does so
Laggards	A group of mainstreamers who are happy and comfortable with the status quo and who resist change as long as they can (until the mainstream changes); hence they are called late adopters
Reactionaries	Those who have vested interests which can be harmed by an innovation (or at least they think so); thus they actively resist the adoption of an innovation
Controllers	The most influential stakeholders who set the rules in the system. They react to how the system evolves but sometimes they actively shape the evolution of this system

Note that the original Amoeba features more roles – as illustrated in Fig. 17.1 (AtKisson 2009)

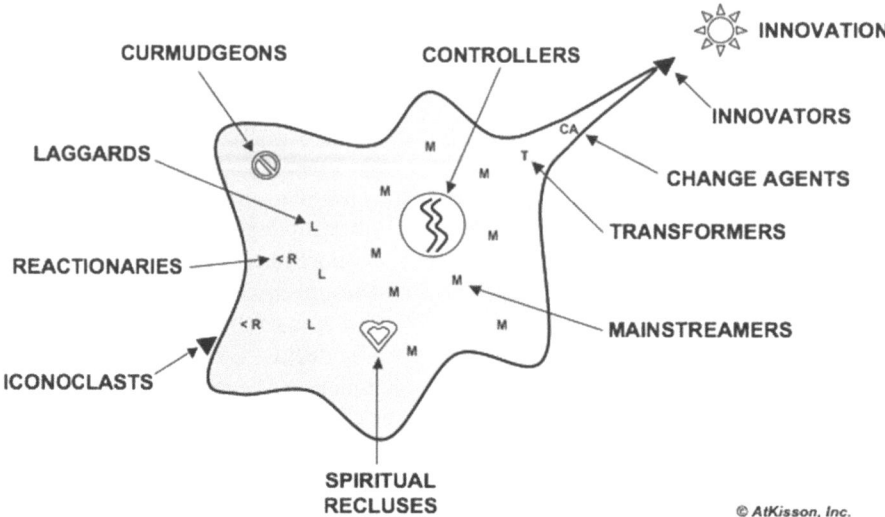

Fig. 17.1 The Amoeba metaphor – roles played by the different stakeholders (Courtesy of Alan AtKisson). For the sake of brevity and simplicity, in this chapter we omitted some of the less important roles. For a full overview, see AtKisson (2009)

alliances. The latter Amoeba's qualities are especially useful when one works on sustainability innovations. The main limitation of this method is its application in the case of a 'fuzzy' situation, especially if one organisation plays different roles in a system (e.g. an NGO is both an innovator and a change agent) – then it could be difficult to decide on how to categorise it.

There are three main sources of information for the Amoeba analyses we conducted in the chapter:

- Own experience and knowledge of the system, gained from several years of research on the diffusion of sustainability innovations in Poland, especially those connected with NBS and green and blue infrastructure, as well as our active participation in some of these processes,
- Desk research, in which we gathered and analysed information on NBS in Poland, mainly from websites and a very limited number of articles and other publications,
- Interviews with stakeholders involved in these processes.

17.3   Nature-Based Solutions in Poland

Although environmental protection was far from being a priority in socialist cities and many green spaces were degraded by polluted water or other by-products of industrial activity, green spaces belonged to the most important aspects of urban planning. Then, the free-market economy brought an overarching focus on satisfying individual needs and freedom to do whatever one wishes on one's land, often coupled with a neglect of public interest. Only in the most recent years, awareness of the broader benefits provided by nature has been rising, and urban green spaces have started to attract increasing attention from the inhabitants and, consequently, from the authorities. Unfortunately, nature is still often seen as a barrier to development, when new investments collide with the remnants of nature, and urban ecosystem disservices seem to be better known than ecosystem services (Kronenberg 2015).

Examples of what we would now call NBS (i.e. conscious use of nature to help urban inhabitants address various environmental, social and economic challenges) that were implemented already in the socialist period included especially a system of ecological corridors. These corridors were planned as green and other open spaces, which were meant to facilitate air exchange in cities. On a smaller scale, green spaces were used for isolation from noise and pollution and to improve health conditions, especially around hospitals and educational facilities. While efforts have been made by urban planners to protect these corridors and other green spaces, they have been under constant pressure from the expansion of built-up areas. In Poland, such pressures have intensified after the fall of socialism and urban spatial planning has become weakened by a number of deregulatory activities.

Nevertheless, environmental degradation which had taken place in the socialist period paved way for new attempts to rehabilitate some urban rivers and green spaces after the fall of socialism. Several 'renaturalisation' projects have been carried out to improve the condition of urban ecosystems. Some of these have been combined with floodwater management, but most focused on recreational opportunities and aesthetics. However, unlike in the socialist period, green spaces have not been seen as solutions to any specific problems related to urban life, rather as an additional aspect of the broader quality of life in cities.

Most recently, discussions on urban nature have intensified, perhaps because its degradation has achieved thresholds that are no longer acceptable to the society or perhaps because of the international trends which make their way to Poland. For example, street trees are increasingly the source of conflict because more and more often urban inhabitants oppose the fact that city authorities uncritically allow for their removal (NIK 2014; Krynicki and Witkoś Gnach 2016). Although the inhabitants are also not always protecting trees, they are in principle in favour of their preservation and are generally aware of the many benefits they provide (Giergiczny and Kronenberg 2014). Other examples of NBS that are increasingly implemented in Poland include urban beekeeping and green roofs, both of which can be linked to the broader initiatives aiming at urban greening.

To sum up the above overview, NBS have been used already in the socialist period, although not fully consistently and without the modern 'hype' that surrounds this concept. There are many examples of NBS in use in Poland, many of which have been developed in the past and still survive (or even thrive), while others have been introduced recently and are usually on the rise – but at a slow pace. Figure 17.2 presents examples of such solutions and their current standing in Poland, depending on their time of origin (by 'old' we mean those developed already in the socialist period and 'new' have only been introduced recently) and implementation dynamics

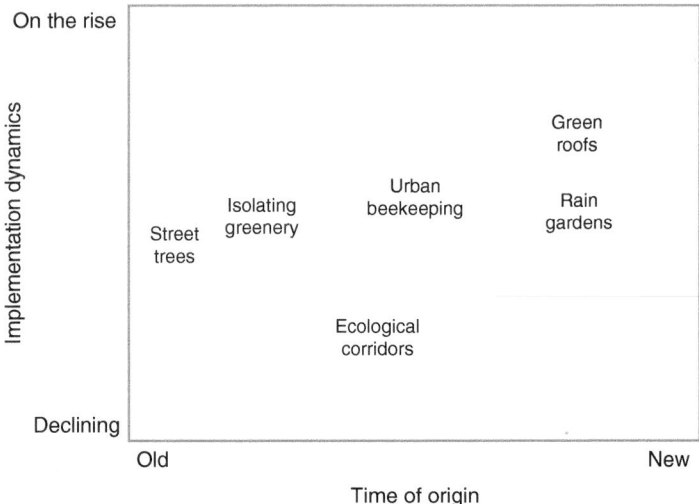

Fig. 17.2 Examples of NBS currently in use in Poland, divided by their time of origin and implementation dynamics (Source: The author's own work)

(whether they are increasingly accepted and used or the opposite). To present the broader context of NBS in Poland, in the following subsections we analyse two extreme cases from Fig. 17.2: an example of a new solution which is on the rise (green roofs) and an example of an old solution which is declining in spite of attempts to restore its importance (ecological corridors). The former example provides an overview on a national level, while the latter concerns an individual city (and a specific concept that is meant to promote ecological corridors in that city). Addressing both examples allows us to see similar mechanisms operating at different scales.

17.3.1 Case Study 1: The Partial Success of Green Roofs

Green roofs provide a broad range of benefits in urban areas which suffer from a significant loss of biologically active surface (Van Mechelen et al. 2015). Besides aesthetic and recreational aspects, they positively influence air quality (removing particles and other pollutants, absorbing CO_2 and producing oxygen), thermal balance (additional insulation, less energy for heating and/or air-conditioning, reducing the urban heat island effect), water balance and flood protection (stormwater retention and evapotranspiration) and biodiversity (connections between urban green areas, especially for birds and insects). They are also more durable and long-lasting than traditional roofs (Bozorg Chenani et al. 2015), also in the case of extreme meteorological events (strong winds, heavy rains, hails, etc.). They are also considered an important tool of urban climate change adaptation (Brenneisen and Gedge 2012).

Even though there are no official statistics, it is visible that green roofs are increasingly popular in Polish cities. Warsaw is the leader in the construction of green roofs (Kania et al. 2013), but they have been installed also in other cities (Energie Cités 2014). Interestingly, there are relatively old examples of green roofs in Polish cities, created in the 1990s, before the idea became widely discussed, e.g. on car parks. In 1999, one of the most inspiring and spectacular examples was created on the Warsaw University Library (Kowalczyk 2011). Other examples followed, such as the Copernicus Science Centre in Warsaw, the National Museum in Krakow, the International Conference Centre in Katowice and shopping malls in different cities. In the case of housing, green roofs are installed very rarely, and if they occur, it is rather on supplementary objects (e.g. garages) or terraces (Kania et al. 2013). Paradoxically, information on green roofs implemented in Poland is rather hard to find outside of trade press. They are still not seen as something that could be used for promotional purposes. Indeed, in the case of housing developers, greening of roofs and other horizontal surfaces is caused by the local regulations preserving the biologically active area, rather than due to the pressure of customers and their awareness of the benefits of green roofs.

The relatively successful diffusion of green roof innovation in Poland is related to the role played by the scientists and NGOs. Polish Green Roof Association (Polskie Stowarzyszenie Dachy Zielone), the partnership of scientists and NGOs, is particularly active and effective in this regard. The Association's main goal is to develop and provide knowledge on the benefits provided by green roofs and the technical guidelines on their design (Kania et al. 2013), as well as collect and promote good examples from Poland and abroad (Energie Cités 2014). A similar role is played by the on line journal *Dachy Zielone* (Green Roofs).

The important force in the dynamics of this innovation's diffusion has been the inspiration and influence of good examples from abroad. However, recently several other supporting mechanisms have been introduced in Poland, for instance, some cities introduced stormwater fees and other restrictions on its release to sewers or surface waters. Furthermore, there is a possibility to use participatory budgets within which citizens decide how to allocate part of a municipal budget to finance pilot installations of NBS, such as green roofs and walls, but also rain gardens and pocket wetlands.

The importance and activities of the actors, influencing the dynamics of green roofs diffusion in Poland, are presented in Table 17.2. The lack of strong and clearly defined innovator is characteristic for most NBS (including both cases analysed in this chapter). Another characteristic phenomenon is the small current activity of controllers, caused probably by the still low implementation rate of this innovation.

Green roofs illustrate an innovation, which is currently in the key turning point in Poland. On the one hand, it is not anymore the avant-garde, the odd novelty, associated with concerns about its durability and safety. On the other hand, it has not reached the status of a widely used and accepted technology, yet, that would be predictable and routinely designed and applied by the representatives of the mainstream construction industry. It is possible that it will pass through the critical phase

Table 17.2 Amoeba roles played by the different stakeholders in the case of green roofs in Poland

Amoeba role	Actor in the case study	Role played in the case study
Innovator	–	It is impossible to define an innovator in this case study. Green roofs are an innovation of fuzzy and unknown origin. However, examples from abroad seem to play the most important role in their diffusion and promotion in Poland
Change agents	NGOs (e.g. Polish Green Roof Association, Energie Cités, Sendzimir Foundation), universities (e.g. Wroclaw University of Environmental and Life Sciences), pioneering investors, designers and architects (e.g. Marek Budzynski)	Development and popularisation of green roofs, scientific discourse, advertising good examples, empowering others, organising conferences and workshops Spectacular investments
Transformers	Selected departments within city offices Mainstream media Big, significant developers and investors, installing green roofs European Union	Mechanisms encouraging investors to install green roofs (e.g. financial and legislation solutions, stormwater management tools) Participation in scientific and pilot projects on NBS (e.g. city offices of Radom, Krakow) Articles and other forms popularising green roofs among construction companies and investors (e.g. *Murator* – the biggest Polish journal on construction of individual houses) Financing the scientific and pilot projects Installing green roofs within municipal investments
Controllers	Ministries Top city authorities responsible for construction regulation Governmental institutions	Institutions capable of introducing regulations, which could enforce green roof installation on a massive scale (both locally and nationally). They could also contribute to the popularisation and credibility of green roofs by installing them on governmental and public buildings. Currently, there are only a few activities of the representatives of this group
Mainstreamers	Private investors Architects and designers Construction companies and developers Residents City officers (spatial planning, local development, municipal investments, etc.)	All individuals and organisations responsible for the design of new buildings, deciding about their technical aspects, both in a scale of individual private buildings, as well as bigger commercial investments, and municipal and public ones, up to the scale of a whole city
Laggards	Construction companies and developers, using traditional roof technologies	Companies specialised in traditional technologies, often not prepared for the transition

(continued)

Table 17.2 (continued)

Amoeba role	Actor in the case study	Role played in the case study
Reactionaries	Conservative architects and designers Traditional roofing industry Conservative private investors, using traditional roofing technology Departments of city halls and municipal institutions	Technical inertia, technical fears and doubts (safety, leakage, higher costs), general fear of novelty and – characteristic for Poles – very conservative approach to building technologies in the construction of houses (especially single-family ones) Attachment of urban policy makers to the technical guidelines and investment, developed through decades (analogous mechanisms as above) These fears are supported by the representatives of traditional – construction companies, which perceive green roof as a threat to their dominant market position These mechanisms are reinforced by the lack of reliable statistics and scientific research, and clear regulations and design guidelines

Source: The author's own work
For a general description of stakeholder roles, see Table 17.1

('rebound' point, after which the acceptance rate is increasing rapidly) and will be accepted by the mainstreamers as a technology widely used to cover roofs in Poland. However, it is difficult to determine when this may occur. Despite the fact that the area of green roofs installed every year in Poland has a growing tendency, the growth rate is still significantly lower than in the leading countries (e.g. Germany). Thus, there is a huge potential to benefit from international cooperation (e.g. good practice, technical know-how, financial and exploitation results), as well as EU support. However, there is also necessity to initiate the national and local actions that could contribute to a more efficient and faster diffusion of green roofs in Poland.

The innovation's diffusion theory points out the particular importance of collaboration between institutions working for this innovation, especially among change agents, and between change agents and transformers (c.f. AtKisson 2009). This is also confirmed by the analysed case study, which highlights the crucial collaboration between NGOs, universities and green roofers (change agents). Such collaboration can lead to practical and in-depth research and publications, as well as reliable statistics concerning green roofs installed in Poland. This in turn could create an opportunity to improve cooperation between change agents and the media (transformers), and widely popularise research results and statistics, as well as design guidelines and/or catalogues of good practices. Conversely, collaboration between change agents and municipal institutions (other transformers) could result in creating the local (municipal) programmes for financing and promoting green roofs, as well as training courses for designers and contractors (improving their technical capacity, competitiveness and business opportunities).

17.3.2   Case Study 2: The Failure of Ecological Corridors

Ecological corridors (plus belts and patches) in cities represent the key structure of urban ecosystems and are now often associated with the concept of urban green and blue infrastructure. As indicated earlier, they were implemented already in the socialist period and have been under pressure from urban growth since the fall of socialism. Similar patterns could have been seen in all Polish cities, but in some cities urban sprawl and pressure on green spaces within urban areas have been especially acute, e.g. the largest cities of Warsaw (Gutry-Korycka 2005), Krakow (Böhm 2007), Lodz (Kronenberg et al. 2017) and Poznan (Kotus 2006). In all of these cities, some stakeholders promoted the ideas of green belts and wedges, often combined with the rehabilitation of some areas that had already been covered by construction, but with little success. The concept of the Blue-Green Network (BGN) put forward in Lodz provides an illustration of such an attempt to consciously use NBS, including through ecosystem restoration, ecohydrology and ecological engineering (Wagner et al. 2013).

The concept of the BGN has been developed as a solution to multiple problems of Lodz, such as stormwater runoff, local flooding and droughts, heat waves, poor air quality and increased prevalence of allergy and asthma, low levels of resilience of urban ecosystems and perceived low quality of public spaces. The BGN encompasses a network of ecological corridors connecting the centre with large green belts surrounding the city. The corridors consist of both existing green spaces (including some that need to be rehabilitated) and the newly constructed ones (including dry reservoirs to increase stormwater retention and infiltration and sedimentation of pollutants, etc.). Ecologically restored (or at least rehabilitated) river valleys are meant to serve as the most important connectors within the BGN.

The concept of the BGN has been built on previous planning documents and scientific analyses, through a participatory process carried out within the EU FP7 SWITCH project (Managing Water for the City of the Future). The initiator of the project, the European Regional Centre for Ecohydrology (ERCE), under the auspices of UNESCO established a Learning Alliance (LA) to promote stakeholder engagement in 2006. The LA was joined by a broad array of stakeholders, from local government institutions, through local media, to schools and NGOs. The roles of the different actors – translated into the roles differentiated in Amoeba – are presented in Table 17.3.

Even though the concept of the BGN has been around for 10 years, it is far from becoming a reality, and in practice only few demonstration projects have been implemented to test potential solutions and to promote the concept. The BGN concept has been incorporated into various strategic and planning documents of the city; it is sometimes discussed as one of the key aspects of future development of Lodz, but still other priorities and interests are favoured over the use of NBS. In particular, like in other Polish cities, the preservation of ecological corridors in Lodz is challenged by poor spatial planning and by numerous other institutional failures that inhibit urban greening in general (Kronenberg 2015). Less than one third of the country's area is covered with local spatial management plans that stipulate the allowed land use patterns (Kowalewski et al. 2013). In the remaining area, decisions regarding land use (construction in particular) are made ad hoc, upon an investor's request, favouring private benefits over public interests.

From the point of view of innovation diffusion which can be captured with Amoeba, the role of the LA has been particularly important in promoting the BGN concept in Lodz (Wagner et al. 2013). The LA has served as a forum where the innovation could be promoted and where it could have been caught on by other stakeholders. The diverse group of LA participants made it possible to exchange

Table 17.3 Amoeba roles played by the different stakeholders in the case of the Blue-Green Network in Lodz

Amoeba role	Actor in the case study	Role played in the case study
Innovator	–	The BGN is not an innovation per se; rather it is a repackaged set of previous ideas, planning documents and analyses in a form that is meant to sell better. Hence, there is no innovator in this system
Change agents	European Regional Centre for Ecohydrology (ERCE) under the auspices of UNESCO Department of Public Utilities (City of Lodz Office) Sendzimir Foundation Other NGOs promoting the BGN concept	ERCE adapted previous ideas regarding ecological corridors, green belts and wedges and urban ecosystem restoration into the BGN concept. ERCE promoted it through its contacts with other stakeholders and implementation of small-scale demonstration projects. For the purposes of promoting the BGN locally, ERCE presents itself as an innovator, which is one of the strategies often adopted by change agents to ensure a stronger outreach for the ideas they are promoting. Other change agents promoted the concept further, through discussions, publications (e.g. Bergier et al. 2014) and lobbying
Transformers	Municipal Planning Office Forward-looking urbanists and researchers City Strategy Office (City of Lodz Office) Researchers (biology, urban planning) Inhabitants concerned with nature conservation Other NGOs and individuals Few investors	Transformers include early adopters who have been keen to translate the concept into practical strategic documents. Examples include featuring the BGN in the Integrated Development Strategy for Lodz 2020+ (City of Lodz Office 2012) and in the city's masterplan (City of Lodz Office 2010), and the establishment of a network of small protected areas dispersed throughout the city (Ratajczyk et al. 2010). Even individuals act as transformers when they protest against the degradation of urban nature and call for its conservation and rehabilitation. Finally, there have been very few investors who actively restored green spaces within and even outside of their investment projects, contributing to the BGN
Controllers	Ministries President of the city	Those in power to ensure that the concept is implemented in practice and that there are legal instruments that require that the creation of the BGN take priority over other issues (this is a potential role only because so far very little has happened)

(continued)

Table 17.3 (continued)

Amoeba role	Actor in the case study	Role played in the case study
Mainstreamers	Lodz City Office Department for Urban Greenery Lodz City Office Department for Environmental Protection and Agriculture Researchers other than those who act as transformers Municipal companies responsible for the provision of basic water-related services	Private and public investors and the relevant departments within the local government that are responsible for the creation and maintenance of green spaces. So far, the authorities responsible for urban green spaces and the environment have had limited opportunities to prevent further degradation of urban green and blue spaces. Municipal companies responsible for sewage systems see their interest in reducing the flow of stormwater into the sewage system, but their activities are not entirely consistent (although they formally endorse the BGN as a way to manage stormwater, they keep investing in large-scale traditional stormwater infrastructure)
Laggards	Investors Public authorities responsible for land management Local politicians and council members Contractors responsible for green space and urban infrastructure maintenance	Most investors are reluctant to protect green spaces on their land and especially to give up private benefits for the sake of public benefits. With no specific regulations that would support the maintenance and creation of ecological corridors, public authorities are unable to prevent further construction on agricultural and forest land and further soil sealing. The authorities keep selling out municipal land to earn profit for the city and fail to buy out private land for conservation purposes. All of the above is reinforced by the inertia of public authorities – it is always easier to maintain the status quo, rather than to prepare for change. Also, laggards include those who are supposed to manage urban green spaces and urban infrastructure but fail to do so in an environmentally friendly way
Reactionaries	Land owners and investors Local government appeal board	Many land owners and investors openly oppose the BGN because they fear that it would reduce the value of their land (e.g. by restricting construction opportunities). They actively resist the implementation of the transformers' prescriptions and general plans. They apply for construction permits on their agricultural and forest land. They seek legal loopholes and benefit from the fact that the legal system in Poland downplays the significance of urban green spaces (Kronenberg 2015). Their right to derive private benefits from their land (as opposed to the delivery of public benefits) is further reinforced by the local government appeal boards to whom private investors can appeal if they are not satisfied with decisions issued by other public institutions. This links to further problems with the overarching idea of freedom, including freedom to build and especially freedom to build on one's land. Reactionaries try to ridicule those who protect nature as outdated, who do not understand the idea of a modern city. They try to discredit environmental NGOs and other groups defending urban nature

Source: The author's own work

For a general description of stakeholder roles, see Table 17.1

and further disseminate knowledge and provide access to the latest examples from abroad. However, the LA required a significant coordination effort, and it was not necessarily composed of those in power to make the relevant decisions. Those in power and interested in promoting change, i.e. transformers, have not been effective in making the change happen. The documents that they prepared (such as the masterplan), which could have translated the BGN into practice, were not internally consistent, and the transformers could not make sure that the prescriptions regarding the BGN were actually implemented. To a large extent, this has been related to institutional failures, such as the fact that the legal system in Poland limits the possibilities to protect urban green spaces. To promote the BGN concept further, and to implement similar concepts in other cities, the controllers would have to endorse this innovation and change the rules of the game to favour its implementation (the first sign of positive change has been the National Urban Policy which explicitly linked to these issues (MIiR 2015)).

17.4   Discussion and Conclusions

The concept of NBS fits well into the neoliberal world where the existence of anything needs to be justified by its ability to solve some problem. Thus, it should sell easily in Poland and other post-socialist and post-transition countries, where neoliberal (economic) ideas have caught on. In fact, the new socio-economic system introduced in Poland as a result of transformation from a socialist country should in theory create a window of opportunity for new solutions. However, in practice NBS are difficult to accept, because modern solutions are usually associated with 'hard' infrastructure, rather than greenery. Furthermore, the ideas of a green city clash with those of a modernist city made of concrete, glass and steel (which are still seen by many as an ideal that Polish cities should finally strive to achieve). As a result, many opportunities to preserve green spaces that are essential from the point of view of ecological corridors are missed, along with opportunities to introduce new components of green and blue infrastructure that would fill the gaps in such corridors. Moreover, NBS have already been known and used for a long time; they are not necessarily seen as an innovation; thus they are not attractive for the mainstreamers. Still, perhaps the NBS framework can help to see nature from a new perspective and convince mainstreamers that we do need nature because it addresses many crucial needs of urban inhabitants and should be seen as an innovation in itself.

Innovators and – in our cases – especially change agents and transformers (i.e. those who promote innovations) are usually relatively less powerful than other stakeholders, especially those who represent well-established solutions. In our case studies, NGOs and research institutes, promoting the use of NBS in Poland, are on the margin of decision-making structures. Transformers who introduce new concepts to the broader public are under pressure from conflicting groups of interests. Meanwhile, those who stand behind well-established solutions often represent these structures. Hence those who promote innovations and their ideas clash against omnipotent structures of laggards and reactionaries and against the barrier of conformism of most mainstreamers. To be more effective in promoting their innovations, they need to collaborate with other stakeholders. In particular, based on the theoretical foundations of Amoeba (AtKisson 2009), change agents need to work with transformers, and the different change agents need to collaborate closely with other change agents. Meanwhile, collaboration in the area of green space management in Polish cities is poor, and few stakeholders are involved in this area (Kronenberg et al. 2016). NGOs, who are particularly strongly involved and are also marginalised, and they may not be able to promote innovations, at least not alone. Indeed, reactionaries often tend to marginalise change agents (discrediting change agents is one of their most effective strategies). Therefore, change agents should work with those who show the potential to accept their ideas and to promote them further, rather than waste their time on talking to reactionaries.

Similar to many other types of sustainable development opportunities, the main driver of potential increased interest in NBS in Poland can be associated with an outside pressure, especially coming from the EU (Kronenberg and Bergier 2012). To some extent, EU institutions act as ultimate controllers who set the general framework for the national social and institutional structures. Their pressure could be the most effective had it been connected with conditional funding (Poland is the largest beneficiary of EU structural funds). Changing the legal framework in Poland to favour NBS could also result from continued pressure from those who have been promoting NBS so far, especially if transformers highlight inconsistencies between the current legal framework and the one necessary for the implementation of NBS.

References

AtKisson A (2009) The sustainability transformation: how to accelerate positive change in challenging times. Routledge, London/New York

Bergier T, Kronenberg J, Wagner I (eds) (2014) Water in the city. Sendzimir Foundation, Krakow

Böhm A (2007) System parków rzecznych w Krakowie. In: Myga-Piątek U (ed) Doliny rzeczne. Przyroda – krajobraz – człowiek. Komisja Krajobrazu Kulturowego PTG, Sosnowiec, pp 277–284

Bozorg Chenani S, Lehvävirta S, Häkkinen T (2015) Life cycle assessment of layers of green roofs. J Clean Prod 90:153–162. doi:10.1016/j.jclepro.2014.11.070

Brenneisen S, Gedge D (2012) Green roof planning in urban areas. In: Meyers RA (ed) Encyclopedia of sustainability science and technology. Springer, New York, pp 4716–4729

City of Lodz Office (2010) Studium uwarunkowań i kierunków zagospodarowania przestrzennego miasta Łodzi. City of Lodz Office, Lodz

City of Lodz Office (2012) Integrated development strategy for Lodz 2020+. City of Lodz Office, Lodz

Energie Cités (2014) Zielone dachy i żyjące ściany. Systemowe rozwiązania i przegląd inwestycji w polskich gminach. Stowarzyszenie Gmin Polska Sieć "Energie Cités", Kraków

European Commission (2015) Towards an EU Research and Innovation policy agenda for Nature-Based Solutions & Re-Naturing Cities. European Commission, Brussels

Giergiczny M, Kronenberg J (2014) From valuation to governance: using choice experiment to value street trees. AMBIO J Hum Environ 43:492–501. doi:10.1007/s13280-014-0516-9

Gutry-Korycka M (ed) (2005) Urban sprawl: Warsaw gglomeration. Warsaw University Press, Warsaw

Kania A, Mioduszewska M, Płonka P, et al (2013) Zasady projektowania i wykonywania zielonych dachów i żyjących ścian. Poradnik dla gmin. Stowarzyszenie Gmin Polska Sieć "Energie Cités", Kraków

Kotus J (2006) Changes in the spatial structure of a large Polish city – The case of Poznań. Cities 23:364–381. doi:10.1016/j.cities.2006.02.002

Kowalczyk A (2011) Green roofs as an opportunity for sustainable development in urban areas. Sustain Dev Appl 2:63–77

Kowalewski A, Mordasewicz J, Osiatyński J et al (2013) Raport o ekonomicznych stratach i społecznych kosztach niekontrolowanej urbanizacji w Polsce. Fundacja Rozwoju Demokracji Lokalnej, Warszawa

Kronenberg J (2015) Why not to green a city? Institutional barriers to preserving urban ecosystem services. Ecosyst Serv 12:218–227. doi:10.1016/j.ecoser.2014.07.002

Kronenberg J (2016) Nature-based solutions. In: Rzeńca A (ed) Ekomiasto#środowisko. Zrównoważony, inteligentny i partycypacyjny rozwój miasta. Wydawnictwo Uniwersytetu Łódzkiego, Łódź, pp 241–256

Kronenberg J, Bergier T (2012) Sustainable development in a transition economy: business case studies from Poland. J Clean Prod 26:18–27. doi:10.1016/j.jclepro.2011.12.010

Kronenberg J, Pietrzyk-Kaszyńska A, Zbieg A, Żak B (2016) Wasting collaboration potential: a study in urban green space governance in a post-transition country. Environ Sci Policy 62:69–78. doi:10.1016/j.envsci.2015.06.018

Kronenberg J, Krauze K, Wagner I (2017) Focusing on ecosystem services in the multiple social-ecological transitions of Lodz. In: Frantzeskaki N, Broto V, Coenen L, Loorbach D (eds) Urban Sustainability Transitions. Routledge, London/New York, chapter 20

Krynicki M, Witkoś Gnach K (2016) Monitoring standardów w zarządzaniu zielenią wysoką w największych miastach Polski. Fundacja Ekorozwoju, Wrocław

MIiR (2015) Krajowa Polityka Miejska 2023. Ministerstwo Infrastruktury i Rozwoju, Warszawa

NIK (2014) Informacja o wynikach kontroli: Ochrona drzew w procesach inwestycyjnych w miastach (P/14/087). Najwyższa Izba Kontroli, Kraków

Ratajczyk N, Wolańska-Kamińska A, Kopeć D (2010) Problemy realizacji systemu przyrodniczego miasta na przykładzie Łodzi. In: Burchard-Dziubińska M, Rzeńca A (eds) Zrównoważony rozwój na poziomie lokalnym i regionalnym: Wyzwania dla miast i obszarów wiejskich. Wydawnictwo Uniwersytetu Łódzkiego, Łódź, pp 78–97

Van Mechelen C, Van Meerbeek K, Dutoit T, Hermy M (2015) Functional diversity as a framework for novel ecosystem design: the example of extensive green roofs. Landsc Urban Plan 136:165–173. doi:10.1016/j.landurbplan.2014.11.022

Wagner I, Krauze K, Zalewski M (2013) Blue aspects of green infrastructure. In: Bergier T, Kronenberg J, Lisicki P (eds) Nature in the city. Solutions. Sendzimir Foundation, Krakow, pp 145–155

Chapter 18
Implementing Nature-Based Solutions in Urban Areas: Financing and Governance Aspects

Nils Droste, Christoph Schröter-Schlaack, Bernd Hansjürgens, and Horst Zimmermann

Abstract Fostering nature-based solutions in urban areas is an issue that receives increasing attention on the political agenda. But in many cases, only insufficient financial resources are available for the implementation of such solutions. A central issue in this context is the structure of municipal revenues, which stem from either municipal tax revenue, fees for municipal services, or fiscal transfers from other governmental levels. Many of these revenues are however absorbed by specific tasks, especially social expenditure; thus there is little room left for autonomous investments, e.g. into nature-based solutions and green infrastructure. In this chapter we elaborate on the structure of the problem such as the corresponding fiscal and constitutional restrictions and analyse which solutions are possible to allow for greater investments into multifunctional urban nature-based solutions.

Keywords Nature-based solutions • Public finance • Municipal revenues • Municipal expenditures • Urban governance

18.1 Introduction: The Nature of Nature-Based Solutions

Nature-based solutions (NBS) in urban areas are receiving increasing attention not only in research but especially on the political agenda. While environmental friendly and ecologically sound practices of agriculture, infrastructure development and human settlements have at least been promoted politically since the Earth Summit

N. Droste • C. Schröter-Schlaack • B. Hansjürgens (✉)
UFZ – Helmholtz Centre for Environmental Research,
Permoserstraße 15, 04318 Leipzig, Germany
e-mail: nils.droste@ufz.de; christoph.schroeter-schlaack@ufz.de;
bernd.hansjuergens@ufz.de

H. Zimmermann
Philipps University of Marburg, Marburg, Germany
e-mail: horstzimmermann1@freenet.de

© The Author(s) 2017   307
N. Kabisch et al. (eds.), *Nature-based Solutions to Climate Change Adaptation in Urban Areas*, Theory and Practice of Urban Sustainability Transitions,
DOI 10.1007/978-3-319-56091-5_18

in Rio 1992, the idea of employing natural elements to substitute and complement man-made infrastructure and production systems is a rather recent development. During the last decades, there has been more and more research focusing on NBS (see as a general introduction ten Brink et al. 2012). At the same time there is an increasing interest on the political level (a prominent example provides the latest EU research strategy; see EC 2015).

Nature-based solutions are multifunctional, i.e. they deliver manifold services at the same time. For example, an urban park is not only a place of recreation for people (and sometimes a touristic factor) but also a corner-stone for (urban) biodiversity. The park is providing cooling effects important for reducing heat stress induced by climate change, relieving public sewerage infrastructures in case of heavy rainfall by providing natural seepage, filtering particular matters and dust and reducing noise of, e.g. traffic, and is contributing to climate change mitigation via carbon sequestration of its vegetation. While on the one hand, the multifunctionality of NBS promises high social return on investments, since they simultaneously address several societal goals, it is on the other hand hard to develop financial arrangements for their realization, since their superiority only becomes visible if all services are considered. Sectorial returns, e.g. in terms of noise reduction, are often falling short of benefits provided by technical and grey infrastructure solutions such as noise-insulating walls.

While there is yet lacking a clear definition (Potschin et al. 2015), the idea of employing nature to satisfy human needs has attracted both conservation experts such as the International Union for Conservation of Nature (IUCN 2016) and major infrastructure investors such as the World Bank (2008). The expert group of the "Nature-Based Solutions and Re-naturing Cities" of the European research programme Horizon 2020 defined NBS as *actions inspired by, supported by or copied from nature; [...] [that] use the features and complex system processes of nature, such as its ability to store carbon and regulate water flows, in order to achieve desired outcomes, such as reduced disaster risk and an environment that improves human well-being and socially inclusive green growth* (EC 2015, p. 24). The concept therefore aims at satisfying human needs by utilizing (cultivated and humanly formed) natural systems. Regarding the urban environment, the idea is to foster multifunctional green spaces, green roofs, roadside greeneries, etc. to improve human health and human well-being.

There are potential similarities and overlaps to concepts such as the "ecosystem approach, ecosystem services, ecosystem-based adaptation/mitigation and green and blue infrastructure" (EC 2015, p. 24). Thus, many different aspects have been addressed from various perspectives; at the same time, however, many details have not been clarified yet.

For an actual implementation, financing aspects are certainly among the issues which are of utmost importance. While we assume that there is no easy blueprint, one-size-fits-all solution, we aim at elaborating on the issue from a perspective of public finance and public choice. Hence, our basic premise is that, in urban areas, public bodies are central actors that provide the required investments for the implementation of NBS. The goal of this chapter is therefore to clarify how cities and municipalities may leverage required investment volumes for NBS. Focussing on municipal actors, the chapter serves three purposes:

1. We provide an overview about potential instruments for implementing NBS and their adoptions at the local and/or regional level (Sect. 18.2).
2. We highlight obstacles for financing NBS regarding the structure and purposes of municipal budgets (Sect. 18.3), for which we will discuss examples from the German institutional setting.
3. We elaborate on selected proposals on how to overcome these barriers and leverage greater investments into urban NBS (Sect. 18.4).

18.2 Policy Instruments for Nature-Based Solutions

Policy instruments for implementing NBS in urban environments are manifold and touch upon almost every single policy area in the urban context. Generally, policy instruments cover a range from rather weak but basic tools such as informational systems like monitoring and accounting, over command-and-control instruments such as municipal green planning, to economic instruments that set incentives via prices or quantity mechanisms. This classification can be extended into the direction of instruments that foster cooperation, since NBS often require interdepartmental teamwork or the support of other governments at the same level (horizontal cooperation) or at upper levels (vertical cooperation) through corresponding programmes.

Figure 18.1 provides an overview of potential instruments, which we briefly introduce. For the practical examples, we refer to the work on incorporating ecosystem

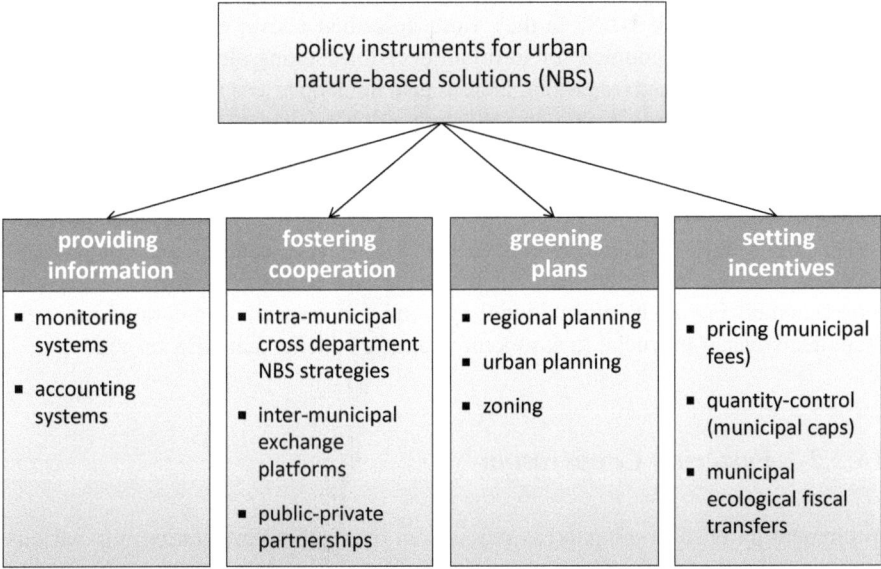

Fig. 18.1 Instruments for implementing nature-based solutions (Source: Adapted from Naturkapital Deutschland – TEEB DE 2016)

services into urban policy-making by the German Natural Capital Project (Naturkapital Deutschland – TEEB DE 2016) which is based on the international TEEB study "The Economics of Ecosystems and Biodiversity" (TEEB 2012).

18.2.1 Informational Tools

At the most basic and fundamental level, green infrastructure planning that implements NBS requires information, for example, about benefits and costs of particular projects or project alternatives. Therefore, the ecosystem services provided by NBS and the natural capital that is accumulated by investing in green infrastructure have to be monitored and translated into accounting systems.

While there are exceptional NBS showcases that are well analysed and researched (see TEEB 2012; Naturkapital Deutschland – TEEB DE 2016), there is no coherent and comprehensive assessment of the natural capital stocks and ecosystem service flows in urban areas. Cooling effects of urban green spaces that increase climate change resilience can contribute to reducing stress factors that stem from overheating and that lead to serious health-related impairments and result in increased morbidity and mortality rates (Gabriel and Endlicher 2011; Heudorf and Meyer 2005; Hoffmann et al. 2008; Schneider et al. 2009). Urban parks and their recreational usage reduce stress, aggression or fears and positively influence concentration and performance (Hartig et al. 2003). Food production in urban gardening raises awareness for local and regional products and for a healthy diet increases by gardening experiences, e.g. in community or school gardens (Lobstein et al. 2015). All the ecosystem services of NBS, as they were described above, are normally neither been assessed nor accounted for in urban decision-making although their value is increasingly recognized (regarding increases in housing prices through proximity to urban green spaces; see Kolbe and Wüstemann 2014). An informed political and administrative decision about which projects to prioritize and implement requires proper evaluation and a sound informational basis. Such basis may consist in both (i) monitoring of NBS, their functioning and the (ecosystem) services they provide, and (ii) the inclusion of natural capital stocks and flows in municipal budgets and bookkeeping systems to account for the value and return of investment in green infrastructure. For both informational systems and respective integrated management decisions, it is crucial to assess the multiple benefits that NBS provide.

18.2.2 Fostering Cooperation

Implementing NBS in urban areas requires an overarching integration into various municipal decision-making processes. The infrastructure department might need to collaborate with the tourism agency, and the climate change and environmental protection administration may be required to cooperate with the social and family department, in order to design coherent policies that minimize trade-offs to other sectors and boost synergies. For supporting and implementing nature-based solution, it is particularly important that urban green spaces do not only receive support by the respective municipal department that is directly responsible but also by other municipal departments that benefit from these NBS. Thus the creation of cross-departmental planning and decision-making procedures within a single municipality's nature-based solution strategy may well enhance the overall performance and speed of implementation.

Furthermore, a single municipality often has limited resources and may not necessarily be able to supply the required budget, manpower and knowledge based for a suitable implementation of NBS. It may also be the case that the benefits of NBS cross municipal borders and affect neighbouring municipalities or that economies of scale, both in terms of increased provision of ecosystem services and in terms of reduced costs of implementation, can be realized when NBS are developed jointly through inter-municipal cooperation. Furthermore, to facilitate access to best-practice examples and learning from mistakes, inter-municipal exchange platforms and associations can serve as a multiplying factor in mainstreaming NBS into urban planning. While there are often existing networks, an installation of a respective working group within and between them would enhance mutual learning processes. This does not only hold for (horizontal) cooperation between municipalities but also for (vertical) cooperation between municipalities and upper governmental levels (state or federal level).

Since public administrations are by far not the only relevant land owners and land users in urban areas, a cooperation of public authorities with private initiative can facilitate an extra aid for the implementation of NBS. Urban gardening, green buildings, or even citizen science, which basically means public participation in scientific processes (cf. Bonney et al. 2009), may be ideas for enhancing monitoring and implementation of green infrastructures and their socio-economic and ecological benefits.

18.2.3 Planning Procedures

Urban development plans and zoning approaches are essential tools for urban decision-making and thus the integration of NBS into the respective procedures. Even in traditional administrative plans, conservation and environmental protection may be integrated already properly. In many instances, however, environmental planning is just one sectorial planning in the course of a development project and sometimes conflicting with other planning goals (such as city compaction, reducing traffic jam or the preservation of historic buildings). The idea of NBS goes further than just conservation and protection and aims at integrating different sectorial planning. The idea of NBS entails approaches where natural ecosystems may provide services that man-made alternatives cannot supply as cheaply or effectively. Since an engineer in the waste-water department may not be aware of nature-based alternatives, integrated planning procedures are required, where feasible and economically viable NBS are streamlined into the various urban planning procedures. However, until now the concept of ecosystem services is not or only implicitly taken into account in planning processes (Hansen et al. 2015). Due to the potential trade-offs and land-use conflicts, elaborated decision support tools such as multi-criteria analysis or environmentally extended cost-benefit analysis can supply the needed information in order to balance and counteract deadweight losses.

18.2.4 Economic Instruments

As has been mentioned, there is a variety of stakeholders that have different interests and ideas for the space available in urban areas. They all will opt for those alternative that offers them the best value for money, given the prevalent rules and conditions of the system in which they operate. Against this background, setting incentives by economic instruments may change these conditions, ultimately stimulating actors to pursue those other project alternative that now seems best to them. For example, a separate charge on rainwater poured into sewage systems can offer incentives to avoid soil sealing (Rüger et al. 2015). The same is true for a wastewater fee that is oriented to the sealed natural ground (Geyler et al. 2014).

Basically, there are three market-based instruments which may help accomplishing NBS, two that address private actors through either prices or quantity mechanisms and one addressing public actors such as municipalities which help in accomplishing NBS.

Price Instruments. A first type of market-based instrument is prices. A possible application of such price-based instruments is either through an incentive-oriented design of existing charges (e.g. municipal fees for water services) or levying new ones (e.g. water charges). In theory, charges shall change the price of using an ecosystem service to reflect the full cost of its provision. Fees for water services usually are based on the principle of cost recovery, often including only technical costs (investment costs, operation and maintenance costs) incurred for the provision of water to users (Gawel 1995). However, there exist margins herein, as the determination of underlying costs allows – at least in principle – to also include environmental and resource costs (Gawel 2016; Rüger et al. 2015).

Quantity Instruments. On the contrary to price-based approaches, quantitative instruments are directly limiting activities impacting natural areas, e.g. by setting a cap on the maximum amount of green fields to be developed (McConnell and Walls 2009). Within the scope of the cap development rights will be auctioned or allocated for free among potential developers. By making development rights tradable, a cost-efficient allocation of development can be assured as those landowners able to realize the highest net benefits from development will buy up rights and develop their land (Mills 1980). However, if such a system is to allow for a targeted protection of specific green infrastructures, it has to be accomplished by land-use zoning (Schröter-Schlaack 2013; Santos et al. 2015).

Fiscal Instruments. While taxes set incentives for private land users, the criteria according to which the levied tax income is distributed among jurisdictions create incentives for decision-making of public authorities. Typically, fiscal transfer is assigned on the basis of comparing population numbers (as a proxy for jurisdiction's fiscal needs) and own tax revenues (as a proxy for jurisdiction's fiscal capacity), thereby stimulating communities to increase the number of inhabitants, e.g. by dedicating land for development to keep property prices low. If a portion of tax revenue would be distributed according to ecological criteria, this may set incentives for providing green infrastructures and NBS. An example may be the ecological fiscal transfers implemented in Brazil or Portugal, where municipalities receive tax revenue for hosting protected areas (Ring 2008; Santos et al. 2012).

18.3 Obstacles and Limits for Financing Nature-Based Solutions on the Local Level

Although there might be a broad variety of obstacles for the successful implementation of NBS in urban areas such as political resistance, path dependencies or dominant interests, we focus on the problem from the perspective of public finance and fiscal federalism (Oates 1972; Boadway and Shah 2009). Therefore, this subsection elaborates on the (missing) recognition of NBS in both the structure of municipal revenues and the competing public functions (and expenditures) that define the municipal spending behaviour. With respect to concrete legal provisions, we refer to the case of Germany. However, similar public finance restrictions can be found in many other developed countries.

18.3.1 Financing NBS from Own Sources: The Structure of Municipal Revenues

If NBS are to be financed, the structure of municipal revenues is decisive. Generally speaking, there are three sources of municipal revenues: (i) fees and charges for publicly provided goods and services, (ii) the revenues from municipal taxes (or the local tax revenue of shared taxes between different government levels) and (iii) the redistribution of tax revenues via fiscal transfers (either vertically from higher governmental levels or horizontally between jurisdictions of the same level). The exact structure depends on the fiscal constitution of a given country (see as a classical example Musgrave 1959) and of the given fiscal transfer system (Zimmermann 2016, Chap. 5).

Municipal Fees and Charges. Fees and charges for public services are often a substantial source of municipal revenues (Wagner 1983) and may play a significant role in financing NBS, e.g. for municipal waste management or water supply. In order to generate an optimal allocation of scarce resources, rates for fees and charges, however, have to follow the benefit principle of public finance: they should make sure that those benefitting from a particular service should also bear the costs (Olson 1969; Hansjürgens 2001). Fees and charges are calculated by municipalities, depending on the underlying costs of providing the service. An example is water pricing. According to Article 9 of the EU Water Framework Directive, water prices could not only be based on water-related investment costs and O&M (operating and maintenance) costs but also on environmental and resource costs (Gawel 2016). Fees and charges can nevertheless only partly serve as instruments for financing NBS, for two reasons: Firstly, a calculation of fees and charges beyond the cost recovery principle is legally forbidden and can be fought via court decisions (Gawel 1995; Hansjürgens 1997). Only if environmental ad resource costs would be included in the calculation in fees and charges, which is currently not the case, these instruments could serve as a financing instrument for NBS. Secondly. financing of NBS through municipal fees and charges (even if environmental and resource costs are included) is always restricted to the underlying service and its costs. Additional revenues that can be spent in policy fields beyond the water or waste management sector are not possible. This clearly limits the scope of fees and charges to finance NBS on the municipal level.

Municipal Tax Revenues. A second source of municipal income is taxes that are completely or partly under the authority of municipalities, such as the property tax in Germany (Zimmermann 2016). Furthermore, municipal tax revenue may also originate from taxes that are collected at other governmental levels and then distributed to the local level, e.g. income tax in Germany.

Three observations are of particular relevance for the possibilities (and limitations) of municipal tax income to finance NBS: (1) Freedom of municipalities to raise revenues from own sources is limited due to a lack of tax competences. Most countries are unitary states with the power to tax residing with the central government – only few nations are federally organized with an independent decentral level with own powers (amongst others, the USA, Canada, Germany, Australia or Switzerland). Moreover, even in federal states it is legally not allowed to tax one and the same tax base several times. As most tax bases are already taxed by upper level governments, there is little possibility left for local levels to introduce own taxes. (2) There are a few taxes with revenue sharing between governmental levels (e.g. in Germany, the income tax of the business tax). In the case of sharing the revenues of joint taxes between different governmental levels, a decentral tax rate setting is – with very few exemptions – not permitted to avoid regional differences in households' or companies' tax rates within one country. There is a high preference for achieving similar living conditions for all citizens within the country; in Germany this is even a constitutional goal. (3) In addition, a loose regulation of taxes under the authority of local governments may lead to an undesired tax competition between the respective authorities. Whether that is beneficial or not desirable is a contested issue: Many fear a "race to the bottom" if tax competences are given to lower governmental levels. Others would welcome this as an element of jurisdictional competition

(Tiebout 1956; Oates and Schwab 1988; Zodrow 1983; Zodrow and Mieszkowski 1986; Sinn 1990; Kenyon and Kincaid 1994; Lenk 2004; Vanberg 2013).

Fiscal Transfers. Depending on the constitutional structure, whether a given country is a federal or a unitary state, the autonomy of local authorities may vary (Boadway and Shah 2009; Zimmermann et al. 2012). However, there are always some public functions that are provided by local governments. Fiscal transfers are an important means to finance these functions on the local level. In general, such transfers between upper and lower governmental levels have two goals: (i) to ensure that the municipal government level has sufficient funds to provide its public functions and (ii) to equalize differences between regions or municipalities, that is, to create a "fairer" revenue distribution. Regarding the first goal, often vertical fiscal transfers from higher government levels to lower tiers are employed – although their share of municipal income is generally higher in non-OECD countries (Shah 2007). In relation to the second goal, there are horizontal fiscal transfers, often also called fiscal equalization schemes that redistribute tax revenue from richer regions or municipalities to poorer ones. Thereby available tax revenue, e.g. per capita, is equalized but rarely to the extent of total equality. Another possibility to realize the second goal is to implement a vertical transfer system that has equalization (and thus horizontal redistribution) effects. As has been mentioned, the criteria according to which the fiscal transfers are distributed may incentivize and thus steer behaviour among (local) governments (see for a German example Baretti et al. 2002). Figure 18.2 shows an overview of fiscal transfers systems.

The scope for financing NBS solutions in such a transfer system is generally limited. Even though the general fiscal situation of a (poor) municipality may improve through such transfers, specific public expenditures for NBS can hardly be financed on this basis. Only if NBS are seen as a specific function that increases fiscal needs that corresponds to the first goal above, an additional financing might be possible.

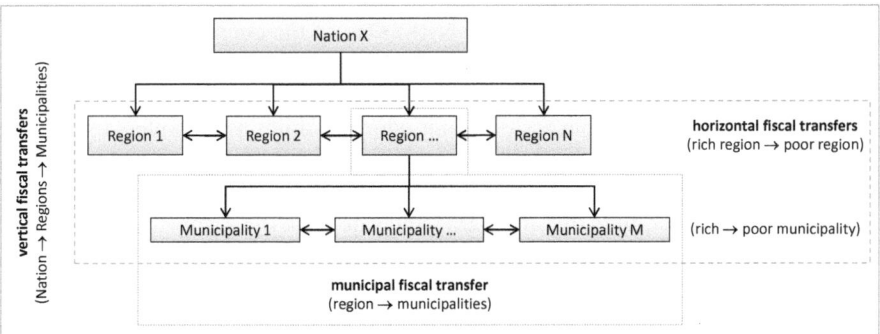

Fig. 18.2 Overview of fiscal transfers in multilevel governments (own representation)

18.3.2   Public Spending for NBS: Limited Municipal Autonomy for Functions and Expenditures

Generally, there is only a limited municipal autonomy in deciding how to allocate expenditures. Most often, public functions are delegated from higher to lower levels, which require that the local governments provide public services that have been defined at higher tiers (Foldvary 1994; Zimmermann 2016).

Municipal governments have numerous functions to fulfil, and for quite a lot of these functions municipalities are (legally or de facto) urged to fulfil these functions and spend the corresponding expenditures (Zimmermann 2016). In this respect, municipal functions (and expenditures) can be classified in:

(i) Those that are fully obligatory or mandatory, because there are legal prescriptions that have to be fulfilled, without any discretionary municipal freedom. Here both the *whether* and *how* to fulfil a function are determined by upper levels.

(ii) Those that have to be taken by the local level and the task as such is prescribed by law. Here, the municipalities have discretionary power to decide *how* to fulfil the prescribed task (water provision or waste-water treatment may serve as an example).

(iii) Those where municipalities have "full freedom" to decide *whether* and *how* to fulfil a function (and the corresponding expenditure) (Zimmermann 2016).

In many nations the share of functions of municipalities that are legally prescribed by upper level legislation (see i) above) is often very high, while the share of "self-determined" decided functions (see iii above) tends to be quite small. Under financial pressures, these self-determined functions are the first that have to be cut down. Urban green spaces and NBS are in most cases falling into this third category of local public functions (e.g. sports and culture). In consequence, although there is discretionary power for municipalities to define whether and how NBS shall be realized, this power is quite limited in practice (at least in poorer communities where freely available money is small).

There is another limit for financing NBS: One of the constitutional principles in Germany (and similarly in other federally organized countries) is connectivity. This principle asks for a match of competencies and expenditure and basically states that who is responsible for deciding upon a public task also has to pay for its fulfilment. It points at the equivalence between public functions and public expenditures as part of the principle of fiscal equivalence (Olson 1969). The connectivity principle requires that if federal states devolve the competence to decide upon NBS to local authorities, the respective financing will also have to be borne by the latter. Given that local authorities have tight budget constraints and various public functions to fulfil, such a delegation without proper cofinancing or respective fiscal transfers will not necessarily help a better implementation of NBS. The point here is that in practice the connectivity principle is often violated: functions are delegated from upper levels to municipalities without fully compensating the corresponding expenditures (Zimmermann 2016). This amplifies the financial problems of local level governments, especially if they belong to the poorer ones.

Summing up the revenue and expenditure considerations of municipal public finance, the central problems for the implementation of NBS are the following: Municipal authorities play a central role in the actual planning and management of (green) infrastructure, environmental protection and nature conservation. Nevertheless, with respect to financing expenditures for investments in NBS, there is not much autonomy left to local authorities: Firstly, the fees and charges for public services such as waste and water management have to cover the costs of the service but may not exceed them. Therefore, they are not suited to generate surplus to cross finance other public functions. Secondly, the powers to raise tax revenues from own sources are rather limited. Thirdly, the income that municipalities levy through their revenues has to be spent on the variety of public functions they are responsible for. And fourthly, current systems of fiscal transfers do not recognize *ecological* public functions (Ring 2002). Therefore, financing NBS faces hard competition with other public functions, cannot be cross-financed from municipal fees and suffers disadvantages in the fiscal transfers system.

18.4 Finding the Appropriate Nature-Based Solutions Policy and Funding Mix

There are three points worth stressing to elaborate on potential approaches for leveraging investments in urban NBS.

First, NBS are multifunctional and require cross-sectoral, cross-departmental planning procedures where different vested interests may be balanced. While originating from the functionality that NBS provide, it means that also different funds have to be acquired and directed towards respective investments. A hypothetical example could be creating an attractive green space with recreational amenity services. Public health studies have shown that these urban green spaces also contribute to positive health impacts (for an overview, see Naturkapital Deutschland 2016, pp. 98). Another example may be the protection against flood risks by urban wetlands (ibid, pp. 86). These NBS could be financed not just by the environmental but also by the health or the municipal water department. Since these are rather intermunicipal decision-making procedures where different interests and power structures create path dependencies, it might be difficult to change current investment patterns through directives for interdepartment cooperation. Information of co-benefits of NBS may nevertheless be a decisive element to create more favourable conditions for corresponding investments by several municipal departments. This is to say, the clearer the "return of investment" for each of the affected sectors, the more likely are respective decision-makers to invest in such "novel" and innovative alternatives to well-known city plans. The required information may either be supplied by science or be included in administrative assessment methodologies, such as cost-benefit analyses that are extended to cover the environmental dimension

(Hanley and Barbier 2009; Hansjürgens 2004) or multi-criteria analyses (Janssen 2001; Tsianou et al. 2013).

Second, public-private partnership may enable urban decision-makers to create alliances that create a favourable climate for investments in NBS (Naturkapital Deutschland – TEEB DE 2016). Citizens, local businesses and potentially even larger enterprises may have an interest in parks, protected areas, urban forests, clean watersheds and a livable city. Especially large enterprises with several locations prefer those with a good living condition, because that helps them to attract high-quality personnel. This opens new opportunities to engage and even support such developments financially. The more stakeholder organize in networks and associations to defend and propagate their interest in, e.g. the conservation of farmlands and forests (Bryant 2006), the more likely a success to balance these land-use pursuits with other land-use interests such as housing development. Arrangements such as Payments for Ecosystem Services (PES) may protect important areas for public services. Several cities have launched such programmes to support their public service provider such as the municipal water utilities to maintain and conserve peri-urban and rural watersheds that supply clean drinking water naturally. It is worth noting that there are, however, critics of such approaches – since they seem to commoditize nature (see, e.g. Fletcher 2014). Nevertheless, there might however be private public partnership approaches that serve the goal of providing NBS without the turmoil of a supposedly sellout of nature (Kallis et al. 2013; Hansjürgens 2015).

Third, the proper integration of ecological public functions within the fiscal constitution may well help and enhance the implementation of NBS in urban areas (Naturkapital Deutschland – TEEB DE 2016). Creating incentives not just for private land users through price mechanisms like taxes and cap-and-trade-based mechanisms for development rights, but essentially incentivizing nature-affine investment behaviour of public authorities, may constitute a well-functioning but not yet well-known addition in the policy mix. It has been shown that the integration of ecological indicators in municipal fiscal transfers incentivizes the respective governments to create additional protected areas (Sauquet et al. 2014; Droste et al. 2015; Ring 2008). Depending on the indicator, also urban green spaces and their ecological public functions could be supported through ecological fiscal transfer mechanisms. Basically, the incentive would function the following way: If a city would receive a portion of fiscal transfers only if it supplied a certain amount of green spaces per capita – to construct an easy example – it might be profitable for the city to invest a certain amount to assure such additional income. By integrating an ecological indicator into the fiscal transfer system, a financial aspect comes into play that may incentivize investments into NBS – whether it actually does might ultimately rather be a question of the amount that can be gained through such investments and the local (opportunity) costs.

Through highlighting three potential routes of how investments into NBS may be supported (inner-municipal cooperation, public-private partnerships and ecological fiscal transfers), it becomes clear that there are different leverage points and that a coherent policy has to be thought in form of a policy mix (cf. Ring and Barton 2015). It also becomes clear that there are different funds available from which such investments may be financed: municipal budgets, public-private funds or fiscal transfer funds. Thus, there is no one-size-fits-all panacea but a toolbox of potentially suitable instruments which may be employed with greater or lesser success in different circumstances.

18.5 Concluding Remarks

We have started by introducing potential instruments for implementing NBS and discussed how difficult it may be to leverage finance to implement them. We ended by presenting three ideas how the required investments may be levied: (i) reorganizing decision-making structure within municipalities to free funds to finance (economic) side benefits of NBS, (ii) organizing alliances and private-public partnership with a vested interest in a clean and green city and (iii) integrating ecological indicators in municipal fiscal transfer systems. These instruments are more likely successful when embedded in a nature-based solution policy mix.

References

Baretti C, Huber B, Lichtblau K (2002) A tax on tax revenue: the incentive effects of equalizing transfers: evidence from Germany. Int Tax Public Financ 9(6):631–649

Boadway R, Shah A (2009) Fiscal federalism: principles and practices of multiorder governance. Cambridge University Press, Cambridge

Breton A (1996) Competitive governments. An economic theory of politics and public finance. Cambridge University Press, Cambridge

Bryant MM (2006) Urban landscape conservation and the role of ecological greenways at local and metropolitan scales. Landsc Urban Plan 76(1-4):23–44

Bonney R, Cooper CB, Dickinson J, Kelling S, Phillips T, Rosenberg KV, Shirk J (2009) Citizen science: a developing tool for expanding science knowledge and scientific literacy. Bioscience 59(11):977–984

Droste N et al (2015) Ecological Fiscal Transfers in Brazil – incentivizing or compensating conservation? In: Paper presented at the 11th international conference of the European Society for Ecological Economics (ESEE). Leeds

EC (2015) Towards an EU Research and Innovation policy agenda for Nature-Based Solutions & Re-Naturing Cities, European Commission. Available at: http://bookshop.europa.eu/en/towards-an-eu-research-and-innovation-policy-agenda-for-nature-based-solutions-re-naturing-cities-pbKI0215162/

Fletcher R (2014) Orchestrating consent: post-politics and Intensification of Nature (TM) Inc. at the 2012 World Conservation Congress. Conserv Soc 12(3):329–342

Foldvary F (1994) Public goods and private communities. Edward Elgar, Cheltenham

Gabriel K, Endlicher W (2011) Urban and rural mortality rates during heat waves in Berlin and Brandenburg, Germany. Environ Pollut 159(8):2044–2050

Gawel E (1995) Die kommunalen Gebühren. Duncker & Humblot, Berlin

Gawel E (2016) Environmental and resource costs under Article 9 water framework directive. Challenges for the implementation of the Principle of Cost Recovery For Water Services. Duncker & Humblot, Berlin

Geyler S, Bedtke N, Gawel E (2014) Nachhaltige Regenwasserbewirtschaftung im Siedlungsbestand – Teil 2: Kommunale Strategien und aktuelle Steuerungstendenzen. gwf - Wasser I Abwasser 2:214–222

Hanley N, Barbier EB (2009) Pricing nature: cost-benefit analysis and environmental policy. Edward Elgar, Cheltenham

Hansen R et al (2015) The uptake of the ecosystem services concept in planning discourses of European and American cities. Ecosyst Serv 12:228–246

Hansjürgens B (1997) Gebührenkalkulation auf Basis volkswirtschaftlicher Kosten. Anwendungsprobleme und Lösungsmöglichkeiten. Archiv für Kommunalwissenschaften 36:233–253

Hansjürgens B (2001) Äquivalenzprinzip und Staatsfinanzierung. Duncker & Humblot, Berlin

Hansjürgens B (2004) Economic valuation through cost-benefit analysis – possibilities and limitations. Toxicology 205(3):241–252

Hansjürgens B (2015) Wider Irrläufer und Fehlinterpretationen. Ökologisches Wirtschaften 30(2):8–9

Hartig T et al (2003) Tracking restoration in natural and urban field settings. J Environ Psychol 23:109–123

Heudorf U, Meyer C (2005) Gesundheitliche Auswirkungen extremer Hitze am Beispiel der Hitzewelle und der Mortalität in Frankfurt am Main im August 2003. Gesundheitswesen 67:369–374

Hoffmann B et al (2008) Increased cause-specific mortality associated with 2003 heat wave in Essen, Germany. J Toxicol Environ Health 71(11/12):759–765

IUCN (2016) Nature-based solutions. Available at: http://www.iucn.org/regions/europe/our-work/nature-based-solutions. Accessed 30 June 2016

Janssen R (2001) On the use of multi-criteria analysis in environmental impact assessment in The Netherlands. J Multi-Criteria Decis Anal 10:101–109

Kallis G, Gómez-Baggethun E, Zografos C (2013) To value or not to value? That is not the question. Ecol Econ 94:97–105

Kenyon DA, Kincaid J (eds) (1994) Competition among states and local governments – efficiency and equity in American Federalism. The Urban Institute Press, Washington, DC

Kolbe J, Wüstemann H (2014) Estimating the value of urban green spaces: a hedonic pricing analysis of the housing market in Cologne, Germany. Folia Oeconomica 5(307):45–61

Lenk T (2004) Mehr Wettbewerb im bundesstaatlichen Finanzausgleich? Eine allokative und distributive Wirkungsanalyse für das Jahr 2005 unter Berücksichtigung der Neuregelungen. Jahrbücher für Nationalökonomie und Statistik 224(3):351–378

Lobstein T et al (2015) Child and adolescent obesity: part of a bigger picture. Lancet 385(9986):2510–2520

McConnell V, Walls M (2009) U.S. experience with transferable development rights. Rev Environ Econ Policy 3:288–303

Mills D (1980) Transferable development rights markets. J Urban Econ 7:63–74

Musgrave RM (1959) The theory of public finance: a study in public economy. McGraw-Hill, New York

Naturkapital Deutschland – TEEB DE (2016) Ökosystemleistungen in der Stadt - Gesundheit schützen und Lebensqualität erhöhen. Technische Universität Berlin, Helmholtz-Zentrum für Umweltforschung – UFZ, Berlin/Leipzig

Oates WE (1972) Fiscal federalism. Harcourt Brace Jovanovich, New York

Oates WE, Schwab RM (1988) Economic competition among jurisdictions: efficiency enhancing or distortion inducing? J Public Econ 35(3):333–354

Olson M (1969) The principle of "fiscal equivalence": the division of responsibilities among different levels of government. Am Econ Rev 59(2):479–487

Potschin M, et al (2015). Nature-based solutions. In: Potschin M, Jax K (eds) OpenNESS Ecosystem Service Reference Book. OpenNESS Synthesis Paper. Available at: http://www.openness-project.eu/library/reference-book/sp-NBS

Ring I (2002) Ecological public functions and fiscal equalisation at the local level in Germany. Ecol Econ 42:415–427

Ring I (2008) Integrating local ecological services into intergovernmental fiscal transfers: the case of the ecological ICMS in Brazil. Land Use Policy 25(4):485–497

Ring I, Barton DN (2015) Economic instrumtents in policymixes for biodiversity conservation and ecosystem governance. In: Martínez-Alier J, Muradian R (eds) Handbook of ecological economics. Edward Edgar, Cheltenham, pp 413–449

Rüger J, Gawel E, Kern K (2015) Reforming the German rain water charge – approaches for an incentive-oriented but still workable design of the charge. GWF – Wasser, Abwasser 156(3):364–372

Santos R et al (2012) Fiscal transfers for biodiversity conservation: the Portuguese local finances law. Land Use Policy 29(2):261–273

Santos R, Schröter-Schlaack C, Antunes P, Ring I, Clemente P (2015) Reviewing the role of habitat banking and tradable development rights in the conservation policy mix. Environ Conserv 42:294–305

Sauquet A, Marchand S, Féres J (2014) Protected areas, local governments, and strategic interactions: the case of the ICMS-Ecológico in the Brazilian state of Paraná. Ecol Econ 107:249–258

Schneider A et al (2009) Ursachenspezifische Mortalität, Herzinfarkt und das Auftreten von Beschwerden bei Herzinfarktüberlebenden in Abhängigkeit von der Lufttemperatur in Bayern (MOHIT). Helmholtz-Zentrum München – Deutsches Forschungszentrum für Gesundheit und Umwelt, Institut für Epidemiologie, München

Schröter-Schlaack C (2013) Steuerung der Flächeninanspruchnahme durch Planung und handelbare Flächenausweisungsrechte. Helmholtz-Zentrum für Umweltforschung – UFZ, Leipzig

Shah A (2007) A practitioner's guide to intergovernmental fiscal transfers. In: Boadway R, Shah A (eds) Intergovernmental fiscal transfers: principles and practices. World Bank, Washington, DC, pp 1–54

Sinn HW (1990) Tax harmonization and tax competition in Europe. Eur Econ Rev 34(2-3):489–504

TEEB (2012) In: Wittmer H, Gundimeda H (eds) The economics of ecosystems and biodiversity for local and regional policy makers. Earthscan, London/Washington

ten Brink P et al (2012) Nature and its role in the transition to a green economy. Institute for European Environmental Policy, Brussels

Tiebout CM (1956) A pure theory of public expenditures. J Polit Econ 64:416–424

Tsianou MA et al (2013) Identifying the criteria underlying the political decision for the prioritization of the Greek Natura 2000 conservation network. Biol Conserv 166:103–110

Vanberg VJ (2013) Föderaler Wettbewerb, Bürgersouveränität und die zwei Rollen des Staates, Freiburg Discussion Papers on Constitutional Economics 13(3). Walter Eucken Institut

Wagner RE (1983) Public finance, revenues and expenditures in a democratic society. Little Brown, Boston

World Bank (2008) Biodiversity, climate change and adaptation – nature-based solutions from the World Bank Portfolio. World Bank, Washington, DC

Zimmermann H (2016) Kommunalfinanzen: Eine Einführung in die finanzwissenschaftliche Analyse der kommunalen Finanzwissenschaft, 3rd edn. Berliner Wissenschaftsverlag, Berlin

Zimmermann H, Henke K-D, Broer M (2012) Finanzwissenschaft: eine Einführung in die Lehre von der öffentlichen Finanzwirtschaft, 11th edn. Vahlen, München

Zodrow GR (ed) (1983) Local provision of public services. The tiebout-model after twenty-five years. Academic Press, New York

Zodrow GR, Mieszkowski P (1986) Pigou, Tiebout, property taxation, and the underprovision of local public goods. J Urban Econ 19(3):356–370

Chapter 19
Nature-Based Solutions for Societal Goals Under Climate Change in Urban Areas – Synthesis and Ways Forward

Nadja Kabisch, Jutta Stadler, Horst Korn, and Aletta Bonn

Abstract Climate change and urbanisation are amongst the greatest global challenges society is facing today. The concept of nature-based solutions has recently been highlighted as key concept in policy and management in achieving alignment of environmental and societal goals (Cohen-Shacham et al. 2016; Kabisch et al. 2016a). The chapters of this volume assess the evidence, debate different concepts, identify policy avenues and display practical management applications that highlight the potential of nature-based solutions in tackling global challenges related to climate change and urbanisation. The authors critically review and present recent findings how urban ecosystem management can be employed to adapt to climate change effects, while at the same time contributing to social and health benefits. The examples discussed for ecosystem services and co-benefits in this volume comprise case studies concerning climate change adaptation by enhancing the regulation of

N. Kabisch (✉)
Department of Ecosystem Services, Helmholtz Centre for Environmental Research – UFZ, Permoserstraße 15, 04318 Leipzig, Germany

Department of Geography, Humboldt-Universität zu Berlin,
Unter den Linden 6, 10099 Berlin, Germany

German Centre for Integrative Biodiversity Research (iDiv) Halle-Jena-Leipzig,
Deutscher Platz 5e, 04103 Leipzig, Germany
e-mail: nadja.kabisch@ufz.de; nadja.kabisch@geo.hu-berlin.de

J. Stadler • H. Korn
Federal Agency of Nature Conservation (BfN), Isle of Vilm, Germany
e-mail: jutta.stadler@bfn.de; horst.korn@bfn.de

A. Bonn
Department of Ecosystem Services, Helmholtz Centre for Environmental Research – UFZ, Permoserstraße 15, 04318 Leipzig, Germany

Institute of Ecology, Friedrich Schiller University Jena,
Dornburger Str. 159, 07743 Jena, Germany

German Centre for Integrative Biodiversity Research (iDiv) Halle-Jena-Leipzig,
Deutscher Platz 5e, 04103 Leipzig, Germany
e-mail: aletta.bonn@ufz.de

© The Author(s) 2017
N. Kabisch et al. (eds.), *Nature-based Solutions to Climate Change Adaptation in Urban Areas*, Theory and Practice of Urban Sustainability Transitions,
DOI 10.1007/978-3-319-56091-5_19

air quality and temperature, contributing to water cycle regulation, provision of recreation potential as well as improvements for human well-being and mental health. In this conclusion chapter, we summarise the main outcomes of the chapters in this book. Considering current European policy, we develop recommendations for putting nature-based solutions into practice and policy and outline outstanding challenges for science and society.

Keywords Conclusions • Synthesis • Nature-based solutions for climate change adaptation • Planning recommendations • Management applications

19.1   Introduction

Climate change and urbanisation are amongst the greatest global challenges society is facing today. The concept of nature-based solutions (NBS) has recently been highlighted as key concept in policy and management in achieving alignment of environmental and societal goals (Cohen-Shacham et al. 2016; Kabisch et al. 2016a). The chapters of this volume assess the evidence, debate different concepts, identify policy avenues and display practical management applications that highlight the potential of NBS in tackling global challenges related to climate change and urbanisation. The authors critically review and present recent findings how urban ecosystem management can be employed to adapt to climate change effects, while at the same time contributing to social and health benefits. The examples discussed for ecosystem services and co-benefits in this volume comprise case studies concerning climate change mitigation via reduction in energy consumption and CO_2 emissions, as well as climate change adaptation by enhancing the regulation of air quality and temperature, contributing to water cycle regulation, provision of recreation potential as well as improvements for human well-being and mental health. In this conclusion chapter, we summarise the main outcomes of the chapters in this book. Considering current European policy as well as recommendations drawn by the ENCA interest group on climate change from the outcomes of the European conference on 'Nature-based solutions to climate change in urban areas and their rural surroundings' (Kabisch et al. 2016b), we develop recommendations for putting NBS into practice and policy and outline outstanding challenges for science and society.

19.2   Climate Change and the Concept of Nature-Based Solutions for Long-Term Sustainability Transition

Challenges from climate change for urban areas include increasing temperatures and changed precipitation dynamics. *Tobias Emilsson* and *Åsa Ode Sang* outline the general impacts and likely direct consequences of climate change for urban areas in Europe, highlighting how NBS through blue and green infrastructure can help to mitigate the effects and to adapt to a changing climate. The authors particularly focus on the cooling potential as well as on hydrological, ecological and social factors of urban green and blue spaces. Based on their review of examples, the authors point out that the current planning practice of urban densification poses threats to urban green spaces and brown fields reducing their capacity to serve as NBS for climate change adaptation. They highlight the approach of dual inner urban development as a solution to combine urban development with the quantitative and qualitative improvement of urban green in order to lead to truly sustainable cities. The authors further suggest that when implementing NBS with a particular focus on heat mitigation, the geographical location should be considered. Heatwaves in northern or central European regions that are usually not regularly affected by heat may have more negative consequences on urban residents in these areas, as they may be less prepared and less adapted. The authors stress that the allocation of NBS projects therefore requires a closer assessment of the specific urban morphology and characteristic of local population to arrive at holistic and targeted solutions. NBS may be most important in those areas where excessive urban heat has largest impact or where local residents have less economic possibilities to adapt. Thus, strategic planning in combination with modelling techniques and collaborative processes with the local population is recommended as a way forward to implement NBS for climate change adaptation while taking into account aspects of environmental justice.

In their chapter, *Stephan Pauleit* and *co-authors* analyse the features of NBS and other related approaches. They focus on the comparison of an issue-specific approach, such as ecosystem-based adaptation with an infrastructure-related approach, i.e. urban green infrastructure, and more general the concept of ecosystem services and their respective linkages to the NBS concept. In comparison to the other concepts, the authors consider NBS as an umbrella concept that covers features from other approaches, but with a distinct focus on deployment of actions on the ground. Still, the authors argue that ecosystem-based adaptation and urban green infrastructure should be considered as complementary and mutually reinforcing concepts related to NBS. The concept of NBS may build bridges between research, society and practice to find a common understanding and improved communication of what nature is beneficial for and may be easier to understand than the ecosystem service concept. A final conclusion from the authors is that there is still the need to operationalise the concepts to arrive from systematisation at implementation.

When implementing NBS in urban areas, its long-term functionality needs to be considered as *Erik Andersson* and *co-authors* point out. The concept of resilience with regard to NBS implies that NBS − once implemented − should not only be considered to be beneficial for current and immediate pressures from climate change but also be able to withstand potential future changes (introduced as a double-insurance value by the authors), both environmental and socio-political changes. Long-term resilience thinking of NBS addressed in urban governance therefore requires the consideration of long-term maintenance and the resources required. This is of particular importance because challenges from climate change will further impact on urban society during the upcoming decades and require long-term adaptation thinking (see also *Yaella Depietri* and *Timon McPhearson*).

To accelerate resilient transition thinking for climate change adaptation and mitigation in planning systems, *Niki Frantzeskaki and co-authors* present a number of European case studies that show how accelerating mechanisms for urban sustainability transitions through NBS unfold in different city contexts. The authors' cross-case comparison resulted in three main recommendations for implementing NBS for accelerating urban sustainability transitions. First, they particularly conclude that it is essential to give support to transition initiatives and urban change agents to mediate and catalyse processes of sustainable transformation. Change agents are central in promoting activities, acting as mediators, translators and networkers, and are further central to identify synergies between departments and sectors. The second important implication is that a proactive collaboration with minimised compartmentalisation is needed as an important precondition for better coordination and policy integration of NBS as part of pathways towards sustainable transition. Third, the authors plea for a policy mix to support long-term stability of projects to enable transition initiatives to spread and scale up. A long-term perspective of local governments on funding is necessary in order to create stability, decrease uncertainty for activities and enable voluntary action for sustainable transition.

19.3 Evidence of Benefits of Nature-Based Solutions Through Ecosystem Service Provision

To support the transition towards using NBS for climate change adaptation and mitigation, evidence of the NBS benefits is needed. This can include complementary approaches that combine the benefits of grey-engineered infrastructure and green and blue infrastructure.

In their synthesis, *Francesc Baró* and *Erik Gómez-Baggethun* present recent evidence on the potential contribution of regulating ecosystem services that are provided by urban green infrastructure to offset carbon emissions, to reduce heat stress and to mitigate air pollution at different urban scales. Using the city of Barcelona as a case study, the authors show that the real potential of NBS to mitigate carbon emissions, heat and air pollution is often limited and is dependent on geographic location and scale of implementation. The impact is particularly pronounced at the local scale, e.g. around specific green spaces, but does not necessarily scale up to city or municipality scale level. The contribution of urban vegetation to greenhouse gas emission reduction may be relatively limited, while urban soils composed of organic materials can act as much more relevant carbon sinks. Based on literature reviewed in this chapter as well as studies undertaken by the authors in Barcelona metropolitan region, they suggest the magnitude of these environmental problems may be too high to be mitigated by NBS only. The authors further highlight the need to provide good practice examples of NBS implementation projects that provide ecosystem services. Orientation on good practices may improve their uptake and upscaling. Finally, they recommend to place attention to the selection of species as well as on design and management of NBS implementation, as these could affect the degree of ecosystem services provision (see also *Vera Enzi and co-authors*).

With regard to disaster risk reduction, *Yaella Depietri* and *Timon McPhearson* also highlight the need for a long-term resilience thinking to tackle impacts from climate change and urbanisation. Reviewing the potentials but also the limitations of green and grey infrastructure for disaster risk reduction in urban areas, they suggest that intermediate 'hybrid' approaches, combining both green and grey approaches, may be the most effective strategy for reducing risk from environmental hazards. This is especially the case when NBS approaches may be insufficient to meet the rising impacts of climate change, in case of space limitation or when resource limitations require cost effectiveness in the context of both climatic and economic uncertainty.

A particular part of green infrastructure in cities – riparian forests and wetlands – are highlighted as NBS with particular benefits for dense urban areas. *Dagmar Haase* introduces the management of these habitats as a NBS with multiple benefits that include risk mitigation and adaptation concerning both climate extremes as well as enhanced flood and drought probabilities, while providing co-benefits such as a buffer against high air temperatures, water availability during heatwaves and recreation potential. The author presents several options that show wetland and riparian forests' functionality as NBS to better face the consequences of climate change in cities and urban regions and stresses that urban riparian forest and wetlands need to not only be conserved but also be restored in case of degradation. In line with the previous authors, *Dagmar Haase* also promotes a hybrid approach of combining natural remnants of wetlands and floodplain forests with technical solutions. To realise 'mini-wetlands' and 'riparian trenches' in areas disconnected from the river may also create new jobs. Following also suggestions by *Yaella Depietri* and *Timon McPhearson*, the complementarity of both NBS and technology may be a clever, pragmatic and at the same time innovative solution to complex socio-ecological problems. In addition, potential disservices from urban riparian forests and wetlands, including vector-borne diseases, e.g. transported via mosquitoes, may need to be considered as well with potential negative effects to human health. Careful consideration including such trade-offs can lead to more informed decision-making.

Complementing the contributions by *Yaella Depietri* and *Timon McPhearson* as well as *Dagmar Haase*, in their chapter *McKenna Davis* and *Sandra Naumann* also focus on disaster risk reduction and introduce sustainable urban drainage systems (SUDS) as a NBS to flood risk management. Based on their case study analysis, including cost-benefit calculations in comparison with conventional piped drainage systems, the authors conclude that SUDS have a high potential as sustainable, cost-effective approaches, which can complement pure 'grey' infrastructure and can be applied within new developments or used to retrofit existing systems. Nevertheless, uncertainties about long-term maintenance, performance and cost (-effectiveness) are main barriers that limit the implementation of SUDS. Authors highlight that making lessons learned and data gathered from existing projects more widely available will support a larger uptake. Maybe most important to support the uptake of SUDS is the fact that NBS such as SUDS can provide multifunctionality, while purely grey and engineered solutions cannot.

Using the example of green roofs and facades as NBS to climate change adaptation and mitigation, *Vera Enzi and co-authors* showcase a number of good practice examples highlighting the multiple benefits provided by them. Green roofs and green walls may contribute to the urban biodiversity network and improve air quality and temperature regulation as well as better management of surface water runoff. The authors provide strong business case arguments and show with concrete values that there is already an active green market in the available technology in Europe. In their conclusion, the authors highlight the vision of a resilient city that adapts to climate change needs to focus first on green and then on grey infrastructure. Nevertheless, the authors showed that hesitation, financial barriers for installation and uncertainty about long-term maintenance (costs) still exist. Knowledge needs to be communicated to planners on the potential of ecologically improved technologies for green roof and living walls, their attractive economic 'return on investment' and the range of associated 'free' benefits to society. In fact, using smart incentives could already speed up NBS implementation when taking into account some quality benchmarks as well as existing informal or even formal implementation guidelines that could be used by policy.

19.4   Health and Social Benefits of Nature-Based Solutions

Nature-based solutions not only provide climate change adaptation potential, they also offer multiple benefits, in particular health and social benefits related to urban green space distribution in cities.

To assess the health benefits of urban green spaces, *Matthias Braubach and co-authors* present an in-depth literature review with specific consideration of disadvantaged groups and health inequalities. They highlight increasing evidence that urban green spaces can have substantial benefits for human health and well-being. The evidence for health benefits appears to be very strong in particular for their potential of relaxation, stress reduction and other psychological effects. However, trade-offs of NBS with societal goals include, e.g. allergies to pollen, infections or potential injuries, e.g. by playing outside. Such detrimental effects can, however, be reduced or prevented by proper urban planning, design and maintenance of urban green areas. The authors therefore conclude that the positive health and social effects of urban green spaces significantly outweigh potential trade-offs. Specific recommendations that help implementing urban green spaces as NBS include the development of harmonised approaches to measuring green space benefits, the production of consistent and comparable data across urban areas and the consistent evaluation and monitoring of health effects of NBS implementation. This knowledge is essential for enabling local governments to assess the effects of planning decisions for interventions on environmental and health outcomes and to identify where future NBS implementation projects are most needed.

As health inequalities are expected to grow with increasing urbanisation and climate change, *Nadja Kabisch* and *Matilda Annerstedt van den Bosch* discuss health benefits from NBS with a particular focus on social justice aspects. They discuss the role of urban green spaces in tackling challenges from both climate change and urbanisation and at the same time counteracting health inequalities across socio-economic status and age scales. Using the case study of Berlin, the authors show that even in cities with comparatively large percentages of urban green spaces, these can be unequally distributed across the city area. Green and blue spaces are not always available in sufficient quantity and may not be easily accessible to all citizens especially vulnerable population groups such as children, elderly or deprived people. In order to provide positive health outcomes for all population groups, an equal distribution of high quality and safe urban natural spaces, adequate for physical activity and play, should therefore be of utmost importance in urban planning for a healthy environment. The authors conclude that quality criteria and standards for urban green spaces and green space implementation projects need to be developed to ensure that they provide the highest number of benefits for most of the potential user groups.

The book section on health and social effects of NBS is complemented by a critical discussion of potential NBS limitations related to social equity, cohesion and inclusion. *Annegret Haase* develops five hypotheses in which she argues comprehensively that NBS are not a solution for social problems on their own, but need complementary (social and political) tools and instruments. Under particular conditions, NBS might even trigger spatial segregation, e.g. when green space establishments lead to gentrification and displacement of population groups. The author concludes that when discussing NBS as contribution to climate change adaptation and mitigation, both synergies and trade-offs between environmental and social developments should be considered, and possible avenues to avoid or minimise negative impacts on communities should be identified. With this, the author highlights the need to 'politicise' the discourse on NBS and to take into account different power structures and social inequalities when dealing with 'greening' interventions.

Ines Cabral and co-authors discuss urban allotment and community gardens as deliberate NBS. They use the case studies of Lisbon, Leipzig, Manchester and Poznan to illustrate the range of ecosystem services that can be provided by urban gardens. Historically, the primary goal of allotment gardens was to mitigate poverty amongst factory workers during the industrial revolution by providing space for recreation and later to grow food. Thus, allotment gardens can be seen as one of the first multifunctional NBS. In some cities urban allotment gardens present a significant share of the community area and provide ecosystem services. Depending on management type and intensity, urban gardens can foster urban biodiversity and can also play an important role as ecological stepping stones within the city greenspace network. Their contribution to climate change mitigation may, however, be limited, since due to allotment codes, they mostly lack large amounts of biomass, e.g. through large trees, to store or sequester carbon. However, they may provide important climate adaptation potential by providing unsealed ground for water infiltration and by offering citizens spaces to escape the urban heat island effect and opportunities for recreation and other health benefits. Importantly, communal urban gardens can form innovative platforms to experiment with change and to promote social cohesion, an important factor for preparing for adaptation.

19.5 Implications for Urban Planning to Implement Nature-Based Solutions

The final section of the book deals with municipal governance and socio-economic aspects of NBS implementation projects. Good practice examples of efficient governance approaches to implement NBS are shown, and multiple actor-networks are highlighted.

Challenges of NBS implementation through urban governance are addressed by *Christine Wamsler and co-authors*. They introduce the concept of adaptation mainstreaming which is understood as the inclusion of climate risk considerations in sector policy and practice. The authors present a holistic framework in which adaptation mainstreaming is suggested to increase sustainability and resilience by not only focussing on preventing or resisting climate hazards but by also fostering a broader systems approach that highlights the importance to learn, to live and to cope with potential environmental risks. The authors argue that at the local level, adaptation mainstreaming requires the active consideration and combination of four approaches to reduce climate risk on the ground. These four key principles of the mitigation hierarchy are to avoid and to reduce exposure, to reduce vulnerability and to prepare an effective response or recovery after impact. Green infrastructure as multipurpose measures (e.g. for cooling and drainage) can help to reduce vulnerability through the inclusion of several elements that reduce dependency on only one system. The authors recommend that whenever possible, NBS should be implemented for both multi-hazard and multipurpose measures, i.e. measures that address both adaptation and other municipal objectives. Especially when addressing vulnerability and preparedness of urban communities, a certain redundancy in measures can be an important element of urban design.

Reflecting on institutional aspects and challenges of the implementation of NBS projects, *Chantal van Ham* and *Helen Klimmek* highlight the need for increased and improved collaboration between sectors and stakeholders as well as for a sound evidence base on the economic, social and environmental benefits of NBS to foster increasing uptake of NBS in urban areas. Examining different case studies across the world, the authors impressively show how multiple partnerships by different actors and sectors have led to climate change adaptation measures while simultaneously providing social, economic and environmental benefits. Demonstrating and sharing these good practice examples can serve as a strong foundation for promoting, investing and inspiring future collaboration to implement NBS. Authors highlight the importance of multidisciplinary and inclusive partnerships in fostering the uptake of NBS. These can result in the creation of synergies between different actors by bringing together resources, skills and knowledge. The involvement of citizens during project planning was shown to create trust during the implementation process and helped to take over ownership and stewardship. As already recommended by *Niki Frantzeskaki and co-authors*, the formation of a common understanding helps to establish and accelerate NBS actions. Developing business cases from NBS implementation projects can result in important arguments to future investments in NBS implementations by the private sector support and through public investment. Therefore, good and cost-effective practice examples that provide a solid and transparent evidence base need to be disseminated. This will particularly increase visibility of the value of a city's natural assets and may promote the uptake of NBS implementation projects in urban governance.

In their chapter, *Jakub Kronenberg and co-authors* provide a detailed analysis of the use, understanding and diffusion of NBS in Poland. Using Poland as a post-socialist and post-transformation case, authors highlight that the acceptance of NBS is relatively low, their visibility limited and subsequently implementation of NBS less spread. They see the main reasons for the low acceptance of NBS in Poland partly in the new socio-economic system that was introduced during the transition period after socialism. This may be explained by the fact that societal expectation of modern development in Poland rather correlates with glass and steel dominated grey infrastructure, than with green and blue developments. Powerful decision-makers do not necessarily work together with NGOs and research institutes that favour the implementation of NBS. To have significant impact, the authors therefore recommend that change agents who promote innovations, as highlighted by *Niki Frantzeskaki and co-authors,* should selectively work closer with those persons that seem to have the highest potential to accept and promote sustainability transitions through NBS. The authors also conclude that external pressure, especially from the EU (Kronenberg and Bergier 2012), may be particularly helpful in stimulating implementation and acceleration of sustainability transitions. By employing conditional funding, e.g. when granting EU structural funds, EU institutions may act as push factors to change the general framework for the national, social and institutional structures. This policy instrument could be an effective measure by funding agencies to overcome barriers and obstacles that are often created by powerful stakeholders that benefit from investments in grey infrastructure.

Finally, *Nils Droste* and *co-authors* focus on fiscal and constitutional restrictions of NBS implementations and analyse solutions to levy greater investments into multifunctional urban NBS. The authors conclude that NBS approaches can provide ecosystem-mediated services that man-made alternatives cannot supply cost-effectively. For leveraging investments in urban NBS, cross-sectoral, cross-departmental planning procedures are required where different interests need to be balanced. The authors suggest that the clearer the 'return of investment' of a NBS implementation is, the more likely it is that respective decision-makers invest in such 'novel' and innovative alternatives to well-known city plans. They further highlight the potential of public-private partnerships to enable urban decision-makers to create alliances that favour a climate for investments in NBS. Finally, the authors suggest the integration of an ecological indicator into the fiscal transfer system to create a financial incentive for investments into NBS.

19.6   Recommendations and Research Challenges to Reach Societal Goals Through Nature-Based Solutions Implementation

In conclusion, the chapters in this volume provide a compelling account of the increasing evidence of the multiple benefits provided by NBS in combatting climate change, in particular for adapting to a changing climate coupled with increasing urbanisation. Case studies across Europe demonstrate successful implementation of the NBS concept and provide important pointers for urban planning and management.

In order to further operationalise the NBS approach as effective instrument in sustainable urban development at a larger scale, we identify key remaining research challenges that can help to foster broad application in practice and policy:

- *Assess effectiveness of NBS at different scales*: The NBS impacts of ecosystem service provisions are highly scale dependent concerning space and time and linked to geographical location in the city as well as other factors such as species selection or management practices. Research should therefore identify at what scale and under which circumstances different NBS are most effective in order to evaluate their potential but also possible limitations. It is important that research is not tailored towards single ecosystem services only but takes into account the multiple benefits possibly generated by a NBS project.
- *Consider NBS effects of urban soil management*: The impact of urban soil management in urban environments, especially the contribution of unsealed soils and high organic soils both to climate mitigation through avoided carbon losses and to adaptation through increased water infiltration and evaporation capacities, needs further scientific attention.
- *Evaluate hybrid approaches of NBS and grey infrastructure combinations*: As several chapters point out, the dichotomy of employing either engineered or NBS solutions may not be useful or effective. Research should also focus on the question how NBS can complement technological solutions. This includes research that combines effects of the building sector (grey infrastructure) with ecosystem management strategies (blue and green infrastructure) in an integrative manner.
- *Analyse cost-benefits of NBS implementation*: In order to evaluate the cost-effectiveness of NBS, cost-benefit analyses are needed to assess the whole range of possible multiple NBS benefits in terms of single project evaluations as well as in terms of a comparison between purely 'grey' and 'green and blue' and "hybrid" options. This includes economic analyses of the costs of inaction as well as the possibility of catastrophic failure of purely technical solutions. The full range of social and economic impacts should be comprehensively taken into account by studying the monetary and non-monetary values of NBS projects.
- *Identify causalities and mechanisms*: In order to assess causalities and mechanisms of NBS effectiveness, research set-ups should include an evaluation of all relevant parameters before and after NBS implementation (pre- and post-assessment). Indicators of efficiency should be selected at the beginning of the project and respective measurements undertaken. In addition, the specific contribution

of biodiversity effects and mechanistic ecosystem functions towards delivery of ecosystem services through NBS need to be disentangled.

- *Identify social and environmental synergies and trade-offs of NBS*: Holistic research approaches are needed that consider both potential synergies and trade-offs between environmental and social developments to assess impacts of, for example, potential gentrification, social displacement or spatial segregation effects. Other potential trade-offs on NBS implementation may concern negative health effects, e.g. through potentially enhanced allergies from transmission of pollen from allergenic plants or increased vector-borne diseases through, e.g. creation of favourable habitats for vectors.
- *Explore efficiency factors in NBS governance and implementation*: Social network analyses and policy analyses may help to assess how successful governance mechanisms can facilitate the participation of relevant institutions and individual actors to arrive at effective decision-making to implement the NBS action. Research should also include the analysis of failure, e.g. why actors do not take decisions in favour of implementing NBS.

19.7 Recommendations to Foster Wider Application of Nature-Based Solutions with Partners from Society and Policy

A number of important conclusions can be drawn from the experiences presented in the case studies. This leads to suggestions for policy and practice that also incorporate the recommendations debated by the ENCA (Network of European Nature Conservation Agencies) interest group on climate change from the outcomes of the European conference on 'Nature-based solutions to climate change in urban areas and their rural surroundings' (Kabisch et al. 2016b). These concern three main areas:

Recommendations to ease future NBS implementation projects:

- *Demonstrating and sharing*: Although a wealth of information on NBS is already available, there is still a need for further collection of case studies and their dissemination through databases and publications. Furthermore, there is also a high demand for synthesis reports, analysing factors of success as well as obstacles encountered and possible ways to overcome them. These need to be produced in accessible formats to urban planners and resource managers. Effective communication of good multipurpose practice examples of NBS implementation projects to planning institutions in other cities will provide orientation and improve NBS uptake and upscaling. Besides information sharing through databases, personal contacts and study visits prove to be important ways of knowledge exchange.
- *Minimise compartmentalisation*: Urban planning needs to minimise compartmentalisation to improve coordination and policy integration of NBS. A change in planning structures that facilitates better communication and cooperation or the integration of intersectoral departments is recommended.

- *Foster participatory processes for co-design, co-creation and co-management of NBS implementation (Co-Co-Co)*: NBS should employ multidisciplinary approaches and inclusive partnerships across sectors and with local communities to create and catalyse synergies between different actors. This brings together resources, skills and knowledge. The involvement of citizens during project planning can create trust during the implementation process and helps citizens to take ownership and stewardship of NBS processes and NBS sites. This is important for further maintenance and sustainable management.
- *Promote change agents*: It has been recognised that individuals, institutions and specific processes that can act as mediators, translators and networkers between the different departments and sectors are central in promoting, accelerating and upscaling NBS implementation projects. These change agents should be actively supported and encouraged to facilitate and promote change through participatory processes.

Recommendations to improve planning processes to NBS implementations:

- *Create long-term stability*: A mix of different policy instruments for implementing NBS is recommended, such as regulation, financial incentives for public-private partnerships, investments as well as participatory community measures. This may support long-term stability of projects to enable to spread and scale up. Importantly, a long-term perspective of local governments on funding is necessary in order to create stability, decrease uncertainty for activities and enable voluntary action for sustainability transitions.
- *Provide monitoring and evaluation*: Detailed monitoring and evaluation of NBS before and beyond the project implementation phase will help to identify benefits and potential trade-offs. This information will help to improve management actions that increase the provision of ecosystem services. This again will increase the value of the site for people.
- *Develop manuals, guidelines and quality criteria*: Based on existing materials, further tools, manuals, guidelines, etc. for practitioners need to be developed in collaboration with science. Evidence and experience-based guidelines about climate change proofing NBS (e.g. species selection) should be developed to ensure that ecological functions and biodiversity gains are resilient to future changes. In some cases, this might mean to be more flexible when considering the provenance of species to be used in a project (e.g. street trees) and give practical advice for the design and management and other relevant aspects of NBS implementation. In addition, quality criteria for urban green spaces as well as their spatial distribution and accessibility need to be developed to ensure that they provide the highest number of benefits for a multitude of potential user groups, especially vulnerable population groups.
- *Consider social trade-offs*: When planning and implementing NBS projects, potential trade-offs with social developments need to be taken into account in order to avoid gentrification developments resulting in spatial segregation and displacement. Planning authorities should aim for a sufficient quantity and easy accessibility of green and blue infrastructure elements by local city residents as well as high-quality and safe urban natural spaces, adequate for phys-

ical activity and mental well-being. Efforts should be highest in those areas where climate change impacts are largest or where local residents have least (economic) possibilities to adapt.

- *Consider environmental trade-offs:* Planning needs to consider the potential disservices when developing, maintaining and managing NBS. Potential trade-offs and ecosystem disservices should be considered, e.g. species selection should consider potential trade-offs to human health such as avoiding the planting of trees or shrubs that cause allergic reactions.

Recommendations to strengthen the business case for NBS implementation:

- *Creating and strengthening the business case*: Strengthening the business case for NBS through promotion of lessons learned as well as data on cost-effectiveness and multiple benefits will result in important arguments to future investments in NBS implementations. Some manuals and tools to evaluate the wider benefits of certain categories of NBS as well as to allow comparisons between 'grey' and 'green' options already exist, while they need further uptake and refinement for other cases.
- *Use incentives and new investments*: Smart incentives included in municipal planning combined with smart funding instruments could speed up NBS implementation. The European Union and other funding agencies should put an emphasis to favour NBS when granting conditional funds, e.g. structural funds.
- *Decrease uncertainty*: Certainty about long-term maintenance costs, performance and overall cost-effectiveness of NBS needs to be developed through practice examples in collaboration between practitioners and researchers. A clear expected return of investment of a NBS implementation helps to convince decision-makers to invest in innovative alternatives for a sustainable city planning.

Over the coming years, it will be important to harness linkages and synergies between science, policy and practice to identify sustainable management of urban development by enhancing the contributions urban ecosystems can provide to society. There are opportunities for scientists, policy advisors and resource managers to engage with the ongoing research and practical urban development programmes of the European Commission and the European Environment Agency on nature-based solution in urban areas (European Commission 2015; European Commission 2016) as well as national developments. In addition, the efforts by the World Health Organization (WHO, Braubach et al., this volume; Romanelli et al. 2015) and the International Union for Conservation of Nature (IUCN, Rizvi et al. 2015; Cohen-Shacham et al. 2016) in fostering NBS for environmental, health and human well-being goals in cities and national programmes should be accompanied by local action. Overall, evidence from research and good practice needs to be made available also to the Intergovernmental Platform on Biodiversity and Ecosystem Services (IPBES) to be included in their assessments to synthesise and to communicate knowledge globally.

With intensifying competition for urban space under a changing climate and coupled ongoing urbanisation processes, it will become increasingly difficult to allocate green or blue spaces for a single purpose only or to rely solely on sectoral solutions to address mitigation and adaptation to climate change. To enhance preparedness and resilience of urban socio-ecological systems, synergies in urban land use and planning need to be found to align environmental and social goals through NBS. Overall, we hope this volume provides a baseline for stimulating discussion on how to address this challenge to foster integrative governance and management approaches that promote healthy, liveable and sustainable cities.

References

Cohen-Shacham E, Walters G, Janzen C, Maginnis S (2016) Nature-based solutions to address societal challenges. IUCN, Gland

European Commission (2015) Towards an EU research and innovation policy agenda for nature-based solutions & re-naturing cities. Final Report of the Horizon 2020 Expert Group on Nature-Based Solutions and Re-Naturing Cities

European Commission (2016) Policy topics: nature-based solutions. https://ec.europa.eu/research/environment/index.cfm?pg=nbs. Accessed 11 Sept 2016

Kabisch N, Frantzeskaki N, Pauleit S, Naumann S, Davis M, Artmann M, Haase D, Knapp S, Korn H, Stadler J, Zaunberger K, Bonn A (2016a) Nature-based solutions to climate change mitigation and adaptation in urban areas: perspectives on indicators, knowledge gaps, barriers, and opportunities for action. Ecol Soc 21:art39. doi: 10.5751/ES-08373-210239

Kabisch N, Stadler J, Korn H, Duffield S, Bonn A (2016b) Proceedings of the European conference on Nature-based solutions to climate change in urban areas and their rural surroundings. BfN-Skripten. German Federal Agency of Conservation, Bonn

Kronenberg J, Bergier T (2012) Sustainable development in a transition economy: business case studies from Poland. J Clean Prod 26:18–27

Rizvi A, Baig S, Verdone M (2015) Ecosystems based adaptation: knowledge gaps in making an economic case for investing in nature based solutions for climate change. Gland, Switzerland: IUCN 48

Romanelli C, Cooper D, Campbell-Lendrum D, Maiero M, Karesh W, Hunter D, Golden C (2015) Connecting global priorities: biodiversity and human health: a state of knowledge review. World Health Organisation and Secretariat of the UN Convention on Biological sity, Geneva

Index